Quantum Information and Consciousness
a gentle introduction

Quantum Information and Consciousness

a gentle introduction

Danko D. Georgiev

CRC Press
Taylor & Francis Group
Boca Raton London New York

CRC Press is an imprint of the
Taylor & Francis Group, an **informa** business

CRC Press
Taylor & Francis Group
6000 Broken Sound Parkway NW, Suite 300
Boca Raton, FL 33487-2742

First issued in paperback 2019

ISBN-13: 978-1-138-10448-8 (hbk)
ISBN-13: 978-0-367-40533-5 (pbk)

Visit the Taylor & Francis Web site at
http://www.taylorandfrancis.com

and the CRC Press Web site at
http://www.crcpress.com

Dedicated to my family

Contents

Foreword

Quantum Information and Consciousness: what a dazzling mosaic of ideas presented by Dr. Danko Georgiev in this book. Here we are escorted on a panoramic excursion of the essential groundwork covering a wealth of intellectually challenging concepts. But for whom is this book intended? Much of its contents, as they unfold chapter by chapter, are to an extent introductory; true for the first part at least. The development of topics seems accessible to the reader who wants to take stock of how logic and classical physics pass over to the quantum realm, and to see what thereafter may spring out, as when biology comes into the picture, for instance. If the reader had been inclined to step into the subject for the first time, without having to enslave a good search engine, then much of what one wants to (or perhaps, should) know is amply stored here under one roof, mainly for getting started upon what may turn out to be a captivating, though often a very perplexing voyage. Having some background in mathematics, physics, and biology, of course helps. So to say this is a 'gentle' introduction depends to some extent on the reader's background, openness of mind, and the willingness to fill out the margins with red ink as she/he makes it to the very end. It may be worth the exercise, as infuriating as that might become for the astute minded reader. So it is hard to think that there are no rewards at stake in committing to a serious study of the various topics, worked through in whatever way with which the reader feels some comfort.

All said and done, the book is actually a very good introduction to the basic theory of quantum systems. But on stepping beyond, we further advance into a world of weirdness riddled with controversy. It may be worth keeping in mind that the problem with a crazy idea is to determine if it is crazy enough, as Niels Bohr would have said it. And Richard Feynman has cautioned us of the dubious credibility of those proclaiming to possess an expert understanding of quantum phenomena. A 'gentle' introduction, or otherwise, Dr. Georgiev's book aptly prepares the reader to confront whatever might be in store later.

I was pleased to have been invited to write this Foreword, and also pleased that the author has deemed it worthwhile to include in his book aspects of our own joint work that spanned several years. I very much thank him in this respect.

James F. Glazebrook

Professor Emeritus
Eastern Illinois University
Charleston, Illinois USA

Preface

Our *minds* are constituted by subjective *conscious experiences* through which we access ourselves and the surrounding world. Examples of conscious experiences are the pain of the toothache, the smell of the rose, or the perceived blueness of the blue sky. Despite the large amount of clinical evidence suggesting an intimate relationship between the brain function and the conscious mind, the nature of this relationship has been subject to a long-lasting controversy. The main culprit in this state of affairs has been the almost exclusive reliance of current neuroscience on classical physics. Thus, some philosophers have promoted flawed theories of consciousness, e.g., we hallucinate that we have conscious experiences (while in fact having none), our consciousness is a causally ineffective epiphenomenon, or our free will is an illusion.

To restore our common sense view of ourselves as conscious minds with free will, we need to adopt a radically new conceptual framework for approaching the physical world such as the one provided by quantum information theory. Because it is impossible to understand the quantum information theory of consciousness without knowing any quantum physics at all, in the first half of the book I gently introduce the reader into the wondrous world of quantum mechanics that was revealed to us by the discovery of the Schrödinger equation in 1926. Only after I derive the differences between classical and quantum information in a set of rigorous theorems, I move forward to discuss the classical origin of seven long-standing problems related to consciousness including the physical boundary problem, the binding problem, the causal potency problem, the free will problem, the inner privacy problem, the mind–brain relationship and the hard problem of consciousness, and then show how these problems can be addressed using the specific tools of quantum information theory. Finally, I discuss the theory-ladenness of experimental observations and highlight the importance of conscious experiences for providing protocol sentences that are used both in the theoretical construction of scientific theories and in the critical assessment of these theories in the light of new experimental data.

Because consciousness is of utmost importance for virtually all forms of human activity, I have written this book with the expectation that it will be of interest to a wide target audience with diverse backgrounds. Consequently, I have attempted to make the exposition self-contained and equally accessible for undergraduate/graduate students and academic professionals. Even though individual chapters are didactically arranged in the order in which they should be read, different readers may proceed with different speeds through the chapters depending on their previous knowledge of the topics discussed. To enhance the overall reading experience and prevent the reader from skipping over essential details, below I provide a general map that displays at a glance the logical relationships between different chapters. This map could be very useful for fast navigation between chapters in a second reading of the book aimed at appreciating the fine details in the presented theories. The specific tasks performed by each chapter for addressing the main problems of consciousness could be summarized as follows.

Chapters 1, 5 and 6 comprise the core of the book that contains a large number of original results in the form of theorems and solved examples. Chapter 1 formulates clearly the seven main problems of consciousness using introspective experiments whose understanding does not require any scientific background. The rest of the book, however, requires mastering of a theoretical minimum of mathematical and physical concepts that are duly introduced in Chapters 2, 3 and 4. Chapter 5 explains the origin of various difficulties encountered when one applies classical information theorems to study consciousness and introduces many of the constraints that a physical theory of consciousness should respect. Two characteristic features of classical physics, namely, determinism and observability of physical states, are singled out as the worst adversaries to consciousness and free will. Chapter 6 constructs an axiomatic quantum information theory of consciousness and shows how fundamental quantum information theorems can be applied to address all of the main problems of consciousness. Chapters 5 and 6 are contrasted to each other thereby enabling the reader to fully grasp what the quantum information theory of consciousness is, what it is not, and why its axioms are not arbitrary due to the large number of constraints that had to be respected. For a complete picture on any one of the seven problems of consciousness, one has to take into consideration a trio of corresponding sections from Chapters 1, 5 and 6.

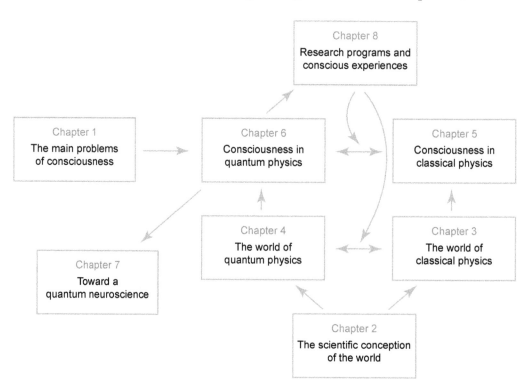

The necessity of axiomatizing science with subsequent proving of valid theorems is introduced in Chapter 2, where the reader is also acquainted with the criteria for rigor expected from any physical theory. Chapters 3 and 4 introduce the basics of the classical or quantum description of the physical world and would be especially valuable for non-physicists. Even though advanced readers could skip over the basics, acquaintance with Sections 3.19 and 4.20 is highly recommended as these provide proofs of the fundamental physical theorems pertaining to classical or quantum information that are subsequently used in Chapters 5 and 6. Chapters 3 and 4 are also contrasted to each other thereby providing a deeper insight into the physical reality of the quantum world.

Chapters 7 and 8 discuss the importance and possible applications of quantum information theory of consciousness. Chapter 7 illustrates how quantum theory could be applied to neurosciences, presents previously published quantum models of synaptic communication between neurons, and provides a number of open questions for future work. Chapter 8 wraps up the presentation by highlighting the importance of conscious experiences for the growth of scientific knowledge through Bayesian inference and assessment of competing scientific research programs based on different scientific theories. It also provides a postponed justification of the methodology for parallel presentation of competing theories and using contrasts for advancing on difficult scientific problems.

My expectations from the reader are minimal and include only a basic familiarity with numbers, mathematical equations, the parts of the human body, and the natural evolution of life on Earth. From there, I take the reader onto a comprehensible journey through logic, mathematics, classical and quantum information theory, neuroscience, and philosophy of mind. The scope of the exposition in each of these disciplines is limited only to a theoretical minimum of concepts, theorems and experimental facts that are essential for understanding and addressing the seven long-standing problems of consciousness. Reducing the number of theoretical concepts, however, leaves room for presenting a detailed explanation of what these concepts really are, why they are needed and how they relate to the problems studied. At the end of our journey, the reader will confidently know that the world of classical information that can be observed, copied, stored or erased is not a complete description of all there is. Instead, the physical world is made of quantum particles that are packets of complex-valued probability amplitudes that evolve in space and time according to the Schrödinger equation until a certain energy threshold is reached for an objective reduction to occur. The quantum information that is carried by the quantum particles cannot be observed, cannot be copied and cannot be erased, and it is exactly this quantum information that is the fabric of which our conscious minds are made. The conscious free will is exercised through the aforementioned objective reductions and the brain is the classical physical record of past mind choices. Thus, the quantum information theory of consciousness provides a physical support for philosophical existentialism according to which you are born free to choose what you want to be within the limits of the physically possible and that ultimately it is you who is responsible for your own decisions and actions.

Acknowledgments

For the realization of this project, first and foremost I would like to thank my family for all their love, caring support and continued encouragement. I am also deeply indebted to Professor James F. Glazebrook for his friendship, valuable advice and inspiring collaboration on the quantum model of SNARE protein zipping that has been prominently featured in our previous research publications and in the pages of this book. In regard to the organization and presentation of the covered material, I thank Lu Han for his editorial guidance and patience throughout all stages of the book preparation.

Danko D. Georgiev

Varna, Bulgaria

About the Author

Danko D. Georgiev was born on April 19, 1980 in Varna, the third largest city in Bulgaria and the largest city and seaside resort on the Bulgarian Black Sea Coast. He earned his MD from Medical University of Varna, Bulgaria, graduating summa cum laude in 2004, and his PhD in Pharmaceutical Sciences from Kanazawa University, Japan, in 2008 for his research in the area of neuronal proliferation and differentiation. From 2004 to 2005, Dr. Georgiev worked as an anesthesiologist at the Department of Anesthesiology and Intensive Care, Naval Hospital, Varna. From 2008 to 2013, he was a postdoctoral researcher at the Department of Psychiatry and Neurobiology, Kanazawa University, where he studied the molecular alterations in the cerebral cortex of subjects with schizophrenia. In the period from 2009 to 2011, he held the prestigious JSPS Postdoctoral Fellowship awarded by the Japan Society for the Promotion of Science. In 2011, Dr. Georgiev was a Short-Term Visiting Scholar at the Department of Psychiatry, University of Pittsburgh, and in 2013, a lecturer at the Biomedical Forum, Annual Program of Continuing Medical Education held at Medical University of Varna. From 2014 to 2015, he held the position of a Postdoctoral Associate at the Department of Environmental and Occupational Health, University of Pittsburgh, where he performed cutting-edge research on the pathogenesis and treatment of Alzheimer's disease. Since 2016, Dr. Georgiev has been a Principal Investigator at the Institute for Advanced Study, Varna, Bulgaria, where he currently employs graph theory and computational linguistics to study the cognitive processes underlying creative problem solving in view of developing technologies for computer-assisted enhancement of human creativity or implementation of creativity in machines endowed with general artificial intelligence. He has published more than 35 research articles, some in world-renowned neuroscience journals such as *American Journal of Psychiatry*, *Schizophrenia Bulletin*, and *Journal of Neuroscience*. His interest in quantum physics, quantum information theory and the problems of consciousness dates back to 2002 and has since led to a number of publications in mathematical and physical journals such as *International Journal of Modern Physics B*, *Axiomathes*, and *Informatica (Slovenia)*, and two book chapters in *Nano and Molecular Electronics Handbook*, CRC Press, 2007 and *Handbook of Nanoscience, Engineering, and Technology*, CRC Press, 2012. Dr. Georgiev has been a member of the Japanese Society of Neuropsychopharmacology since 2007, Japan Neuroscience Society since 2010, Society for Neuroscience since 2010, and an honorary member of the Bulgarian Society for Cell Biology since 2013.

Part I

Introduction

CHAPTER 1

The main problems of consciousness

We are *sentient* beings that are able to *experience* and *feel*. We live our lives from a subjective, first-person point of view. We are *aware* of ourselves and possess a sense of selfhood. We have passions, emotions and desires [268]. We are conscious *minds* that are composed of experiences [279]. Thus, *sentience* is the capacity of a physical entity to be *conscious*.

The term *consciousness* has been used by different authors to refer to a variety of things, including our ability to think logically, to behave rationally, to contemplate upon problems or to solve problems efficiently. None of these features, however, seems to be essential for defining what consciousness is. For example, when we think how to solve a logical problem, we may calculate and assess possibilities similarly to what computer programs do, but we do not consider ourselves unconscious when having a rest without any particular problem in mind. Similarly, some people could behave irrationally when they gamble on roulette or when they self-administer narcotic drugs, but we do not consider gamblers or drug addicts to be unconscious due to lack of rationality. In order to present clearly the main problems of consciousness that need to be solved, we have to strip all irrelevant details from our definition of what consciousness is. Thus, we will adopt a definition proposed by Thomas Nagel [351] and later popularized by David Chalmers [84], according to which the essence of consciousness is *experience*. Such a definition highlights the fact that the only way to access our inner selves and the surrounding world is through our conscious experiences.

Definition 1.1. *(Consciousness) Consciousness refers to the subjective, first-person point of view of our mental states, experiences or feelings. A conscious state is a state of experience. The terms* consciousness, mind *and* experience *will be used interchangeably hereafter.*

When we refer to ourselves as conscious minds we mean the subjective, first-person experiences of whom, what or how we are. Our mental states are states of experience (Fig. 1.1). René Descartes (1596–1650) was the first scientist to explicitly defend the viewpoint that the existence of our conscious minds and our experiences is a thing that we can be absolutely sure of [127, p. 171]. Experiences can have an illusory content in the sense that hearing the voice of an angel may not be reflecting the existence of a real angel outside one's mind, but be due to a dream, a hallucination, or a psychiatric disease. Remarkably, it is possible to hallucinate having had certain experiences in the past without actually having had those experiences. However, since every hallucination is a form of conscious experience [203, 360, 443, 450], *it is nonsense to say that one may be hallucinating having experiences without really having any experiences.*

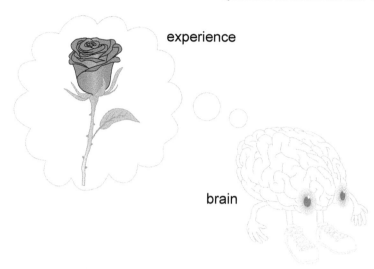

experience

brain

Figure 1.1   Minds consist of subjective, first-person, phenomenal conscious experiences. Thus, consciousness is experience, whereas the brain could only be what the conscious mind looks like from an objective, third-person point of view.

Definition 1.2. *(Qualia) Qualia (the plural form of* quale*) are the subjective, first-person, phenomenal, qualitative properties of conscious experiences. When you experience a red rose, there is something it is like for you to undergo that experience. What it is like to experience the redness of the red rose is something that is subjectively very different from what it is like to experience the whiteness of a white rose. If the* quale *of each conscious experience is what gives the experience its characteristic subjective feel [493], then the redness is the* quale *of experiencing the color red and the whiteness is the* quale *of experiencing the color white.*

To state that we are conscious and possess minds is just an alternative way to affirm that we are sentient creatures with inner feelings, desires and experiences. Thus, the phrase *conscious mind* becomes logically redundant since unconscious entities should not be called minds. Expressions such as *unconscious experience* or *unconscious mind* are also eliminated from our scientific vocabulary, and terms such as *subconsciousness* or *superego* that are frequently used in *psychoanalysis* originated by Sigmund Freud (1856–1939) [12, 197, 198] are relegated to the realm of pseudoscience where most of Freud's work justifiably belongs [94, 102, 146, 513]. *Unconscious brains* do exist, however, and will be discussed extensively hereafter.

Next, we will introduce seven long-standing problems of consciousness that seriously clash with the principles of classical physics and force some philosophers to defend flawed theories of consciousness that contradict our own introspective testimony. The portfolio of seven problems does not aim to be an exhaustive list of all problems related to consciousness, but rather to provide a fertile ground for testing the applicability of classical or quantum information theory to consciousness. For other problems that have been already satisfactorily addressed by classical neuroscience, we will directly present their solutions without further ado.

1.1 The physical boundary problem

Everything that exists is physical and should be subject to physical laws. Since the universe is the collection of all existing things and we do exist inside the universe (Section 2.5), the very first question one may ask is where the physical boundary of one's own conscious mind is. Without making any prior assumption on whether consciousness is a physical thing or a process involving physical things, the physical boundary problem can be stated as follows: What is the rule that sets the boundaries between your mind (or the physical objects that generate your mind) and the rest of the world?

We, as human individuals, possess bodies in addition to our minds. Although we treasure each of our bodily parts, there are some bodily parts that do not appear to be directly related to the generation of our conscious experiences. For example, cutting your hair does not lead to any impairment of your conscious memories or cognitive abilities. Thus, the physical boundary problem can be restated as follows: What is the rule that determines how much of your bodily parts, including pieces of organs such as the heart or the brain, could be discarded or replaced with transplants from a donor before your conscious mind is impaired or suffers a qualitative loss?

In the ancient world, people wrongly believed that the seat of consciousness is the *heart* [228, pp. 25–51]. It was considered that the heart controls our thoughts, emotions and actions. Importantly, this was not irrational guesswork, but based on the available evidence at the time as exemplified by no other but Aristotle, the father of logical reasoning (Section 2.2.1). Aristotle based his arguments on several lines of evidence [225]. First, he considered *physiological* evidence: Strong emotions such as anger lead to increased heartbeat [15, p. 479^b], so there is a direct link between emotions and heart function. In contrast, with the naked eye there were no observable effects of emotions upon the brain. In addition, if the brain of a living animal is exposed, touching the brain produced no sensations in which respect the brain resembles the blood of animals and their excrement [13, p. 652^b]. Second, he considered *anatomical* evidence: The heart is located centrally in the body and is connected with all parts of the body through the veins [15, p. 469^a]. In contrast, the brain appeared to be located to one end and disconnected from the rest of the body, since nerves were not well studied at the time. Thus, for Aristotle, if the soul is to be close to each part of the body, it has to be in the centrally located heart. Third, he considered *embryological* evidence: In sanguineous animals the heart is developed before the brain. Hence the source both of the sensitive and the nutritive soul has to be in the heart [15, pp. 468^b–469^a]. Fourth, he considered *physical* evidence: Living sanguineous animals are warm, whereas dead animals are cold [15, p. 469^b]. Because the living state is associated with consciousness, but not the dead one, Aristotle reasoned that the source of the warmth in the body should be producing consciousness as well. Furthermore, he thought that it is the heart whose fire produces the warmth of the body [15, p. 474^b], since air is needed for sustaining a fire [15, p. 470^a] and the heart is connected with the inhaled air through the lungs. In contrast, the brain appeared to be bloodless, devoid of veins, and naturally cold to the touch [16, p. 495^a]. Thus, for Aristotle, if animal is defined by the possession of sensitive soul, this soul must, in the sanguineous animals, be in the heart. The brain with its watery appearance was assumed to be a cooling device that exerted a chilling effect upon the blood. Consistent with the proposed interpretation of the brain as a cooling device, of all animals, man has the largest brain in proportion to his size; and the brain is larger in men than in women because the region of the heart and of the lung is hotter and richer in blood [13, p. 653^a].

Now we know that Aristotle's interpretation is wrong. Replacing the heart of a human with a mechanical pump or with the heart from a deceased donor does not affect one's mind or replace it with the mind of the diseased donor. The first heart transplantation in which the recipient lived for over two weeks was performed in 1967 by Christiaan Barnard (1922–2001) in Cape Town, South Africa [28]. At present, the number of heart transplantations performed in the world is over 4000 per year [332], and the average life expectancy after a heart transplant is over 10 years. Thus, an overwhelming amount of experimental data shows that the heart is not the seat of consciousness. Instead, modern medicine points to the *brain cortex* as the place where our consciousness is generated.

The *brain* is a part of the nervous system (Fig. 1.2). The brain is connected with the *spinal cord* that gives rise to both motor and sensory nerve fibers. The spinal cord is divided into 31 segments, based on the origin of the spinal nerves. There are 8 pairs of cervical nerves, 12 pairs of thoracic nerves, 5 pairs of lumbar nerves, 5 pairs of sacral nerves, and 1 pair of coccygeal nerves [345, p. 743]. Each spinal nerve is formed from the combination of nerve fibers from its posterior and anterior roots. The posterior root is the afferent sensory root and carries sensory information from the body to the brain. The anterior root is the efferent motor root and carries motor information from the brain to the body. The spinal nerves further branch into peripheral nerves that innervate the corresponding parts of the body. Besides the spinal nerves, there are 12 pairs of cranial nerves that emerge directly from the brain and primarily exchange information between the brain and regions of the head and the neck. The brain and the spinal cord form the *central nervous system*, which is well protected by the bones of the skull and the vertebral column. The *peripheral nervous system* is formed by ganglia and nerves that connect the central nervous system to the rest of the body.

Galen of Pergamon (129–216 AD) was the first physician to use routinely *vivisection* combined with putting ligatures (fine threads made of wool) around the parts in the living animal in order to learn what function is injured. For his experiments, Galen used a wide variety of animals, including dogs, goats, bears, pigs, cows, monkeys and lions [520]. The first public demonstration showing that it is not the heart that consciously controls the body through the veins, but it is the brain that consciously controls the body through the nerves, was performed by Galen in ancient Rome circa 160 AD [226, 228, 520]. The experiment consisted of cutting the *laryngeal nerve* that controls the vocal apparatus of a vivisected pig. Because at the time anesthetics were not used, the pig struggled violently and squealed in pain during the operation. However, at the moment the laryngeal

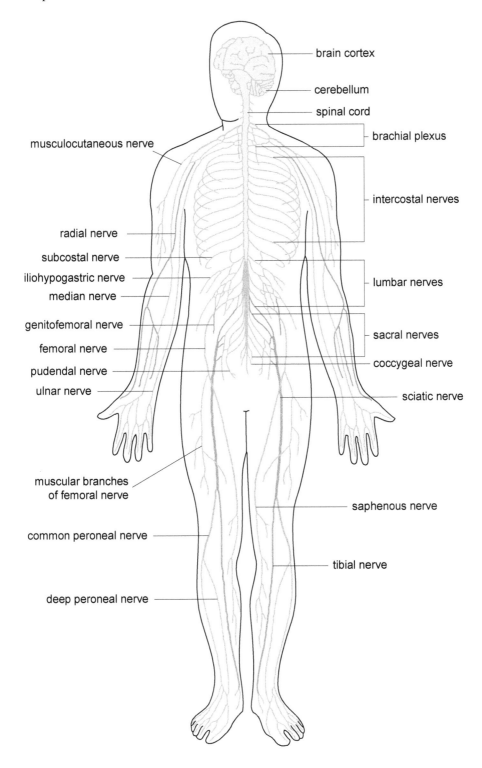

Figure 1.2 Illustration of the human nervous system. The central nervous system consists of the brain and the spinal cord. The peripheral nervous system consists of ganglia and nerves that connect the central nervous system to the rest of the body.

nerve was cut the *squealing pig* fell silent. Galen reasoned that since the *voice* is the most important of all psychic operations and is used to announce the thoughts of the rational soul, then consciousness has to be a product of the brain that controls the vocal apparatus through the nerves. Galen also studied various injuries to the spinal cord and showed that the animals were paralyzed beyond the level of the spinal cord injury.

Medical evidence showing that the seat of consciousness is localized in the *brain cortex* (Fig. 1.2) has steadily accumulated in the past 160 years. In 1855, Bartolomeo Panizza (1785–1867) was the first scientist to produce experimental evidence for the role of the brain cortex in *sensation* [97, 228]. He performed numerous experiments with crows, because they are lively and strong animals with soft skulls that can be cut with a simple knife, thus exposing the brain without damaging any of its functions. Exposing the cerebral hemispheres in crows and cutting through the occipital lobe of the brain cortex produced *blindness* in the opposite eye, even though the movements of the iris were maintained, as were the senses and movements of the body of the animals. The loss of sight was confirmed by allowing the crows to wander and observing that they ran at every step into the wall and other objects placed on the side of the eye opposite to the cortical lesion. The animals were kept alive for two days and then sacrificed in order to confirm the localization of the inflicted lesions. The occipital part of the brain cortex responsible for vision is now referred to as the *visual cortex* [266].

Further experimental evidence that the brain cortex is also involved in the control of body movements was found 15 years after Panizza's observations. In 1870, Gustav T. Fritsch (1838–1927) and Eduard Hitzig (1839–1907) showed that electrical stimulation within the frontal lobe of the brain cortex of a dog produced movements [227, 472]. They applied brief pulses of direct electric current to the brain cortex and observed muscle twitches on the opposite (contralateral) side of the body with respect to the stimulated brain hemisphere. With the use of this method, they found that electric stimulation is able to evoke muscle twitches only if it is applied to a certain portion of the brain cortex, now called the *motor cortex*. They also observed that the stimulation of different parts of the motor cortex activated different groups of muscles. Furthermore, the correspondence between the parts of the motor cortex and the muscle groups formed a topographical map that was the same across all individuals of the same animal species. These findings were soon replicated and extended in macaque monkeys by the Scottish neurologist David Ferrier (1843–1928) [168, 169, 170, 227]. He used alternating electric currents applied for relatively long time periods and observed that the brain cortex is able to elicit complete body movements such as walking, arm retraction, flexion and extension of the wrist, mouth opening, protrusion of the tongue, etc.

Subsequent clinicopathological findings from traumatic injuries, vascular incidents occurring during strokes, or tumors affecting discrete regions of the brain cortex in humans revealed specific losses of certain cognitive abilities such as understanding or articulation of speech, recognition of objects, shapes or faces, planning and execution of tasks, etc., and mapped those losses to specific areas in the brain cortex. With the advancement of neurosurgery, it become possible to apply electric currents to the cortex of awake humans and then obtain their subjective, first-person reports of the experiences elicited. Convincing evidence for the link between the brain cortex and conscious experiences was provided by the Canadian neurosurgeon Wilder G. Penfield (1891–1976) in patients with intractable temporal lobe epilepsy whose seizures failed to come under control with drug treatment [374]. The alternative to drug usage is surgical excision of the epileptogenic region of brain cortex. To avoid unnecessary removal of healthy portions of the brain cortex, the patient has to be conscious during electric stimulations performed in order to differentiate between the source of seizures and the healthy cortical tissue. Penfield found that electric stimulation within the temporal lobe of the brain cortex elicits experiences and one may relive past events, including complete sceneries of places, hearing concert music, seeing other people, etc.

> [A] mother told me she was suddenly aware, as my electrode touched the cortex, of being in her kitchen listening to the voice of her little boy who was playing outside in the yard. She was aware of the neighborhood noises, such as passing motor cars, that might mean danger to him.
>
> A young man stated he was sitting at a baseball game in a small town and watching a little boy crawl under the fence to join the audience. Another was in a concert hall listening to music. "An orchestration," he explained. He could hear the different instruments. All these were unimportant events, but recalled with complete detail.
>
> D.F. could hear instruments playing a melody. I re-stimulated the same point thirty times (!) trying to mislead her, and dictated each response to a stenographer. Each time I re-stimulated, she heard the melody again. It began at the same place and went on from chorus to verse. When she hummed an accompaniment to the music, the tempo was what would have been expected. [374, pp. 21–22]

In 2000, the groundbreaking work of William H. Dobelle (1941–2004) provided direct evidence that the brain cortex does not require the subcortical nerve pathways for producing conscious experiences [141]. He successfully restored the vision of a blind man by implanting *electrodes* that deliver directly the electric signals from a digital camera to the visual cortex [141]. The direct connection between the camera and the visual cortex bypasses the neural pathways that deliver the visual information from the retina through the thalamus to the visual cortex in healthy people. Yet the blind person with the implanted electrodes in the visual cortex consciously sees images of the surrounding world through flashes of light called *phosphenes*. That it is the brain cortex that consciously sees things is also corroborated by the conscious reports of people with injured visual cortex, but intact eyes, retinas and subcortical structures. Because the visual reflexes are controlled by extracortical brain areas, these people exhibit intact visual reflexes and react to visual stimuli, while at the same time insisting that they do not see anything. These conscious reports show that the extracortical brain areas are indeed not involved in the generation of consciousness, whereas the brain cortex is.

Example 1.1. *(Knee jerk reflex) Reflexes are involuntary motor responses that occur in humans as a consequence of applied stimuli. For example, striking the patellar ligament with a reflex hammer just below the patella (knee cap) will invariably cause your leg to kick out. The experience is as if your leg moved on its own without your conscious intention to move it. Indeed, your feeling is correct: the knee jerk reflex was executed at the level of the spinal cord before the electric signals from your knee had any time to reach the brain cortex. Since the brain cortex was not in control of the knee jerk, you feel and know that your conscious mind has not caused the jerk. Thus, testing your knee jerk reflex is an entertaining way to convince yourself that the spinal reflexes are not under your conscious control, hence the spinal cord cannot be the seat of consciousness.*

Some philosophers, including Maxwell Bennett and Peter Hacker, have claimed that the mind should be attributed to the person, not to the brain. They used the expression "mereological fallacy" to argue against the attribution to a part (the brain) that which should only be attributed to the whole (the person) [43, pp. 241–253]. However, limiting the boundary of the mind to the level of a complete human individual has been tested and proven false by the conscious reports of clinical patients: First, the conscious testimony by any person who has lost a limb or an internal organ other than the brain [406, 407] directly contradicts the claim that a mutilated human being cannot have a mind. Second, human patients with severe medical conditions, such as amyotrophic lateral sclerosis, progressive muscular dystrophies or tetraplegia due to cervical spinal cord injury, are very close to being minds locked in the head since the rest of the body is heavily incapacitated. In such cases, brain–machine interfaces could allow for direct mind control of robotic arms [69, 257, 364, 438, 446, 500] for grabbing objects or communication, which would still be the case if the whole body beyond the neck were amputated, leaving the head with the brain under the support of artificial mechanical systems. Third, head transplantation from a donor onto a recipient body has been successfully achieved in mice [412], dogs [120, pp. 139–149], and monkeys [518], with the transplanted head being able to hear, smell, taste, eat and follow objects with its eyes. Head transplantation is currently considered to be surgically possible for human volunteers who have a healthy brain but terminally ill bodies [67]. Surgically, switching the heads of two human beings will also switch their minds, whereas switching the hearts [28], lungs [237], or kidneys [337] of two human beings will leave the minds in their original bodies. From the latter fact, one could further refine the localization of the mind from the head to the brain, and then to the brain cortex in the light of the experimental neurological data collected by Panizza, Fritsch, Hitzig, Ferrier, Penfield, and Dobelle discussed above.

Even though experimental and clinical evidence strongly suggests that the seat of human consciousness is within the brain cortex, it is not known how much of the brain cortex is sufficient for the generation of conscious experiences. A successful theory of consciousness should be able to explain *what is the physical rule that sets the boundary of the conscious mind.* By doing so, the theory would also inform us *what prevents the mind from extending out into the environment where the mind could have exerted paranormal effects upon the surrounding world.*

1.2   The binding problem

Healthy human subjects experience themselves as a *single mind* rather than a *collection of minds*. Introspectively, we could verify that the sensory information obtained from the five sense organs, sight, smell, taste, touch, and hearing, is experienced as a single seamlessly integrated mental picture. For example, if you watch a movie in the cinema, you could be seeing the images projected on the screen, hearing the voices of the actors, smelling the popcorn eaten behind you, tasting the drink that you have bought, feeling the warmth of the theater room, and all that experienced at the same time by you. In order to explain this subjective unity of consciousness, however, one needs to find a mechanism that binds the experiences produced by discrete cortical areas into a single whole [336]. We know that such a mechanism should exist, because different people experience the contents of their own minds but do not have direct access to the experiences inside the minds of other people. In other words, there is nothing that binds your visual with my tactile experiences, but there is something that binds together your visual with your tactile experiences. Presumably, revealing the nature of that something would explain the manifest oneness of personal experience.

Because the existence of *minds within minds* leads to logical inconsistencies as it is not clear who does what, terms such as *collective consciousness, universal mind* or *cosmic mind* are considered to denote physically impossible scenarios. Consequently, the binding problem requires not only an explanation of *what binds your conscious experiences together*, but also *what keeps the experiences of different people with their own minds from binding together into a single global mind.*

Historically, the *unity of mind* has been used as an argument against the possible localization of cognitive functions to discrete areas of the brain cortex [228, p. 94]. Now we know that different functions are indeed localized to specific cortical areas. Localization of motor function to a part of the frontal lobe and vision to the occipital lobe has already been discussed in Section 1.1. To describe anatomical locations more precisely, however, we should go beyond the division of the brain cortex into lobes by identifying various grooves and ridges in the cortex.

Macroscopically, the brain cortex is divided into four lobes: frontal, occipital, temporal and parietal. The *frontal lobe* is located at the front of the head, the *occipital lobe* at the back of the head, the *temporal lobe* on the side of the head above the ear, and the *parietal lobe* in the middle upper part of the head above the temporal lobe. Each cortical lobe contains ridges and grooves. In Latin, the ridge is translated as *gyrus* (plural *gyri*) and the groove as *sulcus* (plural *sulci*). Gyri and sulci create the folded appearance of the brain. The folding of the brain cortex allows a greater surface area to be compacted in the limited volume confined by the skull. Figures 1.3 and 1.4 show the localization of important brain areas whose function will be further elaborated upon in Examples 1.2, 5.4, 5.12, 6.18, and 7.1.

Different cognitive, perceptual, and motor abilities are localized to distinct specialized areas in the brain cortex. A nonexhaustive list of functions includes the following: On the lateral surface (Fig. 1.3), the main motor control of the body is performed by the *precentral gyrus* [285, p. 11]; the main sense of touch from differ-

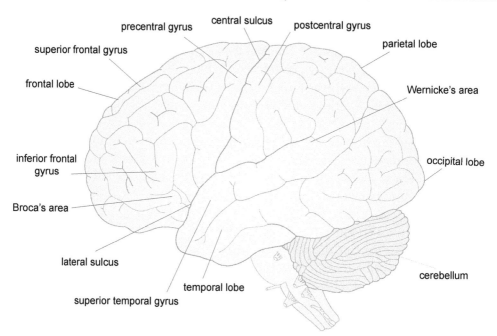

Figure 1.3   Lateral surface of the left brain hemisphere. Highly important are the primary somatosensory cortex in the postcentral gyrus, the primary motor cortex in the precentral gyrus, Wernicke's area for receptive language, and Broca's area for expressive language.

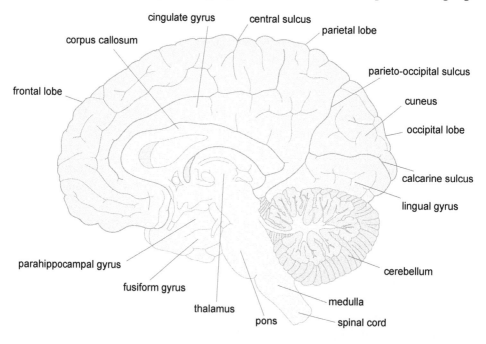

Figure 1.4   Medial surface of the right brain hemisphere. Highly important are the primary visual cortex in the cuneus, the cingulate gyrus controlling emotions, the corpus callosum composed of nerve fibers connecting the left and right brain hemispheres, and the thalamus relaying sensory and motor signals to the cerebral cortex.

ent parts of the body results from the activity of the *postcentral gyrus* [285, p. 11]; speech and expressive language are controlled by Broca's area in the *inferior frontal gyrus* [145]; understanding of written and spoken language is due to the activity of Wernicke's area in the *superior temporal gyrus* [53, 342, 400]; self-awareness and laughter are generated by the *superior frontal gyrus* [199, 219]. On the medial surface (Fig. 1.4), vision and processing of visual information occur in the *cuneus* and the *lingual gyrus* [54, 499]; processing of color information and face recognition result from the activity of the *fusiform gyrus* [354]; emotions form in the *cingulate gyrus* [503]; memories are encoded and retrieved by the *parahippocampal gyrus* [4, 471]. To solve the binding problem of conscious experiences one needs to explain what exactly is the physical mechanism that causes the production of a single mind from the activities of all those distinct brain cortical areas.

Example 1.2. *(Split-brain patients) The unity of consciousness requires integrity of the brain since after certain surgical operations one brain can host more than one individual mind. Clinical observations and psychological tests of split-brain patients, who had their* corpus callosum *(Fig. 1.4) severed as a therapeutic procedure for refractory epilepsy, have shown that the split-brain hosts two minds, one in each of the disconnected hemispheres. Interestingly, each of the two minds is self-aware and readily recognizes and identifies himself or herself in a multiple-choice task, but is blissfully unaware of the existence of the other mind localized in the opposite brain hemisphere [447]. Because the corpus callosum consists of nerve fibers that project from each brain hemisphere to the opposite one, it should be the case that the activity of those nerve fibers somehow glues together and binds the conscious experiences. What remains to be identified is the physical process that enforces this binding.*

1.3   The causal potency problem

Introspectively, it appears to us that our conscious experiences could affect our bodily functions, behavior and future decisions. We, as conscious minds, feel in control of our bodies. Through the conscious control of our skeletal muscles, we are able to transform the surrounding world. In particular, our minds that are hosted in the brain cortex are able to trigger motor electric signals propagated from the *upper motor neurons* located in layer 5 of the motor cortex through the *corticospinal tract* toward the spinal *alpha motor neurons*, which in turn control the action of *skeletal muscles* (Fig. 1.5). Injuries at any level of the corticospinal tract or the alpha motor neurons result in *paralysis* [144, 481]. However, if electrodes are surgically implanted in the motor cortex of either monkeys [69] or paralyzed human patients [257] in such a way that the recorded electric activity from groups of cortical neurons directly controls a robotic arm by a fixed computer algorithm, then after several months of practice the conscious mind is able to train itself to control the robotic arm without actually moving any of the body muscles. Thus, our minds being composed of conscious experiences appear to be causally potent agents within the physical world, because if they were not, conscious control of *brain–machine interfaces* should not have been possible.

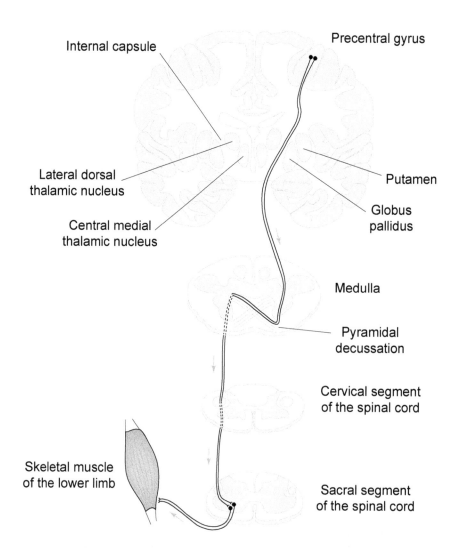

Figure 1.5   A motor neural pathway from the brain cortex. The corticospinal tract transmits motor electric signals from the motor region of the brain cortex to the spinal alpha motor neurons, which in turn control the contraction of target skeletal muscles. The myelinated axons of the upper motor neurons that are pyramidal neurons in layer 5 of the motor cortex pass through the internal capsule and then cross to the opposite side of the body at the level of pyramidal decussation in the medulla. The upper motor neurons synapse onto lower motor neurons that are alpha motor neurons in the anterior horn of the spinal cord. The alpha motor neurons innervate skeletal muscles at the corresponding level of the spinal cord. In the case of a sacral spinal segment, the innervated muscles are located in the lower limbs. Injuries at any level of the corticospinal tract or of the alpha motor neurons result in paralysis. The spinal cord segments and medulla are represented with their transversal sections, whereas thalamus and cortex are shown in frontal slice.

The theory of natural evolution originated by Charles Darwin (1809–1882) [106, 107, 270], and spectacularly corroborated by massive evidence from geology, paleontology, comparative anatomy and genetics [113, 114, 115, 116, 117], also requires that the mind is a causally potent agent in order to be selected for by natural selection. The evolutionary history of life could be outlined as follows:

The life on Earth originated in the oceans $\approx$ 3.5 billion years ago in the form of *prokaryotes* that are single-celled organisms lacking any membrane-bound organelles. In the process of evolution by natural selection, a common prokaryote ancestor diversified and gave rise to all present-day organisms. The shared origin of all living organisms is now clearly seen in the genetic code and the similarities of genes across different species [68].

The second important event in the evolutionary history of life is the origin of *photosynthesis* in cyanobacteria $\approx$ 2.5 billion years ago. Photosynthesis allowed utilization of light energy from the sun for the production of sugars from carbon dioxide and water, releasing oxygen as a waste product. Initially, the oxygen released by the photosynthetic activity of cyanobacteria was used to oxidize and precipitate iron dissolved in the oceans of early Earth [166]. Once all of the dissolved iron was used up, the oxygen could escape from the oceans and accumulate into the atmosphere, eventually reaching the current 21% at sea level.

The third important event is the origin of *eukaryotes* $\approx$ 2 billion years ago. Because eukaryote cells are not enclosed in a rigid cell wall, they are able to change their cell shape easily. To achieve stability of their fragile phospholipid cell membrane, however, eukaryotes had to develop a complex protein *cytoskeleton* composed of *actin, intermediate filaments* and *microtubules*. In addition to the dynamic changes of cellular shape afforded by the cytoskeleton, eukaryotes also developed membrane-bound *organelles* such as the *nucleus* containing the genetic material, *rough endoplasmic reticulum* involved in protein production, *Golgi apparatus* involved in protein packaging and trafficking with the use of membrane-bound vesicles, and *mitochondria* involved in cellular respiration that supplies the cell with ready-to-use biochemical energy in the form of adenosine triphosphate (ATP). Because the cellular respiration performed by mitochondria requires a constant supply of oxygen, the successful reproduction of eukaryote cells depends profoundly on the well-being of photosynthetic organisms.

The fourth important event is the origin of multicellular organisms $\approx$ 1 billion years ago. Diversification of early multicellular organisms gave rise to fish $\approx$ 500 million years ago, land plants $\approx$ 475 million years ago, insects $\approx$ 400 million years ago, amphibians $\approx$ 360 million years ago, reptiles $\approx$ 300 million years ago, mammals $\approx$ 200 million years ago and birds $\approx$ 150 million years ago.

The fifth event, fortunately for us, is the massive asteroid impact $\approx$ 65 million years ago, forming the Chicxulub crater buried underneath the Yucatán Peninsula in Mexico [251]. The crater width of 180 km and depth of 20 km implies that the impacting asteroid was at least 10 km in diameter. Computer simulations also reveal that the impact by such a massive body could produce giant tsunamis, earthquakes, glowing fireballs from the falling rocky debris, and prolonged blockade of sun radiation by dust particles, leading to a global decrease of temperature known as an *impact winter* [489]. The environmental changes following the impact caused mass extinction of numerous plant and animal groups, including *dinosaurs*. Following the death of dinosaurs, the continents of the planet Earth were ready for taking by the survivors. *Mammals*, who at the time were no bigger than a rat and lived in burrows or hibernated, have survived and diversified into many new forms and ecological niches. Rapid mammalian evolution led to the appearance of the first hominins $\approx$ 4 million years ago [526, p. 71], and eventually to anatomically modern *Homo sapiens* $\approx$ 200 thousand years ago [526, p. 104].

Since humans are linked to other animal species through common ancestry [154, 439, 457], it follows that human consciousness should have been subject to natural evolution as well. Thus, evolution theory lends strong support for the causal potency of the conscious mind because if our minds were not causally effective there would have been nothing to select the human mind for by natural processes [278]. Nevertheless, conscious experiences are not mentioned as a fundamental constituent of current physical theories and one is left with the troublesome problem of explaining *where and how exactly the conscious experiences enter the physical picture of the world as causal agents.*

1.4   The free will problem

We can choose what to do in our lives. Daily, we have to make choices and be responsible for our own actions. Our desires influence our actions and it may be the case that it is more likely or less likely to make a given choice due to personal preferences. For example, when offered pizza or pasta for lunch we might be biased to choose pizza in 70% and pasta in 30% of the time. Still, regardless of our biases, we feel that it is up to us to do otherwise if we choose. We feel free to make choices and know that we can be held responsible for what we have chosen. Thus, the major prerequisite for free will is the availability of choices to make. This means that there must be at least two possible alternatives from which we can choose. If there were only a single possible option given to us, then we could not make a choice and could not exercise our free will. Instead, we would feel coerced to act in a way for which it is not up to us to decide to do otherwise.

Definition 1.3. *(Free will) Free will is the capacity of agents to choose a course of action from among various future possibilities. An agent has free will only if there are at least two different alternatives genuinely open to the agent when facing a choice and provided that the outcome of the choice has not been forced upon the agent.*

Performing an action in accordance with your own desires could be regarded as a manifestation of your *will*. Retrospectively, it is always possible to state that an action has been *willed* and performed because there had been some internal desire for performing the action. In order to be *free*, however, the action should not have been the only available option coerced upon the agent. Instead, the agent should have been capable of choosing not to perform the action as well.

Possessing free will does not imply that we should be able to choose everything including choosing what alternatives are available to us. On the contrary, we know from experience that we can only choose among physically possible things. Thus, free will does not grant us the power to perform miracles.

We also cannot choose whether we have free will or not. If we possess free will, this is due to, not our whims, but the fundamental physical laws of the universe. In all cases when we are facing the choice between performing and not performing a given action, we either do the action or do not do it. As a result, if we do have free will, there is no sense in which we can choose to give up our free will. Since we cannot stop the flow of time, it is impossible for us not to use our free will.

The ability to use our free will is remarkable, but for many of our daily choices we hardly even consider that the action is subject to voluntary control. A typical example is our breathing. We can choose to inhale, temporarily stop breathing, then exhale, but our attention is rarely focused upon the fact that we are breathing and that we may use our free will to control our breathing. In other words, we are making choices and using our free will virtually all the time, yet we are so busy that we almost never contemplate the fact that we are making all those choices.

The concept of free will is intimately connected to the concept of *moral responsibility*. For agents with free will, not doing an action is equivalent to choosing not to do the action, hence such agents are always *morally responsible* for the things that they have chosen not to do. Acting with free will could be considered as the sole requirement that makes agents responsible for things that they have done. Agents with free will are always morally responsible for their choices regardless of the emotional status they were in while making the choices or whether they were intelligent enough to foresee how their actions may have caused suffering to others.

Moral responsibility is ultimately connected with the questions of what exactly is the *meaning of life* and whether we are born in this world for fulfilling a definite *purpose*. After witnessing the horrors performed by human beings in World War II, Jean-Paul Sartre (1905–1980) realized that we are born free and we are able to decide ourselves what the purpose and the meaning of our own lives are. Thus, there are no excuses for our actions. In philosophy, Sartre's ideas are known as *existentialism* due to the claim that "existence precedes essence," where under the *essence* of something is understood its meaning or its intended purpose. For example, an airplane is made to fly; that is its essence. Humans, however, do not have an essence because there is no greater purpose, no predetermined plan and no ultimate meaning of our lives set by something outside us. We are simply here, and it is up to us to define ourselves and choose our own purpose of life.

> Dostoyevsky once wrote: "If God does not exist, everything is permissible." This is the starting point of existentialism. Indeed, everything is permissible if God does not exist, and man is consequently abandoned, for he cannot find anything to rely on—neither within nor without. First, he finds there are no excuses. For if it is true that existence precedes essence, we can never explain our actions by reference to a given and immutable human nature.

In other words, there is no determinism—man is free, man is freedom. If, however, God does not exist, we will encounter no values or orders that can legitimize our conduct. Thus, we have neither behind us, nor before us, in the luminous realm of values, any means of justification or excuse. We are left alone and without excuse. That is what I mean when I say that man is condemned to be free: condemned, because he did not create himself, yet nonetheless free, because once cast into the world, he is responsible for everything he does. [420, pp. 28–29]

Existentialism is humanism, because man is confronted with the power of his free will and the responsibility for his own life. Because people can choose to be either good or evil, they are essentially neither of these things until they make their choice. The world can only determine the available choices, but ultimately it is up to the human agent to make the choice. Therefore, if people stop viewing themselves as victims of the circumstances, they can use their free will to make the world a better place.

Man is not only that which he conceives himself to be, but that which he wills himself to be, and since he conceives of himself only after he exists, just as he wills himself to be after being thrown into existence, man is nothing other than what he makes of himself. [420, p. 22]

Being rational or intelligent enough to assess the consequences of your own actions is not a prerequisite for exercising your free will. We are all born as ignorant babies without any knowledge about ourselves or the surrounding world. Without knowledge, we are unable to reason and cannot be rational. We learn as we grow, however, and with the enhancement of our *knowledge, intelligence* and *rationality*, we become able to predict what the consequences of our actions would be under certain circumstances. Still, *ignorance* cannot be claimed as an excuse for disregarding our moral obligations. Regardless of whether a choice is rational or irrational, informed or uninformed, the free will is exercised in the very act of choice making. Thus, morality is dependent not on how knowledgeable or clever we are, but on whether we were capable of acting otherwise than we actually acted.

Due to the profound importance of free will for human morality, it would be very disturbing if science denies our ability to make choices and precludes the existence of free will. In fact, the intuitive notion of free will seriously clashes with deterministic physical theories such as classical mechanics (Section 3.12) or Einstein's relativity (Section 3.18). In such theories, the future is uniquely determined by the physical laws in combination with the present physical state of affairs, and there is no sense in which an agent such as a conscious human being is given alternative choices to make. Interestingly, many physicists are still prejudiced to believe that physical theories should necessarily be deterministic. Whereas such a desire might be psychologically pleasing because an ultimate control over nature would appear to be achievable, it also makes humans susceptible to such an external control by others. Fortunately, an accumulating amount of experimental evidence shows that the physical processes are inherently indeterministic, hence not necessarily incompatible with the existence of free will.

In order to dispel skepticism in regard to the possibility of a physical theory of free will, it would be instructive to explicitly describe what a physical theory of agents endowed with free will should look like. In fact, with the use of straightforward theoretical arguments based on our common sense view of what we are and what we can do, two basic features of the putative physical theory could be derived.

First, the physical theory should be indeterministic in regard to the long-term time evolution of free agents. This follows from the definition of free will as the capacity of the agent to make choices among at least two available future courses of action. In particular, the existence of genuine choices implies the existence of genuine bifurcation points in the time evolution of free agents.

Second, the physical theory should have at least one, but possibly more, physical laws. Thus, the existence of free will does not itself imply physical *lawlessness*. On the contrary, many kinds of physical laws are compatible with the existence of free will:

(1) There could be a physical law that attributes free will to some agents, and thus makes it clear whose free will is operating in the physical world. Preferably, the agents should be identified as minds, because it is our conscious minds that possess free will.

(2) There could be a physical law that determines what the agent with free will can do and what he/she cannot do. Such a law will provide a set of available courses of future actions, but will leave it up to the agent to choose which course of action will be actualized at a given instance.

(3) There could be a physical law that determines the propensity or the probability with which the free agent could actualize any of the available choices provided that the choosing can be repeated multiple times. Such a law will essentially introduce inherent biases in the agent's choosing. The law does not need to imply time reversibility or that the same choosing can be actually repeated multiple times.

(4) There could be a physical law that determines the frequency of the instances in time at which the free agent has to choose. Such a law will introduce points of bifurcation in the time evolution of the free agent and will make it impossible for the agent to give up its free will. Simply, the further time evolution could not be done without choosing one of the available physical trajectories.

(5) For ensuring the logical consistency of the physical theory, the physical laws should forbid occurrence of minds within minds, otherwise it would not be clear whose free will is operating. In addition, methods for obtaining absolutely certain foreknowledge by the mind of which future choices will be actualized such as backwards-in-time traveling should not be physically available. Thus, a physical theory of agents with free will is meaningful only if there is an irreversible flow of time in the universe.

In Chapter 6, we will present a physical theory that supports free will and provides physical laws addressing all of the points listed above.

1.5 The inner privacy problem

Conscious experiences are *private* and accessible through *introspection* from a first-person, subjective, phenomenal perspective, but remain *unobservable* from a third-person, objective perspective [351]. As a consequence, there is no objective scientific way to determine if any other person, animal or object is conscious or not, because we do not have direct access to someone else's experiences through *observation*. We may observe someone else's brain but the process of observation alone does not make us experience what that brain is experiencing (Fig. 1.6). Thus, there are some things that exist in the universe, such as one's own experiences, but which are fundamentally *unobservable*. Furthermore, the phenomenal nature of individual conscious experiences is *incommunicable*. We may define the subject whose experiences we are interested in or the situation under which certain experiences are elicited; however, we are unable to describe in words what it is like to have those experiences. Therefore, we are unable to communicate what it is like to have any of our experiences to others and those others are unable to communicate what it is like to have their experiences to us.

Example 1.3. *(What it is like to hear music) Suppose that a person who was born deaf and never knew what it is like to hear sounds is given the opportunity to observe the electric impulses in someone else's brain during the performance of a philharmonic orchestra. If, from the act of observation, the born deaf person was able to deduce that "the person under observation is experiencing the sounds of musical instruments," such an expression could be true but still the born deaf person would have no way whatsoever to imagine what hearing the sounds of the musical instruments would be like. Thus, for the born deaf person the expression that someone "is experiencing sounds" is close to meaningless, whereas for people with unimpaired hearing the expression that someone "is experiencing sounds" is meaningful due to the fact they had previously experienced sounds and know what it is like to hear sounds.*

Example 1.4. *(What it is like to be a bat) Consider bat navigation in the real world through sonar. No human is able to imagine what it is like to be a bat or what it is like to experience the surrounding world through reflected ultrasound waves. To say that from recording the electric activity of a bat's brain we have observed the bat's sonar experience could be made true by definition but the actual bat's experience could have been subjectively very different, and yet, we still would have called it "bat sonar experience." The expression "bat sonar experience" is well-defined regardless of what the actual bat experience is and whether we can imagine what it is like to have it or not. Thus, without being logically inconsistent, it is possible to define and refer to things that we do not understand, things that we do not know, or things that we cannot experience.*

One may try to redefine the term "observation" and insist that experiences are observable in the sense that they are "deducible" from the observed data. For example, one may observe the physical electric impulses triggered in someone else's brain and deduce that the observed brain is experiencing pain, even though the observer himself is not experiencing pain. The "deducibility" of the pain, however, is not really an "observation" because it presupposes that the observer already

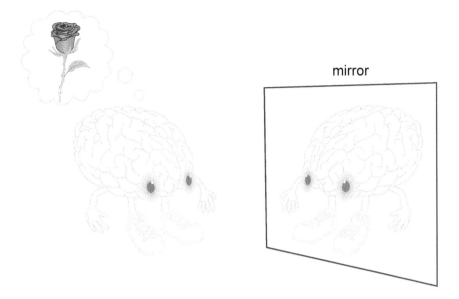

Figure 1.6   The conscious experiences are unobservable from a third-person perspective. If we were experiencing the mental image of a red rose and we were able to take a look at ourselves in a mirror, we would see the brain and not the conscious experience. Because the brain looks quite different from what we experience, we are able to learn new things about ourselves with the use of a mirror.

has a kind of personal incommunicable knowledge of what it is like to feel pain, even before the deduction has been made. If the observer does not already know subjectively what it is like to experience pain, then the word "pain" would be either a label devoid of meaning or just a shorthand notation for the observed brain electric impulses. Redefinition of the term "observation" cannot show that experiences are observable, it can only corrupt the meaning of the word "observation" to "utterance of word labels whose meaning we really do not understand." Thus, the *inner privacy* of conscious experiences is a problem that cannot be avoided and needs to be properly addressed by any physical theory. Insisting that "the universe contains only observable entities" is either false in the usual meaning of the term "observation" or meaningless if one attaches by definition the label "observable" to all existing things regardless of whether they can really be observed or not.

The inner privacy of consciousness implies that in the real world, if we happen to look at ourselves in a mirror we will observe something different from what we are already experiencing (Fig. 1.6). If a part of our skull is surgically removed, we will see our brain in the mirror, not our mind. Thus, by taking a look at ourselves in a mirror we are able to learn new things that we cannot learn solely from an introspective analysis of our own conscious experiences. Through accumulated observations of other brains, we will discover that the brain has similar anatomical structure and organization across all individuals of the same animal species, providing evidence for inherited genetic information and common ancestry.

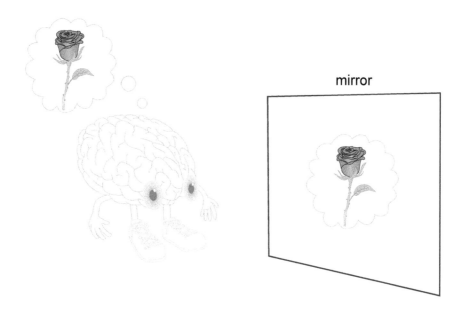

Figure 1.7   Illustration of an imaginary world in which conscious experiences happen to be observable from a third-person perspective. If we were experiencing the mental image of a red rose and we were able to take a look at ourselves in a mirror, we would see just the conscious experience of the red rose.

In an imaginary world in which conscious experiences happen to be observable, looking at ourselves in a mirror would not be remarkable for us since we already experience ourselves and know what we are experiencing (Fig. 1.7). What would be remarkable, however, is that we would have had direct access to the experiences of others by just taking a look at them. Such a world could be just one global mind, if it is a single connected network where everybody observes somebody and is observed by somebody.

The inner privacy problem is directly related to the fact that the subjective, first-person point of view of our conscious experiences is *not communicable* to others. If our conscious experiences were communicable to others through our communications, we would have been able to make a blind person see the world as we see it or a deaf person hear the world as we hear it. Unfortunately, we are unable to restore missing senses through communication alone. For example, we may communicate to a blind person that there is an obstacle in front of him, but this will not make the blind person see the obstacle. The blind person could understand the meaning of what an obstacle is due to the fact that already through another sense, such as touch, he has explored the world. The exploration of the world through touch allows one to form the abstract notion of *location* that does not depend on the exact sense through which a localized object is experienced. Thus, there may be some regularities within our conscious experiences that we may communicate to others. What we are unable to communicate to other people is the phenomenal nature of qualia, namely, what it is like to feel what we experience.

The inner privacy of consciousness is related to the binding problem presented in Section 1.2 through the fact that neither communication nor observation lead to binding of conscious experiences. This is evident from all kinds of everyday life scenarios where you communicate with other people and observe them without actually becoming a single social mind and experiencing what others do experience. Thus, conscious experiences are incommunicable and unobservable, but because the consciousness is endowed with free will, it is possible to observe the conscious choices that were already made, and communicate the choices that could be made or could have been made. In Section 1.4, we have listed some of the physical laws that a physical theory of agents endowed with free will could have. Here, we further characterize the kinds of physical laws that may be present in the physical theory of a subjective, private and unobservable consciousness, the nature of whose conscious experiences is incommunicable.

(1) There could be physical laws that postulate the existence of unobservable conscious minds. Thus, to each conscious mind could be attached a *label* such as ψ and questions related to that mind could be meaningfully asked. Different conscious minds and different conscious experiences could have different labels.

(2) There could be no psychophysical laws that explain what exactly it is like to feel the conscious experiences of a mind labeled as ψ. Indeed, if it were possible to describe what the subjective, first-person point of view of conscious experiences is for the purposes of a physical law, then since the physical laws in a theory are communicable, conscious experiences would have been communicable too.

(3) Because different conscious states within the physical theory cannot be theoretically differentiated by the incommunicable subjective nature of their content, they could be characterized and differentiated from one another by something communicable such as the probability distributions for different possible courses of action from which the conscious mind is able to choose. Because within deterministic theories the probability distributions are reduced to a single course of action, it follows that indeterministic theories of consciousness are much richer and have the capacity to differentiate between a larger number of conscious states.

1.6   The mind–brain relationship

Our conscious experiences are real and there is no doubt that they do exist [127]. However, our mental life is not publicly observable by others and each of us individually is the only agent that has privileged introspective access to his or her own *mind*. In contrast, our brains are publicly observable but because we do not have direct introspective access to the anatomical or physiological description of our brain, we may doubt whether we have a brain at all. For example, if we were raised in isolation or not educated, we could have lived perfectly well without knowing that we have a brain. Thus, there is a need to explain what exactly is the physical relationship between the mind and the brain: Do the mind and the brain interact with each other? Does one generate or cause the other? Or, are the mind and the brain just two words that refer to the same physical thing in reality?

Example 1.5. *(Descartes' argument for mind–brain distinction) Descartes tried to prove that the mind is not the brain. As a start, he decided to doubt everything and then search for anything that could be true. His reasoning went as follows: The fact that I am doubting is impervious to doubt, because were I to doubt that I was doubting, I would still be doubting. The fact that I am thinking is also something that I cannot be deceived about. Were something to deceive me into merely thinking that I am thinking, I would still be thinking. Thus, my doubting and thinking imply that I exist and have a mind. Therefore, I cannot doubt that I have a mind. The same cannot be said for the brain. Perhaps there is an evil demon that is trying to deceive me by feeding false information directly to my mind. Perhaps I have an ethereal mind resembling a cloud of light rather than a brain. Therefore, I can doubt that I have a brain. At the end, Descartes concluded that the mind and the brain should be two different things because if they were identical everything true of one must be true of the other, which apparently is not the case as he can doubt that he has a brain but cannot doubt that he has a mind [224].*

Theoretical terms in physics are often used to signify existing physical objects of which we may have only an incomplete knowledge. Consequently, the theoretical concepts in a scientific theory should not be confused with the corresponding existing physical realities. Namely, the map is not the territory [299, p. 58]. Similarly, the word "mind" is not the existing mind. When we say that we do not doubt the existence of our minds, we do not mean that we do not doubt the existence of the word "mind." Instead, we mean that because we know by definition what the word "mind" signifies in reality, we do not doubt that the expression "my mind exists" is true within the framework of the scientific theory. Thus, the scientifically relevant question is whether the two words "mind" and "brain" signify the same physical reality or not. Scientific theories are communicable and contain only a third-person perspective of existing things. Because when we look inside the skull of other people we do not see their minds but we see their brains, we may hypothesize that the brain is what the mind looks like from a third-person perspective. Hence, even though the "mind" and the "brain" are not the same as words, they could refer to the same thing in reality and be equivalent within the scientific theory. Because doubting whether two words refer to the same thing in reality has no bearing on the actual relation between the two words, Descartes' argument is deficient and does not show that the mind is different from the brain.

Example 1.6. *(Doubting an identity does not disprove it) Let us define the value of x with the integral equation $x = \int_0^\infty \frac{1}{t^2+1}\,dt$. We are absolutely certain of what x is in terms of the integral equation because we have defined it so. (Similarly, we are absolutely certain that we are conscious minds because the word "mind" has been defined in terms of our conscious experiences.) Suppose now that we are given the numbers π and $\frac{\pi}{2}$, and we are asked whether x is the same as π or $\frac{\pi}{2}$. (Such questions would be equivalent to asking whether the "mind" is the same as the "brain.") Without having knowledge of how to solve integral equations we may doubt whether $x = \pi$ or $x = \frac{\pi}{2}$. Our doubts will not change the truth-values of the latter two equations. Indeed, $x = \pi$ is false, whereas $x = \frac{\pi}{2}$ is true. (Hence, our doubting of the mind–brain identity cannot change the truth-value of the identity; it is either true or false even though we may not know the answer.)*

While we may have no a priori reason to believe that the mind is different from the brain, modern medicine has provided us with overwhelming experimental evidence that the human brain *is not* the human mind.

Example 1.7. *(Dead brain) The existence of* dead brains *that do not produce conscious experiences implies that dead brains do not have minds. After the death of a person, the brain remains in existence and can be studied using pathoanatomical analysis. If the brain were the mind, it would have been impossible to* separate *the mind from the brain.*

Example 1.8. *(Anesthetized brain) The possibility of general anesthesia together with the existence of* anesthetized brains *implies that the brain can exist in a state in which human consciousness is temporarily turned off. If the brain were the mind, it would have been impossible to turn off the mind without annihilating the brain out of existence. Thus, while general anesthesia provides strong evidence for the claim that the brain is not the mind, the two related questions of how general anesthesia works and how the mind relates to the brain need to be answered.*

1.7   The hard problem

At the microscopic level, the brain is composed of nerve cells called *neurons* (Fig. 1.8). Individual neurons within the brain are interconnected through synaptic contacts and assembled into neural networks [285]. Each neuron has all of the organelles possessed by eukaryotic cells and is able to perform all the basic biochemical processes that lead to synthesis of biomolecules, secretion, chemical signaling, energy production in the form of adenosine triphosphate (ATP), etc. All neurons are also highly specialized in order to input, process, and output information in the form of electric signals in the neural network. The neuronal specializations are both structural and functional.

Structurally, neurons have a cell body called *soma*, and numerous cable-like projections collectively called *neurites* that could extend from millimeters up to tens of centimeters away from the neuronal soma. Neurites that input electric information toward the soma are called *dendrites* [255, 408, 409]. Dendrites of pyramidal neurons inside the brain cortex could be further classified into *apical dendrites* that project upwards toward the first cortical layer, and *basal dendrites* that project laterally from the soma (Fig. 1.8). Neurites that output electric information from the soma toward other target cells are called *axons* [408, 409, 507]. The axons of cortical pyramidal neurons typically project downwards from the soma toward the subcortical white matter [343, 344] but there are also lateral axonal branches that innervate the nearby gray matter of the brain cortex [59]. Neurons are dynamically polarized cells since within each neuron the transmission of information is from the dendrites through the soma toward the axon. Neurons transmit information between each other at special places of contact called *synapses* [191].

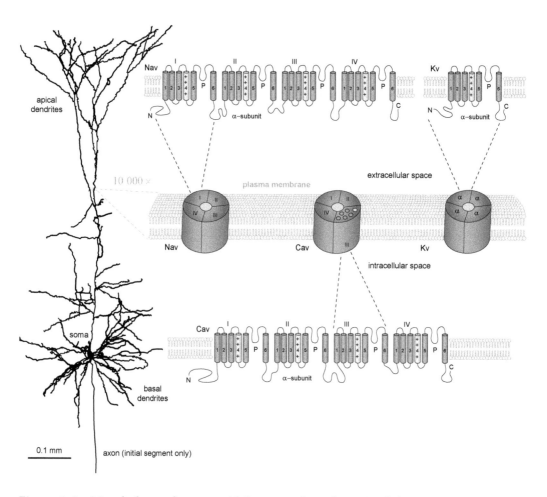

Figure 1.8 Morphology of a pyramidal neuron from layer 5 of the motor cortex in rat (NeuroMorpho.org NMO_09566) and common structure of voltage-gated ion channels. Apical and basal dendrites receive synaptic inputs in the form of excitatory or inhibitory electric currents that summate spatially and temporally at the soma. If the transmembrane voltage at the axon initial segment reaches a certain threshold of depolarization around −55 mV, the neuron fires an action potential (spike) that propagates along the axon down to terminal axonal arborizations that form synapses mainly onto the dendrites (and rarely onto soma or axon) of target neurons. Neuronal electric properties are due to opening and closing of sodium (Nav), potassium (Kv) and calcium (Cav) voltage-gated ion channels. Structurally, each channel is built of four protein domains I–IV, each of which contains six transmembrane α-helices (numbered from 1 to 6). The channel pore is formed by protein loops (P) located between the 5th and 6th α-helices, whereas the voltage sensing is performed by the 4th electrically charged α-helix within each domain. The 0.1 millimeter scale bar applies only to the neuron reconstruction in the left panel. Since the thickness of the plasma membrane is only 10 nanometers, the right panel with the ion channels has an additional 10,000 × magnification.

Functionally, neurons have excitable plasma membranes due to the opening or closing of excitatory or inhibitory ion channels incorporated in the plasma membrane (Fig. 1.8). Physiologically most important are three groups of *voltage-gated ion channels*: sodium (Nav), potassium (Kv) and calcium (Cav) ion channels, respectively conductive for Na^+, K^+ or Ca^{2+} ions [223, 315]. All three families of ion channels share a common evolutionary conserved structure (Fig. 1.8). The pore of the ion channel is formed by a protein α-subunit. The α-subunit of sodium (Nav) and calcium (Cav) channels is composed of four protein domains I–IV, each of which contains six transmembrane α-helices. There is a minor difference in the structure of the potassium (Kv) channels in which the protein domains I–IV are disconnected from each other, giving rise to four α-subunits instead of a single one (Fig. 1.8). The channel pore is formed by four protein loops (P) located between the 5th and the 6th α-helices of the protein domains I–IV. The voltage sensing is performed by a charged 4th α-helix of each domain [49, 50, 159].

Macroscopic electric currents produced by voltage-gated ion channels flow across the neuronal plasma membrane depending on the *transmembrane voltage* and the density of the ion channels [167, 282, 331, 492]. As a result, the transmembrane voltage of neurons undergoes dynamical changes in time. For most of the time, the neuron stays at rest and the transmembrane voltage in the soma and axon does not exceed a threshold value of −55 mV [34, 258]. At certain instances in time, however, the transmembrane voltage reaches the threshold value of −55 mV and the neuron fires a brief *electric spike* called *action potential* that propagates down the axon in order to activate synapses innervating other target neurons [413]. With the use of electric spikes propagating within the neural network, the brain is able to perform a large number of computational tasks such as acquiring knowledge, learning, memorization, memory recall, problem solving, prediction, optimization, class identification, categorization, pattern recognition, and error correction. Unfortunately, the electrophysiological processes occurring in the nerve cells of the human brain do not always generate consciousness, as shown in general anesthesia [272, 307, 308].

Computational neuroscience has been successful in regard to modeling and explaining most of the comparatively straightforward problems concerning input, processing, storage, and output of classical information [118, 200, 413]. Even though these achievements may seem quite impressive, the philosopher David Chalmers refers to such computational problems as the *easy problems of consciousness* [83, 84]. For him, the really *hard problem of consciousness* is to explain why neurons in the brain generate any conscious experiences at all, and why it is the case that the brain does not operate in a *mindless brain* mode (Fig. 1.9) where the neurons perform the computational functions that they are supposed to do but without generation of conscious experiences [83, 84, 85, 293].

It would have been easy to just say that consciousness is what brains do, but there are multiple counterexamples showing that a human brain could temporarily lose consciousness. Loss of consciousness could occur due to a sudden drop of the blood glucose level, inhalation of anesthetic gases, or mechanical brain trauma. Dead brains fixed with paraformaldehyde and stored in a glass jar do not give rise

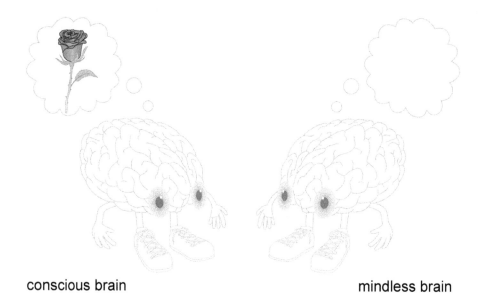

conscious brain   mindless brain

Figure 1.9   The hard problem of consciousness is to explain why the brain generates conscious experiences and why it is not possible for the brain to do what it does but in a mindless brain mode that lacks any conscious experience whatsoever.

to conscious experiences too. Therefore, in general the brain states are not identical with states of conscious experience. Since the mind is not the brain, in order to solve the hard problem one needs to have a precise physical law that correctly predicts which brain states give rise to conscious experiences and which brain states do not. The same physical law should be able to also inform us whether minds can exist within other minds and whether conscious experiences need to be produced by brains only or they could also occur in machines.

In essence, the hard problem of consciousness is to explain why qualia exist, not what the phenomenal nature of qualia is. In Section 1.5, we have argued that the phenomenal nature of qualia is incommunicable. For example, you need to be a bat in order to experience what it is like to see the world through a bat's sonar system. Because in the scientific description of the world enters only what is communicable, science cannot tell you what it is like to be a bat. There is no reason to think, however, that science is incapable of solving the hard problem too. Explaining why the bat has any experience when operating with its sonar is something that can be communicated if it can be summarized in a physical law specifying the conditions under which conscious experiences arise. Then, the physical law together with a given particular initial condition could be used to deduce whether any conscious experience at all will occur or not.

Part II

Some background for beginners

CHAPTER 2

The scientific conception of the world

When we are born, we have no memories and know nothing about ourselves or the surrounding world. This state of ignorance, however, does not persist for long. As we grow, we explore the world and learn facts about existing physical objects including other people and ourselves. We also learn a language in order to express ourselves and to communicate with others. And, sooner or later, having collected a certain amount of knowledge, we start asking questions whose answers may not have immediate practical applications, but are needed for the sake of satisfying our own human curiosity. The most important of these questions are those that concern ourselves in relation to the world we live in: Where are we? Who are we? What are we? Why do we exist? What is the purpose of our existence?

Through introspection, we know that we exist and that we have experiences. We, as *conscious minds*, are composed of *experiences*. Our experiences are the most intimate aspects of our lives and our only way to access the surrounding world. We learn *how things are* by accumulating memories about our past experiences. We are able to learn new things about ourselves and the rest of the world because of two facts: we have direct conscious control over our bodies and we have only indirect control over the rest of the world. Indeed, if we were unable to directly control our bodies or if we were directly controlling the whole universe, it would have been impossible to learn anything new. Henri Poincaré (1854–1912) was the first to appreciate the importance of trial and error for learning new things: We try to do something, it fails to be done due to existing physical laws, and from the error we learn that there are some objective reasons that prevent us from achieving what we wanted to achieve. Thus, if we were unable to perform any actions or if we were controlling the whole universe, there would have been no errors and no learning [389, p. 59].

From our experiences of how things are, we are able to conceive abstract concepts which go beyond appearances. Thus, we become capable of contemplating questions not of how things are, but *what* things are and *why* they are. Fortunately, we do not have to learn everything from newborns to adults in complete isolation. Human society has made it possible for each individual to learn from the accumulated knowledge of previous generations. For example, humans no longer aimlessly explore the world by trial and error until accidentally rediscovering fire, but learn from others what fire is, how to make it, and how to use it [48, 277]. Acquiring knowledge and learning from others gives us a chance to use our limited lifetime in a more efficient way, exploring new horizons and discovering new things that were inaccessible to our ancestors.

2.1   Subjective and objective knowledge

We often say that we *know* what the sensations of joy or pain are. This kind of knowledge, however, is subjective and incommunicable because you must be or have been in a given state of experience in order to know what that experience is.

Definition 2.1. *(Subjective knowledge)* Subjective knowledge *refers to our conscious experiences including our conscious recollections of past experiences.*

We also say that we *know* what trees or birds are. This kind of knowledge is objective and communicable because we can provide others with explicit examples of what a tree or a bird is. Once the knowledge can be shared, it becomes criticizable [396].

Definition 2.2. *(Objective knowledge)* Objective knowledge *refers to information that can be shared with others.*

Accumulation and sharing of knowledge between people is probably the single most important factor that makes humans what they are. With the current unprecedented access to computers and other digital devices, students are able to obtain knowledge not only in schools or libraries, but also on the internet, where an abundance of free information is available. Online discussion forums also facilitate the exchange and critical evaluation of new ideas. We all live in the same *physical world* and we are part of it. We can share knowledge about the world because there are immutable *physical laws* that govern the world. And because the physical world is one, regardless of how we divide our knowledge into scientific subjects, ultimately all knowledge, including the *theory of consciousness*, should be unified and derivable from *physics* alone (cf. [339, 355, 470]).

Since our minds exist in the universe, they should be described by physics and governed by physical laws [207]. Unfortunately, even though physics is supposed to study the universe and all existing things in it, current physical theories have nothing to say on consciousness [381]. Two historical reasons contribute to this state of affairs. First, philosophers of mind have persistently used the terms "physical" for the brain and "non-physical" for the mind [240, 274, 293, 318, 348], without recognizing the paradox in defining physics as the natural science studying the universe and then branding all conscious minds in the universe as "non-physical" as if they were outside the scope of physics. Second, physicists have habitually shied away from discussing consciousness and the difficult problems associated with it. While it is true that classical physics is poorly suited to accommodate consciousness, as we will show in Chapter 5, this does not imply that quantum physics should also fail in providing a physical theory of consciousness. On the contrary, in Chapter 6 we will demonstrate how quantum information theory could be used to address the main problems of consciousness presented in Chapter 1.

Before we proceed further, it would be useful to discuss in some detail what is *science* and what is a *scientific theory*.

2.2   Science and scientific theories

Science is a systematically organized body of *objective knowledge* about the *world* we live in. To qualify science as a kind of objective knowledge, two important requirements should be met. First, any knowledge should be *logically consistent* (Section 2.2.1). Second, objective knowledge should be *communicable* to others (Section 2.2.2). Thus, by definition, something that is not logically consistent or is not communicable to other people cannot be classified as objective knowledge or science. Providing a consistent communicable description of the world we live in, however, is not sufficient to define science. Science should also be able to *answer* all questions that we may have about ourselves or the surrounding world and *explain why* the world we live in appears to be the way it is (Section 2.2.3). Furthermore, in order to be credible, the scientific explanation needs to be *empirically corroborated* (Section 2.2.4).

2.2.1   Logical consistency

Scientific theories should produce statements about the world that are either *true* or *false*, but never *both*. An inconsistent theory is one that is able to simultaneously prove both some specific statement P and the negation of that same statement $\neg P$. The conjunction of a statement with its own negation, $P \wedge \neg P$, is referred to as *logical contradiction*. Therefore, the inconsistency of a theory implies the existence of a contradiction.

In Section 2.3, we shall see that every inconsistency is explosive: From a contradiction follows everything [129, 265, 401]. To be able to prove everything, however, is equivalent to knowing nothing. A theory that contains a contradiction is able to prove everything and as a consequence provides no knowledge. Contradictory theories could be viewed as meaningless, vacuous, self-defeating or nonsense. The requirement for logical consistency demarcates science from other products of the human mind, including art, poetry and religion.

Example 2.1. *(Knowledge is prohibition) Every knowledge has to be consistent. Because logical consistency is prohibitive, knowledge is prohibitive too. To know that some specific statement P is true means that it definitely cannot be false, $P \rightarrow \neg(\neg P)$, and to know that the statement P is false means that it definitely cannot be true, $\neg P \rightarrow \neg(P)$.*

The Greek philosopher Aristotle (384–322 BC) was the first to explicitly postulate the need for logical consistency as a universal principle that requires no proof [14, 1005b].

Definition 2.3. *(Axiom) An axiom is a true statement, first principle, postulate or a law that is accepted as true without proof. The word comes from the Greek axioma meaning "that which is thought worthy or fit."*

Definition 2.4. *(Theorem) A theorem is a statement that has been proven with the use of logical reasoning from a set of axioms. If the proof of the theorem is correct, it is guaranteed that the theorem is true given that all axioms within the set are true.*

Theorem 1. *(Aristotle's theorem) Scientific theories should be formulated, provided or recast in the form of a list of axioms because to provide a proof of every statement is impossible* [14, 997a].

Proof. Because any proof consists of statements, S_i, if one requires every statement to be proved, there will be only two kinds of such proofs: an *infinitely long proof* in which every statement is proved by another statement that has not been used in the proof before

$$\ldots \to S_6 \to S_5 \to S_4 \to S_3 \to S_2 \to S_1 \tag{2.1}$$

or a finitely long *circular proof* in which a statement is proved based on premises that are subsequently proved using the very statement that had to be proved

$$S_1 \to S_n \to S_{n-1} \to \ldots \to S_3 \to S_2 \to S_1 \tag{2.2}$$

Neither of these two kinds of proofs is valid. An infinitely long proof composed of an ever growing list of novel statements, also known as an *infinite regress*, cannot be actually completed, written down and communicated to others. A circular proof, also known as a *vicious circle*, is a kind of cheating, presenting as a proof something that is not. Therefore, any proof should be finite in length and based on a list of axioms that are true statements accepted as true without further proof. □

Since we cannot prove every statement, we can accept a small number of statements as axioms. Axioms are accepted as true without proof, but we do not cheat about them and do not pretend that we can prove them. Instead, we explicitly label the axioms as such [128, 164, 249, 250, 388]. Whether the axioms are good or bad, we judge from their results, which are the theorems that we can prove from the axioms. If we can prove a contradiction or derive a prediction not supported by experimental observations, then the axioms are not good and we need to revisit and replace one or more of them with new, hopefully better axioms.

For Aristotle, the principle of non-contradiction is the most certain of all principles, because it is impossible for anyone to believe the same thing to be and not to be [14, 1005b]. If the statements "X is Y" and "X is not Y" are indistinguishable, then no argumentation is possible [265]. Aristotle's argument could be stated as follows: if X is a word, and Y is its meaning, one may think of all meanings $\{A, B, C, D, \ldots\}$ that are not Y. Together with Y, one obtains all possible meanings $\{Y, A, B, C, D, \ldots\}$. Thus, from "$X$ is Y" and "X is not Y" follows that the word X has all possible meanings, which is the same as having no meaning. If X can stand for any word, then all words will be meaningless. If words have no meaning, however, our reasoning with one another or with ourselves would be annihilated and it would be impossible to think [14, 1006b]. So, without the principle of non-contradiction we could not know anything that we do know [221].

The German mathematician Gottfried Wilhelm Leibniz (1646–1716) argued in a similar fashion: The principle of non-contradiction is needed, since otherwise there would be no difference between truth and falsehood, and all investigation would cease at once, if to say yes or no were a matter of indifference [312, p. 14].

If not convinced by logical arguments alone, the skeptics of the principle of non-contradiction could be confronted with extreme experimental tests proposed by the Arab scholar Ibn Sina (980–1037): He lets the skeptics be plunged into fire, since fire and non-fire are identical; beaten, since suffering and not suffering are the same; and deprived of food and drink, since eating and drinking are identical to abstaining [265].

Noteworthy, with the use of the axioms of logic presented in Section 2.3, the principle of non-contradiction becomes provable as a theorem [519, p. 111]. The proof starts from Axiom 2.3.4, and goes

$$\neg P \vee P \rightarrow \neg P \vee \neg\neg P \rightarrow \neg(P) \vee \neg(\neg P) \rightarrow \neg(P \wedge \neg P) \tag{2.3}$$

2.2.2 Communicability

Scientific theories, being a kind of objective knowledge, should be communicable to others. For a theory to be communicable means that it could be recorded, copied, transmitted to and understood by others. Historically, oral transmission of knowledge precedes the development of writing. Nevertheless, communicability implies that one could in principle record what is going to be communicated. Since the record itself is objective, one may say that science is objective.

Every scientific record can be read, copied, multiplied, or, if necessary, erased. Communicable knowledge is *classical information* and as such can be quantified in the form of classical bits [435, 436].

Definition 2.5. *(Bit) A classical bit is the basic unit of information contained in the answer of a single yes-or-no question.*

Information stored on optical media such as compact disks is in the form of a string of classical bits. Each bit may have one of two values, typically labeled as 0 and 1. Letters in an alphabet contain much more information than a single bit. Assuming that all letters have an equal information content, for the amount of information I contained by a single letter of an alphabet containing N letters, we have

$$I = \log_2 N \tag{2.4}$$

In the Icelandic alphabet, there are 32 letters [253], which means that a single Icelandic letter contains $\log_2 32 = \log_2 2^5 = 5$ bits of information.

Example 2.2. *(Quantifying classical information in bits) Suppose that we are asked to find how many bits of information are contained by a single letter A of an alphabet composed of eight letters $\{A, B, C, D, E, F, G, H\}$. To solve the problem, we need to evaluate how many yes-or-no questions are needed in order to find out what the single letter is if it is drawn by chance from the set of eight letters. From Eq. (2.4) we can calculate three yes-or-no questions. The following algorithm explains why this is indeed the correct answer. First, we divide all letters of the alphabet into two equally large sets with four letters each and ask whether the letter is in the first set. If the answer is yes,* *we keep the first set and if no, we keep the second set. Next, we divide the remaining letters into two equally large sets with two letters each. The second question asks again whether the letter is in the first set. If the answer is yes, we keep the first set and if no, we keep the second set. Lastly, we divide the remaining letters into two equally large sets with one letter each. A final third question asks again whether the letter is in the first set and from the answer we deduce with certainty what the single letter is. Noteworthy, the quantity of classical information that we learn from a single letter A does not come from A itself, but from the knowledge that the letter is not any of the remaining letters $\{B, C, D, E, F, G, H\}$. If the letter A were a part of a shorter alphabet composed of only four letters $\{A, B, C, D\}$, the letter A would have delivered only 2 bits of information instead of 3 bits.*

Since scientists do not have direct access to each other's minds, they need to explicitly list all axioms of their scientific theories in order to ensure proper communication. Without knowing all of the axioms of a scientific theory, the theory becomes incommunicable and uncriticizable. Thus, communicability of scientific theories is only possible when each theory is explicitly formalized into a list of axioms. Once all of the axioms are known, one is able to derive theorems, thereby understanding what these axioms really say. If the axioms lead to a contradiction, then the theory is flawed and useless.

2.2.3   Explanatory power

Collecting observational data about the world we live in is an important scientific activity. However, a list of collected data is not itself a good *scientific theory*. We may know that certain statements about the world are true, that other statements are false, but still we may not have any idea *why* this happens to be so. The scientific theory is expected to *explain* why the world appears to be the way it is. *Explanations* are logically valid inferences of observed facts from a *theory* that includes a set of axioms and rules of inference. Good explanations are those in which the complexity of the theory does not exceed the complexity of the observed facts.

Explanations need to be logically valid. Every logically valid chain of reasoning is *deductive* in nature and proves something. To explain an experimental observation or a natural phenomenon means to *prove* that the occurrence of the observation or the phenomenon *was to be expected* given certain premises. The given premises include one or more *physical laws*, L_1, L_2, ..., L_k, and a set of *initial conditions*, C_1, C_2, ..., C_n. If we refer to what is to be explained as *explanandum* and to the set of premises that do the explaining as *explanans*, we may say that the scientific explanation is adequate if its explanans could have served as a basis for predicting the explanandum [248, p. 138]. To be able to *explain* means to be able to *predict* the occurrence of the phenomenon under consideration. In mathematical notation, we write

$$L_1, L_2, \ldots, L_{k-1}, L_k, C_1, C_2, \ldots, C_n \vdash E \tag{2.5}$$

where the symbol $\vdash$ means *prove* and E is the description of the phenomenon that needs to be explained. Thus, *explanations* are *proofs*.

The set of initial conditions, C_1, C_2, ..., C_n, characterizes the particular circumstances in which the experimental observations are performed. Changing the initial conditions changes the nature of the experiment. Thus, it is meaningless to ask why the initial conditions are the way they are. Instead, the set of initial conditions, C_1, C_2, ..., C_n, should be viewed as an *input* that is transformed according to the set of *physical laws*, L_1, L_2, ..., L_k, into an *output*, E.

The *physical laws*, L_1, L_2, ..., L_k, are the core of the scientific theory. To search for an explanation why a certain physical law L_k holds is not necessarily meaningless. In some cases, we may discover a simpler law L_k' that explains the law L_k. Since $L_k' \vdash L_k$, we can provide a better explanation of E

$$L_1, L_2, ..., L_{k-1}, L_k', C_1, C_2, ..., C_n \vdash E \tag{2.6}$$

It is also possible, however, that there is no deeper reason why the physical law L_k holds. Indeed, Aristotle's theorem shows that it is impossible to prove, and therefore explain, every statement. As a consequence some statements should be accepted as axioms without further explanation. If L_k happens to be true for no other simpler reason, we refer to it as a *fundamental physical law*. Fundamental physical laws simply *are* the way they are. Discovering the fundamental physical laws is the ultimate goal of science.

Most of the physical laws are *universal statements* that contain the *universal quantifier* $\forall$ meaning *for all*. A universal statement is "All swans are white." If X is the set of swans, the symbol $\in$ denotes set membership, and $P(x)$ is a propositional function denoting the property of being white, symbolically we write

$$\forall x \in X \, P(x) \tag{2.7}$$

meaning "For all x in the set X the property $P(x)$ is true." Universal statements could be alternatively rewritten with the use of the negation $\neg$ of the *existential quantifier* $\exists$ meaning *exists*. Thus, $\forall x \in X \, P(x)$ is equivalent to

$$\neg \exists x \in X \, \neg P(x) \tag{2.8}$$

meaning "Does not exist x in the set X for which the property non-$P(x)$ is true." In the swan example, the statement becomes "No swan is non-white."

The negation of a propositional function's existential quantification is a universal quantification of that propositional function's negation

$$\neg \exists x \in X \, P(x) \leftrightarrow \forall x \in X \, \neg P(x) \tag{2.9}$$

Since universal statements can be rewritten using only the existential quantifier, and vice versa, it is possible to dispense with one of the two quantifiers and still express logical statements just as accurately. To preserve human readability, however, we will keep both quantifiers.

Universal statements cannot be proved from a finite number of singular statements alone. That is why universal statements cannot be *verified* experimentally, but they can be *falsified*. For example, observing a large number of white swans does not verify the statement "All swans are white," but observing even a single black swan falsifies it. In Chapter 8, we will show that falsifiability could be used to assess competing theories but only if the scientific theories are taken as a whole and not broken down into individual statements.

Some physical laws are *singular existential statements*. A singular existential statement is "Exists a black swan." If X is the set of swans and $Q(x)$ is a propositional function denoting the property of being black, symbolically we write

$$\exists x \in X \, Q(x) \tag{2.10}$$

meaning "There is at least one x in the set X for which the property $Q(x)$ is true." Singular existential statements could be alternatively rewritten using negation of the universal quantifier $\forall$. Thus, $\exists x \in X \, Q(x)$ is equivalent to

$$\neg \forall x \in X \, \neg Q(x) \tag{2.11}$$

meaning "Not for all x in X the property non-$Q(x)$ is true." In other words, "Not all swans are non-black." With the use of the two identities $\neg\neg x \leftrightarrow x$ and $(x \leftrightarrow y) \leftrightarrow (\neg x \leftrightarrow \neg y)$, Eq. (2.9) indeed implies that the negation of a propositional function's universal quantification is an existential quantification of that propositional function's negation

$$\neg\neg \forall x \in X \, Q(x) \quad \leftrightarrow \quad \neg \exists x \in X \, \neg Q(x) \tag{2.12}$$
$$\neg \forall x \in X \, Q(x) \quad \leftrightarrow \quad \exists x \in X \, \neg Q(x) \tag{2.13}$$

Noteworthy, singular existential statements cannot be *falsified* by experiment, but they can be *verified*. For example, observing a large number of white swans cannot falsify the statement "Exists a black swan," whereas the observation of a single black swan verifies it.

Universal statements have a greater explanatory power compared to singular existential statements because from a universal statement could always be proved a singular existential statement

$$\forall x \in X \, P(x) \vdash \exists x \in X \, P(x) \tag{2.14}$$

meaning that "For all x in the set X the property $P(x)$ is true" proves "Exists at least one x in the set X for which the property $P(x)$ is true." Actually, a universal statement is able to prove an infinite number of singular statements. From the statement "All swans are white" follows "This swan is white," "That swan is white," "That swan is white, too," etc.

Mathematically proving things from a few axioms is less risky than assuming many experimentally suggested assertions. Furthermore, proving things from a list with fewer axioms is less risky than proving things from a list with a larger number of axioms. Because each axiom can be either correctly guessed or not [81], a theory T with n axioms has a *prior probability* $p(T)$ of being correct given by

$$p(T) = \left(\frac{1}{2}\right)^n = 2^{-n} \tag{2.15}$$

Since a long list of axioms could always be joined into a single axiom using logical *and* conjunction operators in the form "Axiom 1 *and* Axiom 2 *and* Axiom 3 *and* ...," the number of axioms in a theory T is to be taken as the number of independent claims postulated to be true.

The *complexity* of the axioms, in addition to their number, is also important when discussing the explanatory power of a scientific theory. A theory is of value only to the extent that it compresses a great many bits of data into a much smaller number of bits of theory [80]. Thus, for quantifying the explanatory power of a scientific theory we can use the *compression ratio* defined as the ratio between the uncompressed size of the data and the compressed size of the theory. The theory is useless if it is more complicated than the facts that it is trying to explain, or equivalently, if the compression ratio achieved is ≤ 1.

According to the information-theoretic point of view advocated by the mathematician Gregory Chaitin [81], every scientific theory is equivalent to a computer program that can be used to calculate the outcome of observations. The smaller the computer program the better the scientific theory is. Thus, science is data compression. If a set of collected observational data cannot be calculated by any computer program that is smaller than the data itself, then the data is algorithmically random, structureless, incomprehensible, irreducible and not subject to explanation by a scientific theory.

Example 2.3. *(Compression of classical information) Suppose that we are asked to find a way to explain the occurrence of the data sequence*

$$4, 6, 8, 12, 14, 18, 20, 24, \ldots$$

Explaining the data requires finding a pattern that neatly encompasses all the numbers together. If the numbers are randomly selected, finding a pattern may be impossible and there would be no way to compress the data. In the given data sequence, however, compression seems to be possible. One way to compress the data is "Take all even numbers greater than 2 with the exception of 10, 16 and 22." This description indeed reproduces the given numbers, but it does not seem to be very good since it still takes a lot of classical bits to explain the data. Moreover, adding exceptions ad hoc is rarely considered enlightening. Instead, a better data compression is "Take all odd primes plus 1." Since the latter compression of data is greater, it provides a better explanation.

Definition 2.6. *(Bad explanation) A scientific theory that is more complex or as complex as the data it is supposed to explain is a* bad *theory and the explanation provided by the bad theory is a* bad *explanation.*

Definition 2.7. *(Good explanation) A scientific theory that is less complex than the data it is supposed to explain is a* good *theory and the explanation provided by the good theory is a* good *explanation.*

Computation is equivalent to a mathematical proof. The operation of every *computer* is based on a set of *rules of inference*. Any *computer program* running on a computer is equivalent to a set of *axioms* written in a language understandable by the computer. The *computer outputs* that are produced while running the program can be viewed as *theorems* proved with the use of the given axioms. Thus, formal axiomatic systems could be thought of as computer programs [79, p. 409].

Thinking about formal scientific theories in terms of computer programs and data compression highlights the importance of universal statements for expressing fundamental physical laws, because universal statements are able to compress an infinite number of singular existential statements. Still, even though singular existential statements have lower explanatory potential than universal statements, physical laws that are in the form of singular existential statements are *indispensable* for the construction of scientific theories.

Physical laws that are singular existential statements include the laws specifying the existence and number of dimensions of the physical world, the existence and number of fundamental physical forces, the existence and number of fundamental physical particles, the existence of a maximal velocity for transfer of classical information that happens to be the speed of light in vacuum c, and so on. Furthermore, the fundament of mathematics itself is built upon the set theoretic *axiom of infinity (3.9.5)* stating that there exists a set with infinitely many members such as the natural numbers [442, p. 13]. Without the axiom of infinity, we would not be able to use irrational numbers such as π or $\sqrt{2}$ (Section 3.4), and much of the mathematics used for solving differential (Section 3.6) or integral (Section 3.8) equations would be incomprehensible.

2.2.4   Empirical corroboration

Successful scientific theories need to provide *predictions* that are confirmed by experiments. There is, however, an important distinction between two kinds of scientific prediction: the prediction of *events of a kind which is known* and the prediction of *new kinds of events that are previously unknown* [395, p. 117]. Scientific theories that predict correctly the outcome of performed experiments are *empirically adequate*. Only those scientific theories that predict correctly *novel* phenomenal regularities are *empirically corroborated*. Thus, an empirically adequate theory need not be also empirically corroborated, but an empirically corroborated theory has to be both empirically adequate and able to predict the outcome of novel physical experiments.

Since all scientific theories are originally built upon some experimental data, it is not surprising that a theory is able to reproduce the experimental data used in the construction of the theory [349, 350]. Indeed, even a false scientific theory may correctly predict, or rather *postdict*, the experimental data used for the construction of the theory. An example of such a false theory is the *geocentric model* created by Claudius Ptolemy (90–168) in which the Earth stands still at the orbital center of all celestial bodies [402]. In the Ptolemaic model, the apparent motions of the celestial bodies were accounted for by treating them as embedded in rotating spheres made of a transparent substance called quintessence. From the belief that the stars did not change their positions relative to one another, it was argued that they must be on the surface of a single outermost sphere containing *fixed stars*.

Thus, the Ptolemaic order of the rotating spheres with embedded celestial bodies from the Earth outward is Moon, Mercury, Venus, Sun, Mars, Jupiter, Saturn, fixed stars [10, 402]. Because astronomical observations have shown that sometimes the planets apparently slow down, stop and reverse their motion around the Earth, Ptolemy postulated that in addition to their circular motion around the Earth, the planets move along smaller circles called *epicycles*. Even though the geocentric model is *false*, it predicts correctly the apparent motion as observed from Earth of those celestial bodies whose data was used to construct the model.

It would be quite surprising, however, if a scientific theory designed to accommodate one phenomenal regularity (or a set of regularities) successfully predicts a quite different regularity (or a set of regularities) [349]. The English philosopher William Whewell (1794–1866) argued that no false supposition could, after being adjusted to one class of phenomena, exactly represent a different class where the agreement was unforeseen and uncontemplated [517, p. 65]. Thus, if one does not believe in extraordinary accidents or miracles, it would be rational to explain the successful novel predictions of the theory with the supposition that the theory is *true*, or that the theory *captures truthfully* at least some of the properties of the real world that are relevant for making the correct predictions.

Science is evolving because the amount of data about the world we live in is increasing. New data invariably makes us reassess and reconsider the scientific theories that we currently have. Scientific theories that are proved unsatisfactory are either modified or discarded. In Example 2.3, we have provided data in the form of a partial sequence of numbers and two different explanations of the data. Suppose that we perform a new experiment and find that the next number in the sequence is 30. This will empirically corroborate the shorter explanation "Take all odd primes plus 1" and will falsify the longer explanation "Take all even numbers greater than 2 with the exception of 10, 16 and 22." Alternatively, if we find that the next number in the sequence is 26, this will empirically corroborate the longer explanation and will falsify the shorter one since $26 - 1 = 25$ is not a prime number. Thus, collection of new empirical data is always beneficial for science. If our old theories are empirically corroborated by the new data, we find out that the old theories actually compress more data than we originally thought. On the other hand, if the new data falsifies the old theories, we find out that our understanding of the world is incomplete and novel explanations need to be found.

2.3   Axioms of logic

The axioms of logic could be used to determine the truth value of propositions that are formed by other propositions with the use of logical operators. The axioms are accepted as true without proof. The small number of axioms, however, does not represent *all* true statements. With the use of a couple of *rules of inference* (Section 2.4), from the axioms we are able to prove many more true statements as *theorems*. Since mathematical proofs are truth preserving, we can be sure that the theorems are true whenever the axioms are true.

Axiom 2.3.1. *The implication operator, $\rightarrow$, is defined by the following two axiom schemas*

$$F \rightarrow (G \rightarrow F) \tag{2.16}$$

$$(F \rightarrow (G \rightarrow H)) \rightarrow ((F \rightarrow G) \rightarrow (F \rightarrow H)) \tag{2.17}$$

Axiom 2.3.2. *The and operator, $\wedge$, is defined by the following three axiom schemas*

$$F \wedge G \rightarrow F \tag{2.18}$$

$$F \wedge G \rightarrow G \tag{2.19}$$

$$F \rightarrow (G \rightarrow (F \wedge G)) \tag{2.20}$$

Axiom 2.3.3. *The non-exclusive or operator, $\vee$, is defined by the following three axiom schemas*

$$F \rightarrow F \vee G \tag{2.21}$$

$$G \rightarrow F \vee G \tag{2.22}$$

$$(F \rightarrow G) \rightarrow ((H \rightarrow G) \rightarrow (F \vee H \rightarrow G)) \tag{2.23}$$

Axiom 2.3.4. *The not operator, $\neg$, is defined by the following three axiom schemas*
Refutation by deriving a contradiction

$$(F \rightarrow G) \rightarrow ((F \rightarrow \neg G) \rightarrow \neg F) \tag{2.24}$$

From contradiction follows everything

$$F \rightarrow (\neg F \rightarrow G) \tag{2.25}$$

The law of excluded middle

$$F \vee \neg F \tag{2.26}$$

Axiom 2.3.5. *The equality operator, $\leftrightarrow$, is defined by the following three axiom schemas*

$$(F \leftrightarrow G) \rightarrow (F \rightarrow G) \tag{2.27}$$

$$(F \leftrightarrow G) \rightarrow (G \rightarrow F) \tag{2.28}$$

$$(F \rightarrow G) \rightarrow ((G \rightarrow F) \rightarrow (F \leftrightarrow G)) \tag{2.29}$$

Axiom 2.3.6. *If $F(x)$ is any formula in which x is a free variable (not inside the scope of a quantifier) and y is a term such that substituting y for the free occurrences of x in $F(x)$ is admissible (in particular, y may be x itself), then*

$$\forall x F(x) \rightarrow F(y) \tag{2.30}$$

$$F(y) \rightarrow \exists x F(x) \tag{2.31}$$

Axiom 2.3.7. *If F is any formula, and G is a formula that does not contain x as a free variable, then*

$$\forall x (G \rightarrow F(x)) \rightarrow (G \rightarrow \forall x F(x)) \tag{2.32}$$

$$\forall x (F(x) \rightarrow G) \rightarrow (\exists x F(x) \rightarrow G) \tag{2.33}$$

2.4 Rules of inference

The rules of inference allow the derivation of theorems from the list of axioms. Only two rules of inference, namely, modus ponens and universal generalization, are sufficient to grant us full deductive power [129].

Modus ponens states that if both $F \rightarrow G$ and F are true, then G is also true

$$F \rightarrow G; F \vdash G \tag{2.34}$$

Example 2.4. *An argument that employs modus ponens*

> *If the tomato is red, then it is ripe.*
> *The tomato is red.*
> *Therefore, the tomato is ripe.*

Universal generalization states that if $F(y)$ is true for any y, namely, no assumptions have been imposed on which particular y is selected, then $\forall x F(x)$ is also true

$$F(y) \vdash \forall x F(x) \tag{2.35}$$

Example 2.5. *An argument that employs universal generalization*

> *Any bat is a mammal.*
> *Therefore, all bats are mammals.*

We will also add one extra rule of inference that can be derived from the list of logical axioms (Section 2.3) together with modus ponens. This rule will not add an extra deductive power, but will explicate the structure of a common type of argument that involves logical contraposition.

Modus tollens states that if both $F \rightarrow G$ and $\neg G$ are true, then $\neg F$ is also true

$$F \rightarrow G; \neg G \vdash \neg F \tag{2.36}$$

Modus tollens can be converted into modus ponens by replacing the conditional statement $F \rightarrow G$ with its logically equivalent contrapositive

$$(F \rightarrow G) \leftrightarrow (\neg G \rightarrow \neg F) \tag{2.37}$$

which results in

$$\neg G \rightarrow \neg F; \neg G \vdash \neg F \tag{2.38}$$

Example 2.6. *An argument that employs modus tollens*

> *If the tomato is red, then it is ripe.*
> *The tomato is not ripe.*
> *Therefore, the tomato is not red.*

2.5 Axioms of natural science

Logical axioms and rules of inference tell us how to reason correctly, but they can be applied to both existing and non-existing things in nature. For example, given the premises that all unicorns have a single horn and a given creature is a unicorn, we are able to conclude that the given creature has a single horn. The conclusion is perfectly logical but it says nothing on whether unicorns do actually exist in the real world. Since science is knowledge about the existing physical world, we need to introduce several basic axioms that define what exists.

Axiom 2.5.1. *There exists a real world. The totality of existence is the universe.*

Axiom 2.5.2. *I am a conscious mind that exists in the universe.*

Axiom 2.5.3. *I am not the whole universe.*

If there were no real world, it would have been impossible to answer what our scientific knowledge is about. Introspectively, each of us is able to verify his or her own existence as a conscious mind [127, p. 171]. Therefore, there must exist something rather than nothing. Also, since we are unable to directly control all existing things, we are able to conclude that there must be something else outside of us. Thus, the existence of ourselves and the surrounding world has to be postulated in a set of basic axioms that lay the foundations of all natural sciences. It needs to be stressed, however, that the existence of an objective world should be accepted without any preconceived expectations of what the *reality* of the world *should be* [390]. Rather, we have to explore the world and be sufficiently open-minded to accept the reality, whatever it may be.

CHAPTER 3

The world of classical physics

The world according to classical physics is built up from *matter, light, space* and *time*. The dynamics of the classical world resembles the working of a ticking clockwork mechanism whose future behavior is completely *determined* by the physical laws and the actual physical state of the universe at any single moment of time. In the classical world everything that exists is *observable*, and as a result, the existence of things can be verified using physical measurements with measuring devices.

Classical physics includes classical mechanics [326, 473], classical electrodynamics [536], and Einstein's special and general theories of relativity [155, 157]. In essence, the term *classical* stands for *non-quantum*. *Determinism* and complete *observability* are the two characteristic features of all classical physical theories. Determinism implies that the knowledge of the mathematical structure of the physical laws and the current state of a physical system is sufficient for calculating, at least in principle, the state of the system at any other future time. Complete observability implies that all physical *observables* describing the state of the physical system can be measured at the same time with arbitrarily high precision, and conversely, that if something is not observable, then it is not physical. These classical features may look pretty innocent, but lead to insurmountable paradoxes in the theory of consciousness, as we shall see in Chapter 5.

3.1 Matter

Matter is made of massive elementary particles such as the *proton, neutron* and *electron*. Each massive particle occupies a certain volume of *space* at a given moment of *time*. The classical size of these particles is of the order of a femtometer (10^{-15} m). Some of the massive particles may also carry an electric charge. In units of *elementary electric charge*, the proton has a positive charge of +1, the electron has a negative charge of −1, whereas the neutron has no electric charge. The particles are able to interact with each other with electromagnetic forces due to the particle electric charges, and with gravitational forces due to the particle masses.

Protons, neutrons and electrons are able to bind together, thereby forming *atoms* of the existing *chemical elements*. Inside atoms, the protons and the neutrons accumulate in the center, forming a positively charged *nucleus*, whereas the electrons orbit around the nucleus, forming negatively charged *electron shells*. Because the atoms as a whole are electrically neutral, it follows that within each atom the number of protons is equal to the number of electrons.

Different chemical elements have unique chemical names and chemical symbols. The *identity* of each chemical element is determined by its *atomic number*, which is the number of protons inside the nucleus [346, 347]. Atoms of the same

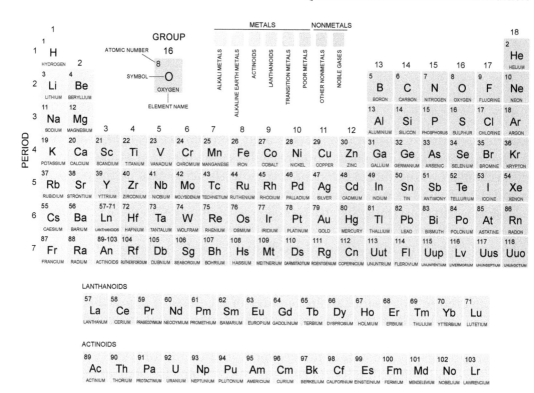

Figure 3.1   Periodic table of the chemical elements. Each element has its own chemical symbol and is identified by its atomic number. The rows of the table are called periods; the columns are called groups. Elements within the same group have the same configuration in their outermost electron shells and exhibit similar chemical properties. Group names are given according to the classification endorsed by the International Union of Pure and Applied Chemistry.

element, however, can have different numbers of neutrons, resulting in different relative atomic masses. Atoms of the same chemical element with different numbers of neutrons are called *isotopes*. For example, there are three hydrogen isotopes: *protium*, ^{1}H, containing one proton and no protons; *deuterium*, ^{2}H, containing one proton and one neutron; and *tritium*, ^{3}H, containing one proton and two neutrons. Because tritium is unstable and undergoes radioactive decay, its natural occurrence is extremely rare.

The chemical properties of the elements are primarily dependent on their electron configurations, particularly in their outermost electron shells. Since the same electron configurations in the outermost electron shells occur periodically in different elements, it is possible to arrange the chemical elements in a periodic table that groups elements with similar chemical properties (Fig. 3.1). The rows in the periodic table represent *periods*. The number of each period indicates the number of electron shells possessed by an element. The columns in the periodic table represent *groups*. The number of electrons in the outermost shell is the same for all

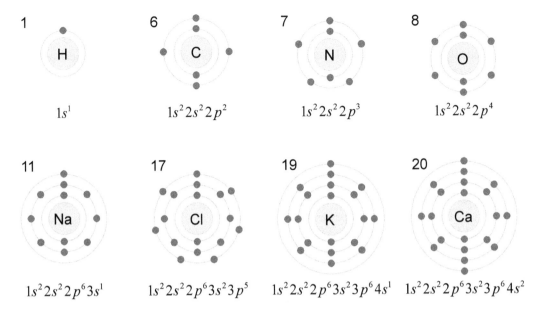

Figure 3.2 Atomic numbers and models of the atoms for some of the main chemical elements occurring in biological systems: H, hydrogen; C, carbon; N, nitrogen; O, oxygen; Na, sodium; Cl, chlorine; K, potassium; Ca, calcium. The electrons are shown as balls and the electron shells are shown as concentric circles. The electron configuration formulas show the distribution of electrons in each subshell, even though subshells are not explicitly shown in the atomic models.

elements within a group. Because the number of electrons in the outermost shell determines the *valence*, which is the maximal number of chemical bonds that an element can form, the outermost electron shell is called the *valence shell*.

The electron shells n are usually labeled from 1 to 7 going from the innermost shell outwards. This numbering also corresponds to the number of subshells possessed by the electron shell. For example, the first shell has 1 subshell, the second shell has 2 subshells, and so on. The subshells are named consecutively as s, p, d, f, g, h, i and are usually numbered with a number ℓ that varies from 0 to 6, respectively. The order in which the electrons fill in the different subshells is determined by their energies, namely, subshells with lowest energy are filled in first. According to the $n + \ell$ ordering rule, the subshell energies increase with increasing $n + \ell$ values. If $n + \ell$ values are equal, higher n has higher energy: $1s$, $2s$, $2p$, $3s$, $3p$, $4s$, $3d$, $4p$, $5s$, $4d$, $5p$, $6s$, $4f$, $5d$, $6p$, $5f$, $6d$, ... The maximal number of electrons that a subshell can contain is $4\ell + 2$. Thus, an s subshell can contain at most 2 electrons, a p subshell can contain at most 6 electrons, a d subshell can contain at most 10 electrons, and so on. The electron configurations of chemical elements are typically written using the subshell names ordered with increasing energies and with the number of electrons in each subshell given as a superscript. The atomic models together with the corresponding electron configuration formulas for some of the main chemical elements occurring in biological systems are shown in Figure 3.2.

Because the energy of the inner subshells is lower than the energy of the outer subshells, the electrons in the outer subshells feel a lesser pull from the positively charged nucleus and can be lost. If an atom loses some of its electrons, it becomes a *positively charged ion*; conversely, if an atom gains some electrons, it becomes a *negatively charged ion*. Oppositely charged ions can attract each other and bind together with the use of ionic bonds, forming *ionic compounds*. Typically, ionic compounds are easily dissolved in solvents such as water and individual ions may be floating around in the solution.

Alternatively, it is possible that electrons are not lost, but donated to another atom in the form of *covalent bonds*, forming a *molecule*. In such a case, the atoms donating the electrons in the covalent bonds still partially owe them and the atomic nuclei exert sufficiently strong control over the electrons in order to avoid their loss. The molecules formed by covalent bonds are much more stable compared to ionic compounds and do not disintegrate down to ions when dissolved in a solvent. Nevertheless, it should be kept in mind that even though the main backbone of a large molecule may be stable in solution, it is possible that some portions of the molecule may be ionized, as is the case with amino acids (Section 7.1).

Different molecules, composed from atoms of the known chemical elements, can interact with each other using gravitational and electromagnetic forces, and such interactions are responsible for the formation of material bodies, including our planet and the living organisms on it. Biologically, of paramount importance are the electromagnetic interactions that occur through the exchange of *light*, or more precisely, exchange of electromagnetic waves (Section 3.17).

3.2   Determinism

Determinism is the main characteristic feature of classical physics. The French astronomer and mathematician Pierre-Simon Laplace (1749–1827) wrote:

> We may regard the present state of the universe as the effect of its past and the cause of its future. An intellect which at a certain moment would know all forces that set nature in motion, and all positions of all items of which nature is composed, if this intellect were also vast enough to submit these data to analysis, it would embrace in a single formula the movements of the greatest bodies of the universe and those of the tiniest atom; for such an intellect nothing would be uncertain and the future just like the past would be present before its eyes. [310, p. 4]

Definition 3.1. *(Determinism) A physical theory is* deterministic *if knowing the physical laws of the universe and the state of a closed system at a single moment of time is sufficient for one to calculate in principle with absolute certainty and arbitrarily high precision the state of the closed system at any other future moment of time.*

An important part of the definition of a deterministic system is played by the possibility to calculate *in principle* the physical state of the closed system at any future moment of time with arbitrarily high precision. Such a possibility does not imply that the calculation should be actually carried out by someone or something, especially when studying enormously complex, nonlinear biological systems such as the human brain. The behavior of the human brain may appear to be indeterministic due to our lack of sufficient computational power to calculate the dynamics of $\approx 10^{26}$ atoms that constitute the brain or our lack of sophisticated instruments that can measure the initial physical state of all these constituent atoms. Classically, however, the appearance of the nonlinear brain as an indeterministic system is considered to be misleading, because deterministic classical physical laws lead to deterministic *chaos* [72, 425, 444]. In Sections 5.3.1, 5.4.3, 6.4.2 and 6.5.1, we will argue that there are no good reasons for why the world has to be deterministic and will provide strong support for indeterminism.

3.3   Observability

All physical quantities in classical physics are *observable* or *measurable* at least in principle. Basing scientific theories on such a postulate has been very effective in fighting *superstition*. For example, demons, ghosts or spirits that remain hidden when you are actively trying to detect them with a measuring device could be safely proclaimed as non-existent, non-physical or fictitious. If all physical quantities in a deterministic physical theory are observable, you will be able to predict in advance the behavior of any physical system with arbitrarily high precision provided that you possess good measurement tools and enough computational power. On the other hand, if there are unobservable physical entities in a deterministic theory, you could only make probabilistic predictions based on assumptions about the actual values and dynamics of these unobservable or hidden variables.

Being able to understand and master the physical reality requires knowledge of some basic mathematics, which we are going to concisely introduce next.

3.4   Real numbers

We all learn how to count objects using the set of *natural numbers* $\mathbb{N} = \{0, 1, 2, 3, \ldots\}$. To solve mathematical *equations*, however, only natural numbers are not sufficient to accomplish the task, which forces us to expand our number system.

Solving equations that involve *addition* like $x + 1 = 0$ requires the introduction of negative numbers such as -1. Taking together all negative whole numbers and all natural numbers gives us the set of *integers* $\mathbb{Z} = \{\ldots, -2, -1, 0, 1, 2, \ldots\}$.

Solving equations that involve *multiplication* like $3x - 2 = 0$ requires the introduction of *fractions* such as $\frac{2}{3}$. The set of *rational numbers* $\mathbb{Q}$ is formed as a collection of all numbers that can be expressed as a fraction of two integers. Thus, the rational numbers include the integers. All rational numbers have infinite periodic decimal expansions, for example, $\frac{2}{3} = 0.666\ldots$ This also holds for rational numbers that have finite decimal expansions, for example, $\frac{1}{2} = 0.5 = 0.5000\ldots = 0.4999\ldots$

Figure 3.3 The real number line consists of points that represent the real numbers. Shown are all integers from −4 to 4, a rational fraction $\frac{2}{3}$, and several irrational numbers such as $\sqrt{2}$, e and π.

Solving equations that involve *exponentiation* like $x^2 - 2 = 0$ requires the introduction of *irrational numbers* such as $\sqrt{2} = 1.41421\ldots$ All irrational numbers are represented by infinite aperiodic decimal expansions. The collection of all rational and all irrational numbers forms the set of *real numbers* $\mathbb{R}$. A convenient way to geometrically represent the real numbers is to use *points* on the *real line* (Fig. 3.3). Thus, real numbers are a 1-dimensional collection of numbers. Real numbers also form a mathematical *field*, which means that they satisfy the set of field axioms.

Definition 3.2. *(Field) A mathematical field F is any set of elements equipped with two operations referred to as "addition" + and "multiplication" · that satisfies the following five field axioms [11, p. 18].*

Axiom 3.4.1. *Addition and multiplication are commutative*

$$x_1 + x_2 = x_2 + x_1 \tag{3.1}$$

$$x_1 \cdot x_2 = x_2 \cdot x_1 \tag{3.2}$$

Axiom 3.4.2. *Addition and multiplication are associative*

$$(x_1 + x_2) + x_3 = x_1 + (x_2 + x_3) \tag{3.3}$$

$$(x_1 \cdot x_2) \cdot x_3 = x_1 \cdot (x_2 \cdot x_3) \tag{3.4}$$

Axiom 3.4.3. *Multiplication is distributive over addition*

$$x_1 \cdot (x_2 + x_3) = x_1 \cdot x_2 + x_1 \cdot x_3 \tag{3.5}$$

Axiom 3.4.4. *There exist two distinct identity elements, denoted as 0 and 1, respectively, for addition and multiplication such that*

$$x + 0 = x \tag{3.6}$$

$$1 \cdot x = x \tag{3.7}$$

Axiom 3.4.5. *For each number $x \neq 0$, there exist inverse elements, negatives $(-x)$ for addition and reciprocals (x^{-1}) for multiplication, such that the corresponding identity elements are returned*

$$x + (-x) = 0 \tag{3.8}$$

$$x \cdot x^{-1} = 1 \tag{3.9}$$

Definition 3.3. *(Scalar) The elements of a mathematical field are called scalars. It should be noted that defining the scalar as an object with a magnitude only is less general because complex numbers (Section 4.1) are scalars too.*

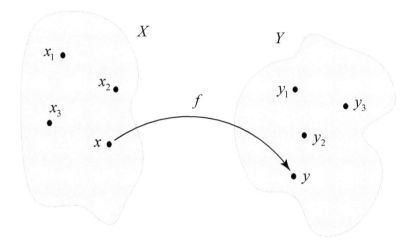

Figure 3.4   Set theoretic representation of a function $f : X \to Y, y = f(x)$.

3.5   Functions

A mathematical *function* associates one quantity, the *argument of the function*, also known as the *input*, with another quantity, the *value of the function*, also known as the *output*. Suppose X and Y are two sets of numbers. If we have a *correspondence rule* according to which for each element $x \in X$ there is at most one corresponding element $y \in Y$, we call this rule a *function* $f : X \to Y, y = f(x)$ [8, pp. 15–16]. Briefly, we read "y is a function of x." Set theoretic representation of a function $f : X \to Y, y = f(x)$ is shown in Figure 3.4.

Definition 3.4. *(Function) A function is an ordered triple of sets, written* (X, Y, F), *where X is the* domain, *Y is the* codomain, *and F is a set of ordered pairs* (x, y). *In each of the ordered pairs, the first element x is from the domain, the second element y is from the codomain, and a necessary condition is that every element in the domain is the first element in exactly one ordered pair. The set of all* y_i *is known as the* range of the function.

Depending on the form in which the correspondence rule is provided, the functions could be classified as *analytic functions, graphic functions, tabular functions*, and so on. Most common are the *analytic functions*, which are given by analytic expressions. A nice feature of analytic functions is that we can easily plot their graphs using currently available computer programs.

If the argument x of a function is a single number, the function $f(x)$ is referred to as a *function of one variable*. If the argument x of a function is an ordered set of numbers $(x_1, x_2, \ldots, x_n)$ the function $f(x_1, x_2, \ldots, x_n)$ is referred to as a *function of several variables*. Graphs of functions of one or several variables are shown in Figure 3.5.

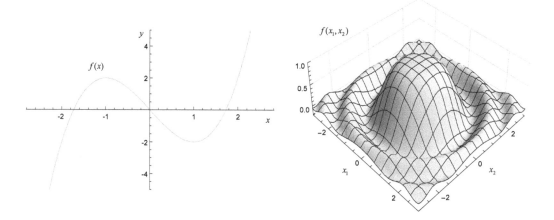

Figure 3.5   Examples of function graphs. The left panel shows a graph of a function of one variable $f : X \to Y, y = x^3 - 3x$. The right panel shows a graph of a function of two variables $f : X_1 \times X_2 \to Y, y = \sin\left(x_1^2 + x_2^2\right)/\left(x_1^2 + x_2^2\right)$.

3.6   Derivatives

Differentiation of mathematical functions is indispensable for the mathematical formulation of physical laws. For example, Hamilton's equations (3.72) and (3.73) that govern the behavior of classical physical systems and the Schrödinger equation (4.157) that governs the behavior of quantum physical systems are differential equations.

Definition 3.5. *(Derivative) Differentiation is the process of finding the derivative* $f'(x) = \frac{df(x)}{dx}$ *that is the* instantaneous *rate of change or* slope *of a mathematical function* $f(x)$ *at a given point* x. *The derivative can be calculated as a limit of the ratio between the rise of the function* $\Delta y = f(x + \Delta x) - f(x)$ *and the run* Δx, *when the run tends to zero* $\Delta x \to 0$. *Symbolically, we write*

$$f'(x) = \frac{df(x)}{dx} = \lim_{\Delta x \to 0} \frac{\Delta y}{\Delta x} = \lim_{\Delta x \to 0} \frac{f(x + \Delta x) - f(x)}{\Delta x} \tag{3.10}$$

The value of the derivative at any point x_0 *is written as*

$$f'(x_0) = \frac{df(x)}{dx}\bigg|_{x=x_0} \tag{3.11}$$

The instantaneous rate of change (slope) of $f(x)$ at a point x_0 is equal to the slope of the tangent line at the point x_0 (Fig. 3.6). Generally, a function $f(x)$ can have varying slopes at different points x_1 and x_2. That is why the derivative $f'(x)$ is also a mathematical function that gives the slope of $f(x)$ at any x. At point x_1 the slope of $f(x)$ is equal to $f'(x_1)$, at point x_2 the slope of $f(x)$ is equal to $f'(x_2)$, and so on.

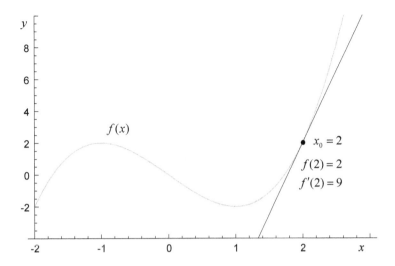

Figure 3.6   The function $f(x) = x^3 - 3x$ has a derivative $f'(x) = 3x^2 - 3$. The value of $f'(x)$ at the point $x_0 = 2$ is equal to the slope of the tangent line to $f(x)$ at $x_0 = 2$.

If a function $f(x)$ has a derivative at a point x_0, it is referred to as a *differentiable function* at the point x_0. A function $f(x)$ may not be differentiable at a point x_0 if the function $f(x)$ is discontinuous at the point x_0 (Fig. 3.7). Importantly, the differentiability of a function $f(x)$ at a point x_0 implies the *continuity* of the function $f(x)$ at the point x_0. The continuity of a function $f(x)$ at a point x_0, however, does not imply the differentiability of the function $f(x)$ at that point since it is possible for the function $f(x)$ to have a sharp bend at x_0 (Fig. 3.7).

Differentiation is a linear operation. Thus, if $f(x)$ and $g(x)$ are two functions, and if a and b are two *scalars*, the following equality holds

$$\frac{d}{dx}[af(x) + bg(x)] = a\frac{df(x)}{dx} + b\frac{dg(x)}{dx} = af'(x) + bg'(x) \tag{3.12}$$

The derivative of a product $f(x)g(x)$ is given by the *product rule*

$$\frac{d}{dx}[f(x)g(x)] = g(x)\frac{df(x)}{dx} + f(x)\frac{dg(x)}{dx} = g(x)f'(x) + f(x)g'(x) \tag{3.13}$$

Noteworthy, the multiplication by $\frac{d}{dx}$ is noncommutative, meaning that the order in which the differential operator $\frac{d}{dx}$ occurs in mathematical expressions is important. In general, $g(x)\frac{d}{dx}f(x) \neq f(x)\frac{d}{dx}g(x)$.

The derivative of a composite function $h(x) = f[g(x)]$ is given by the *chain rule*

$$\frac{d}{dx}h(x) = \frac{df[g(x)]}{dg(x)}\frac{dg(x)}{dx} = f'[g(x)]g'(x) \tag{3.14}$$

Many physically interesting quantities are represented by functions of several variables $f(x_1, x_2, \ldots, x_n)$. In the presence of several variables one can calculate two different kinds of derivatives: *partial derivatives* or *total derivatives*.

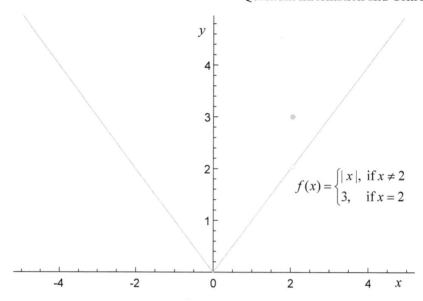

$$f(x) = \begin{cases} |x|, & \text{if } x \neq 2 \\ 3, & \text{if } x = 2 \end{cases}$$

Figure 3.7   Graph of the discontinuous function $f(x)$ which is equal to $|x|$ almost every-where except at $x = 2$ where $f(x) = 3$. The function $f(x)$ does not have a derivative at the point $x = 2$ due to discontinuity and at the point $x = 0$ due to a sharp bend.

Definition 3.6. *(Partial derivative) The derivative of $f(x_1, x_2, \ldots, x_n)$ with respect to one of those variables while the remaining variables are held constant is called a* partial derivative *and is denoted with the symbol ∂. For the function $f(x_1, x_2, \ldots, x_n)$ there will be n such partial derivatives given by $\frac{\partial f(x_1, x_2, \ldots, x_n)}{\partial x_1}$, $\frac{\partial f(x_1, x_2, \ldots, x_n)}{\partial x_2}$, $\ldots$, $\frac{\partial f(x_1, x_2, \ldots, x_n)}{\partial x_n}$. Holding the variables $x_1, x_2, \ldots, x_{i-1}, x_{i+1}, \ldots, x_n$ constant when calculating the partial derivative $\frac{\partial f(x_1, x_2, \ldots, x_n)}{\partial x_i}$ means that one is allowed to mathematically manipulate any of $x_1, x_2, \ldots, x_{i-1}, x_{i+1}, \ldots, x_n$ as if it were a scalar constant, not a variable.*

Definition 3.7. *(Total derivative) The derivative of $f(x_1, x_2, \ldots, x_n)$ with respect to one of those variables without the assumption that the remaining variables are held constant is called a* total derivative *and is denoted with the symbol d. The total derivative of $f(x_1, x_2, \ldots, x_n)$ with respect to x_i is given by*

$$\frac{df}{dx_i} = \frac{\partial f}{\partial x_1}\frac{dx_1}{dx_i} + \frac{\partial f}{\partial x_2}\frac{dx_2}{dx_i} + \ldots + \frac{\partial f}{\partial x_i}\frac{dx_i}{dx_i} + \ldots \frac{\partial f}{\partial x_n}\frac{dx_n}{dx_i} \tag{3.15}$$

The total derivative $\frac{df}{dx_i}$ will be equal to the partial derivative $\frac{\partial f}{\partial x_i}$ only if the variables $x_1, x_2, \ldots, x_{i-1}, x_{i+1}, \ldots, x_n$ do not depend on x_i, because in such a case the following equations will hold: $\frac{dx_1}{dx_i} = 0$, $\frac{dx_2}{dx_i} = 0$, $\ldots$, $\frac{dx_{i-1}}{dx_i} = 0$, $\frac{dx_{i+1}}{dx_i} = 0$, $\ldots$, $\frac{dx_n}{dx_i} = 0$.

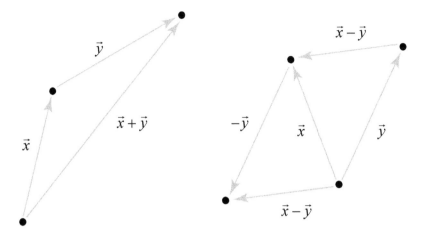

Figure 3.8 Vector addition and subtraction of two vectors $\vec{x}$ and $\vec{y}$. The negative of a vector $-\vec{y}$ has the same length but opposite direction of $\vec{y}$. Vector subtraction could be viewed as vector addition either of $\vec{x}$ and $-\vec{y}$ to get $\vec{x}-\vec{y}$ or of $\vec{y}$ and $\vec{x}-\vec{y}$ to get $\vec{x}$.

3.7 Vectors

Line segments in which there is an initial point p_1 and a terminal point p_2 are called directed line segments and can be written as ordered pairs (p_1, p_2). The set of all directed line segments of the same length and direction is called a *vector*. Vectors are usually indicated with a small overhead arrow, for example, $\vec{x}$, $\vec{y}$, $\vec{z}$, etc. Vectors can be added together into a vector sum. The so-called parallelogram law gives the rule for vector addition of two or more vectors. For two vectors $\vec{x}$ and $\vec{y}$, the vector sum $\vec{x}+\vec{y}$ is obtained by placing them head to tail and drawing the vector from the free tail to the free head. Vector subtraction can be obtained from the parallelogram law using the fact that the negative of a vector $-\vec{x}$ has the same length but opposite direction to the vector $\vec{x}$. Examples of vector addition and subtraction are shown in Figure 3.8.

A set of orthogonal unit vectors is often used for the introduction of a coordinate system that parametrizes the space of the surrounding world. In three dimensions, it is possible to define right-handed or left-handed coordinate systems that cannot be transformed into each other through translation and rotation alone. Instead, transformation of a right-handed into a left-handed coordinate system requires translation, rotation and mirror reflection.

Definition 3.8. *(Right-handed coordinate system)* Right-handed coordinate system $\mathcal{O}(x, y, z)$ *is any system such that if the z-axis points toward your face, the counterclockwise rotation of the x-axis to the y-axis has the shortest possible path (Fig. 3.9). The positive normal +n of a given surface s closed by contour ℓ is collinear with the z-axis of a right-handed coordinate system $\mathcal{O}(x, y, z)$ whose x-axis and y-axis are within the plane of the surface. The positive direction of the contour ℓ is the direction in which the rotation of the x-axis to the y-axis has the shortest possible path* [536].

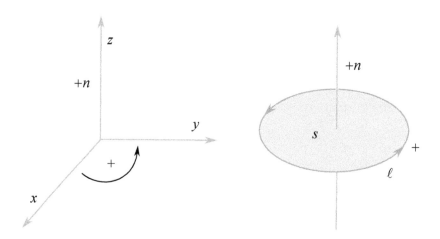

Figure 3.9   Right-handed coordinate system $\mathcal{O}(x, y, z)$ and direction of the positive normal $+n$ of a surface s depending on the positive direction of the contour ℓ.

Multiplication of vectors could be performed in two different ways: the dot product and the cross product, each producing a different type of output.

Definition 3.9. *(Dot product) The* dot product *of two vectors $\vec{a}$ and $\vec{b}$ is a scalar whose magnitude is equal to the length of the first vector times the projection of the second vector along the first. The symbol used to represent this operation is a small dot at middle height (·), which is where the name* dot product *comes from. Since this product has magnitude only, it is also known as the scalar product*

$$\vec{a} \cdot \vec{b} = ab \cos \theta \qquad (3.16)$$

where $a = |\vec{a}|$ is the length of the first vector, $b = |\vec{b}|$ is the length of the second vector, and θ is the angle between the two vectors $\vec{a}$ and $\vec{b}$.

Definition 3.10. *(Cross product) The* cross product *of two vectors $\vec{a}$ and $\vec{b}$ is a vector whose magnitude is equal to the area of the parallelogram between $\vec{a}$ and $\vec{b}$, and whose direction is along the unit vector $\vec{n}$ that is normal to the plane formed by $\vec{a}$ and $\vec{b}$. The symbol used to represent this operation is a large diagonal cross (×), which is where the name* cross product *comes from. Since this product has magnitude and direction, it is also known as the vector product*

$$\vec{a} \times \vec{b} = ab \sin \theta \, \vec{n} \qquad (3.17)$$

The direction of $\vec{n}$ is easily determined by the right hand rule, *according to which if you hold your right hand out flat with your fingers pointing in the direction of the first vector and orient your palm so that you can fold your fingers in the direction of the second vector, then your thumb will point in the direction of the cross product. In Section 3.14, we shall see that vector cross products are extremely useful for the formulation of the laws of electrodynamics.*

In physical theories, *forces* acting on physical systems are represented by vectors. Typically, forces are generated by scalar potential energy fields $\varphi(x,y,z)$ that are mathematical functions of several variables. The force $\vec{F}(x,y,z)$ resulting from a scalar potential energy field $\varphi(x,y,z)$ is given by partial derivatives in the form

$$\vec{F}(x,y,z) = -\frac{\partial\varphi(x,y,z)}{\partial x}\vec{e}_x - \frac{\partial\varphi(x,y,z)}{\partial y}\vec{e}_y - \frac{\partial\varphi(x,y,z)}{\partial z}\vec{e}_z \tag{3.18}$$

where $\vec{e}_x$, $\vec{e}_y$ and $\vec{e}_z$ are unit vectors pointing in the x, y and z directions, respectively. The assignment of a vector to each point in space produces a *vector field*.

Definition 3.11. *(Gradient) The gradient vector field of a scalar function $\varphi(x_1,x_2,\ldots,x_n)$ is denoted $\nabla\varphi$, where ∇ is the gradient vector operator, also known as del or nabla. The components of the gradient vector field are given by the partial derivatives of $\varphi(x_1,x_2,\ldots,x_n)$ in the form*

$$\nabla\varphi = \frac{\partial\varphi(x_1,x_2,\ldots,x_n)}{\partial x_1}\vec{e}_1 + \frac{\partial\varphi(x_1,x_2,\ldots,x_n)}{\partial x_2}\vec{e}_2 + \ldots + \frac{\partial\varphi(x_1,x_2,\ldots,x_n)}{\partial x_n}\vec{e}_n \tag{3.19}$$

where $\vec{e}_i$ are orthogonal unit vectors pointing in the corresponding directions x_i.

The force $\vec{F}(x,y,z)$ resulting from a scalar potential energy field $\varphi(x,y,z)$ can be concisely expressed with the use of the del operator as

$$\vec{F}(x,y,z) = -\nabla\varphi(x,y,z) \tag{3.20}$$

An illustration of a 2-dimensional scalar function $\varphi(x,y)$ and its gradient vector field $\nabla\varphi(x,y)$ is shown in Figure 3.10. The resulting forces act in the direction opposite to the one of the gradient vector field.

In biological systems, gradients determine the direction in which passive processes occur. For example, the diffusion of chemical molecules occurs from regions with higher concentration toward regions with lower concentration, the electrotonic propagation of electric currents occurs from regions with higher electric potential toward regions with lower electric potential, etc.

3.8 Integrals

Since most physical laws are formulated with the use of differential equations, one needs to know how to solve differential equations. For example, given a mathematical function $f(x)$ one is often asked to find an *antiderivative* $F(x)$ such that

$$F'(x) = \frac{dF(x)}{dx} = f(x) \tag{3.21}$$

To approach the problem, consider a function $g(x) = C$, where C is a scalar *constant*. Because the graph of the function $g(x)$ is just a horizontal line at a height C, the slope of $g(x)$ will be *zero* for all x, hence the derivative will be identically zero,

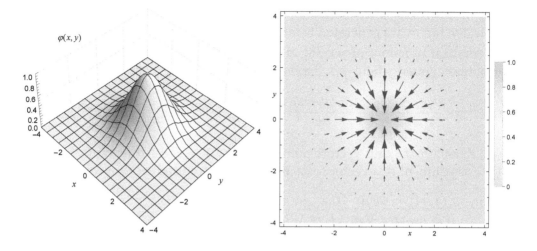

Figure 3.10   An example of a 2-dimensional scalar field $\varphi(x,y) = e^{-\frac{1}{2}(x^2+y^2)}$ is shown in the left panel. The gradient vector field $\nabla\varphi(x,y)$ is shown superposed on a density plot of the scalar field $\varphi(x,y)$ in the right panel. The arrows of the gradient vector field point toward higher values of the scalar field $\varphi(x,y)$. Typically, physical forces act in the direction opposite to the one of the gradient.

$g'(x) = \frac{dg(x)}{dx} = 0$. Suppose now that $F'(x) = f(x)$ and we calculate the derivative of the function $F(x) + g(x)$. From the linearity of differentiation (Eq. 3.12), we have

$$\frac{d}{dx}[F(x) + g(x)] = F'(x) + g'(x) = f(x) + 0 = f(x) \tag{3.22}$$

Thus, there is a set of antiderivatives $F(x) + C$ whose derivative is equal to $f(x)$.

Definition 3.12. *(Indefinite integral) The set of all antiderivative functions that have derivative $f(x)$ is called the* **indefinite integral** *of a function $f(x)$ and is written as*

$$\int f(x)dx = F(x) + C \tag{3.23}$$

where C is a scalar constant.

From the definition of indefinite integral it follows that

$$\frac{d}{dx}\int f(x)dx = \left[\int f(x)dx\right]' = f(x) \tag{3.24}$$

$$\int \frac{d}{dx}f(x)dx = \int f'(x)dx = f(x) \tag{3.25}$$

$$\int df(x) = f(x) \tag{3.26}$$

$$d\left[\int f(x)dx\right] = f(x)dx \tag{3.27}$$

Integration is also a *linear operation*. Thus, if $f(x)$ and $g(x)$ are two functions, and if a and b are two *scalars*, the following equality holds

$$\int [af(x) + bg(x)]\, dx = \int af(x)\, dx + \int bg(x)\, dx = a \int f(x)\, dx + b \int g(x)\, dx \quad (3.28)$$

Integration is closely related to summation and can be defined with the use of infinite sums, as we will show next.

Definition 3.13. *(Summation operator) When we add a large number of terms, it is very convenient to use the* summation operator $\sum$ *defined as*

$$\sum_{i=m}^{n} a_i = a_m + a_{m+1} + a_{m+2} + \ldots + a_{n-1} + a_n \qquad (3.29)$$

If there is an infinite number of terms in the series, the summation notation becomes

$$\sum_{i=m}^{\infty} a_i = a_m + a_{m+1} + a_{m+2} + a_{m+3} + \ldots \qquad (3.30)$$

Theorem 2. *The summation operator $\sum$ has the following mathematical properties*

$$\sum_{i=m}^{n} a_i = a_m + \sum_{i=m+1}^{n} a_i = \sum_{i=m}^{n-1} a_i + a_n \qquad (3.31)$$

$$\sum_{i=m}^{n} c \cdot a_i = c \sum_{i=m}^{n} a_i \qquad (3.32)$$

$$\sum_{i=m}^{n} (a_i \pm b_i) = \sum_{i=m}^{n} a_i \pm \sum_{i=m}^{n} b_i \qquad (3.33)$$

$$\sum_{i=m}^{n} a_i = \sum_{k=m}^{n} a_k \qquad (3.34)$$

$$\sum_{i=m}^{n} 1 = n - (m-1) = n - m + 1 \qquad (3.35)$$

$$\sum_{i=1}^{p} \sum_{j=1}^{q} a_{ij} = \sum_{i=1}^{p} \left(a_{i1} + a_{i2} + \ldots + a_{iq} \right) \qquad (3.36)$$

$$\sum_{i=m}^{n} \sum_{j=p}^{q} a_{ij} = \sum_{j=p}^{q} \sum_{i=m}^{n} a_{ij} \qquad (3.37)$$

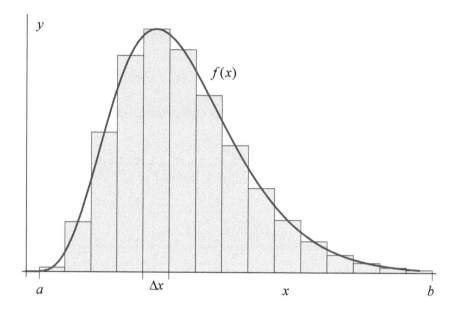

Figure 3.11   Illustration of the Riemann summation method for approximating the area under a curve $f(x)$. The error of the approximation vanishes as the width of the rectangles becomes infinitesimally small, $\Delta x \to 0$.

Whereas differentiation can be used to calculate the instantaneous rate of change of a function $f(x)$, integration can be used to calculate the area enclosed between a function $f(x)$ and the x-axis. Suppose that one wants to calculate the area enclosed by a function $f(x)$ for x-values varying from a to b. It is possible to approximate the area using a Riemann sum of rectangle areas (Fig. 3.11) as follows: First, partition the interval from a to b into n subintervals with width $\Delta x = \frac{b-a}{n}$. Then, for each subinterval i draw a rectangle with height equal to the value of the function at the midpoint of the subinterval, namely, $f\left[a + \left(i - \frac{1}{2}\right)\Delta x\right]$. Finally, sum the areas of all rectangles to get

$$\sum_{i=1}^{n} f\left[a + \left(i - \frac{1}{2}\right)\Delta x\right]\Delta x \tag{3.38}$$

The approximation of the area enclosed between the function $f(x)$ and the x-axis becomes better as the number of partitions n increases. In the limit $n \to \infty$, the error from the approximation becomes *zero*. Thus, the calculated area enclosed by $f(x)$ becomes

$$\lim_{n \to \infty} \sum_{i=1}^{n} f\left[a + \left(i - \frac{1}{2}\right)\Delta x\right]\Delta x = \int_{a}^{b} f(x)dx \tag{3.39}$$

Definition 3.14. *(Definite integral) The mathematical expression*

$$\int_a^b f(x)dx = \lim_{n \to \infty} \sum_{i=1}^{n} f\left[a + \left(i - \frac{1}{2}\right)\Delta x\right]\Delta x$$

is called a definite integral *of* $f(x)$ *from a to b. Since the definite integral is represented by an infinite Riemann sum, it follows that whenever the function* $f(x)$ *has a negative value by going under the x-axis, the area enclosed between the x-axis and* $f(x)$ *also has a negative sign.*

Theorem 3. *(The fundamental theorem of calculus) The definite integral of the function* $f(x)$ *from a to b is given by*

$$\int_a^b f(x)dx = F(b) - F(a) \tag{3.40}$$

where $F(x)$ *is any antiderivative of the function* $f(x)$*.*

Definite integrals are also linear. The linearity of integration is directly related to the linearity of summation, since integrals could be thought of as infinite sums. From Theorem 3, we have the following immediate results

$$\int_a^a f(x)dx = 0 \tag{3.41}$$

$$\int_b^a f(x)dx = -\int_a^b f(x)dx \tag{3.42}$$

$$\int_a^b f(x)dx = \int_a^c f(x)dx + \int_c^b f(x)dx \tag{3.43}$$

3.9   Sets

A *set* is a *collection of objects* that is considered as an object in its own right. Each object in the set is called an *element* of the set. The elements of the set can be anything: numbers, people, other sets, etc. Sets are usually denoted with capital letters $A, B, C, D \ldots$ whereas their elements are denoted with small letters $a, b, c, d \ldots$ If a is an element of the set B, we write $a \in B$, and read a belongs to B. Conversely, if a is not an element of B, we write $a \notin B$, and read a does not belong to B. Sets can be geometrically represented with the use of *Venn diagrams* as enclosed parts of the plane (Fig. 3.12), whereas their elements can be represented as points within such enclosed parts of the plane.

Definition 3.15. *(Subset) If we have two sets A and B for which we know that each element of A is also an element of B, we say that A is a* subset *of B and we write* $A \subseteq B$*. If A is a subset of B, but A is not equal to B, then we say that A is a* proper subset *of B and we write* $A \subset B$*.*

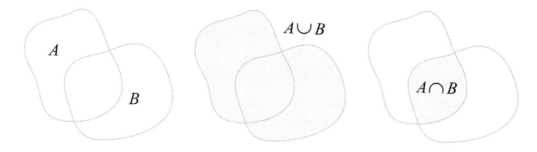

Figure 3.12   Venn diagrams of sets A and B including the results from the corresponding set union $A \cup B$ and set intersection $A \cap B$ operations.

Within set theory, the logical *non-exclusive or operator* $\vee$ becomes the *set union operator* $\cup$, while the logical *and operator* $\wedge$ becomes the *set intersection operator* $\cap$.

Definition 3.16. *(Set union) If A and B are two sets, their* union *is the set $A \cup B$ whose elements belong to A, to B, or to both A and B, namely, $A \cup B = \{x | x \in A \vee x \in B\}$. For example, if $A = \{x_1, x_2, x_3\}$ and $B = \{x_3, x_4, x_5\}$, then $A \cup B = \{x_1, x_2, x_3, x_4, x_5\}$.*

Definition 3.17. *(Set intersection) If A and B are two sets, their* intersection *is the set $A \cap B$ whose elements belong to both A and B, namely, $A \cap B = \{x | x \in A \wedge x \in B\}$. For example, if $A = \{x_1, x_2, x_3\}$ and $B = \{x_3, x_4, x_5\}$, then $A \cap B = \{x_3\}$.*

Definition 3.18. *(Ordered pair) An* ordered pair *(a, b) is defined to be $\{\{a\}, \{a, b\}\}$. Because sets do not have a notion of order, namely, $\{a, b\} = \{b, a\}$, one needs to encode the order within the ordered pair (a, b) with the use of a* singleton *$\{a\}$ for specifying the first element and an* unordered pair *$\{a, b\}$ for introducing the second element.*

The existence of certain sets, as well as their mathematical properties and relations to other existing sets, can be derived from the following list of nine axioms proposed by the mathematicians Ernst Zermelo (1871–1953) [533] and Abraham Fraenkel (1891–1965) [192].

Axiom 3.9.1. *(Axiom of extensionality) Given any set A and any set B, A is equal to B if given any set C, C is a member of A if and only if C is a member of B [442, p. 12]*

$$\forall A \forall B (\forall C (C \in A \leftrightarrow C \in B) \rightarrow A = B) \tag{3.44}$$

In other words, each set is determined uniquely by its members and the repetition or change in the order of elements in a set does not change its identity. For example, the set $A = \{z, x, x, y, y, y\}$ is the same as the set $A = \{x, y, z\}$.

Axiom 3.9.2. *(Axiom of empty set) There exists an empty set $\emptyset$ such that no set is a member of it [442, p. 12]*

$$\exists \emptyset \forall A (A \notin \emptyset) \tag{3.45}$$

Axiom 3.9.3. *(Axiom of pairing) Given any set A and any set B, there is a set C such that, given any set D, D is a member of C if and only if D is equal to A or D is equal to B [533, p. 30]*

$$\forall A \forall B \exists C \forall D (D \in C \leftrightarrow (D = A \lor D = B)) \tag{3.46}$$

In other words, given two sets A and B, we can always find a set C = {A, B} whose members are precisely A and B. Pairing of a set A with itself produces its singleton {A}.

Axiom 3.9.4. *(Axiom of union) Given any set A, there is a set B such that, given any set C, C is a member of B if and only if there is a set D such that D is a member of A and C is a member of D [442, p. 13]*

$$\forall A \exists B \forall C (C \in B \leftrightarrow (\exists D (D \in A \land C \in D))) \tag{3.47}$$

In other words, for any set A, there is a union set $\bigcup A$ which consists of just the elements of the elements of A. For example, if $A = \{A_1, A_2\}$, where $A_1 = \{a_1, a_2, a_3\}$ and $A_2 = \{a_4, a_5\}$, the union of the set A is given by $\bigcup A = A_1 \cup A_2 = \{a_1, a_2, a_3, a_4, a_5\}$.

Axiom 3.9.5. *(Axiom of infinity) There exists an infinite set ω, such that the empty set $\emptyset$ is in ω and such that whenever A is a member of ω, the set formed by taking the union of A with its singleton {A} is also a member of ω [442, p. 13]*

$$\exists \omega (\emptyset \in \omega \land \forall A (A \in \omega \rightarrow (A \cup \{A\}) \in \omega)) \tag{3.48}$$

If we denote $A \cup \{A\}$ as the successor of A, we can define recursively all natural numbers with the use of sets, where the zero is represented by the empty set and each number $n + 1 = n \cup \{n\}$ is represented by the successor of n. Explicitly written, we get $0 = \emptyset$, $1 = 0 \cup \{0\} = \{0\}$, $2 = 1 \cup \{1\} = \{0, 1\}$, $3 = 2 \cup \{2\} = \{0, 1, 2\}$, ... Hence, it could be said that the infinite set ω contains all natural numbers $\omega = \{0, 1, 2, 3, ...\}$.

Axiom 3.9.6. *(Axiom schema of separation) Given any set A, there is a set B such that, given any set C, C is a member of B if and only if C is a member of A and the property P holds for C [442, p. 12]*

$$\forall A \exists B \forall C (C \in B \leftrightarrow (C \in A \land P(C))) \tag{3.49}$$

In other words, given a set A and a property P, we can always find a subset B of A whose members are precisely the members of A that satisfy P. If no member of A satisfies P, then the subset B is equal to the empty set $\emptyset$.

Axiom 3.9.7. *(Axiom schema of replacement) Given any set A, there is a set B such that, given any set C, C is a member of B if and only if there is a set D such that D is a member of A and C is equal to the value of the function F at D [192, p. 231]*

$$\forall A \exists B \forall C (C \in B \leftrightarrow \exists D (D \in A \land C = F(D))) \tag{3.50}$$

In other words, given a set A, we can find a set B whose members are precisely the values of the function F at the members of A.

Axiom 3.9.8. *(Axiom of power set) Given any set A, there is a set B such that, given any set C, C is a member of B if and only if, given any set D, if D is a member of C, then D is a member of A [442, p. 13]*

$$\forall A \exists B \forall C (C \in B \leftrightarrow (\forall D (D \in C \rightarrow D \in A))) \tag{3.51}$$

The set B is called the power set of A and consists precisely of all subsets of A. Because the number of different subsets of the set A is equal to $2^{|A|}$, where $|A|$ is the number of elements in the set A, the power set of A is usually written as 2^A, hence $B = 2^A$.

Axiom 3.9.9. *(Axiom of foundation) Every non-empty set A contains an element x which is disjoint from A [533, p. 31]*

$$\forall A (A \neq \emptyset \rightarrow \exists x (x \in A \wedge (x \cap A) = \emptyset)) \tag{3.52}$$

Since for every set A we can always obtain its singleton {A} and the axiom of foundation requires $A \cap \{A\} = \emptyset$, it follows that no set can be an element of itself, hence $A \notin A$.

3.10   Classical probability theory

Classical probabilities obey the set theoretic axioms introduced by the Russian mathematician Andrey Kolmogorov (1903–1987) [296]. If we have a defined sample space S that is the set of all possible values the physical events may assume, we can write

$$S = \bigcup_{i=1}^{n} E_i = E_1 \cup E_2 \cup \ldots \cup E_n \tag{3.53}$$

The number n of all physical events E_i can be either finite or countably infinite. The probability $p(E_i)$ of each event E_i satisfies the following three axioms:

Axiom 3.10.1. *The probability of each event is non-negative and not greater than 1*

$$0 \leq p(E_i) \leq 1 \tag{3.54}$$

Axiom 3.10.2. *The probability of an event E_i and the probability of its complement $\bar{E}_i$ in S sum up to 1*

$$p(E_i) + p(\bar{E}_i) = p(S) = 1 \tag{3.55}$$

Axiom 3.10.3. *For k mutually exclusive subsets of S the probability for the occurrence of at least one of these events is given by the sum of the probabilities for the individual events*

$$p\left(\bigcup_{i=1}^{k} E_i\right) = \sum_{i=1}^{k} p(E_i) \tag{3.56}$$

Definition 3.19. *(Conditional probability) The conditional probability $p(A|B)$ for the occurrence of an event A provided that event B has occurred is*

$$p(A|B) = \frac{p(A \cap B)}{p(B)}, \qquad p(B) \neq 0 \tag{3.57}$$

where $p(A)$ is the probability of event A, $p(B)$ is the probability of event B, and $p(A \cap B)$ is the probability of the joint occurrence of events A and B.

Conversely, the conditional probability $p(B|A)$ for the occurrence of an event B provided that event A has occurred is

$$p(B|A) = \frac{p(A \cap B)}{p(A)}, \qquad p(A) \neq 0 \tag{3.58}$$

The probability $p(A \cap B)$ for the joint occurrence of events A and B could be expressed through Eqs. (3.57) and (3.58) as follows

$$p(A \cap B) = p(A|B)p(B) = p(B|A)p(A) \tag{3.59}$$

The probability $p(A \cup B)$ for the occurrence of at least one of the two events A or B is given by

$$p(A \cup B) = p(A) + p(B) - p(A \cap B) \tag{3.60}$$

Conversely, $p(A \cap B)$ is expressible through $p(A \cup B)$ as

$$p(A \cap B) = p(A) + p(B) - p(A \cup B) \tag{3.61}$$

Theorem 4. *The probability $p(A \cap B)$ for the joint occurrence of events A and B is bounded by*

$$\max\left[0, p(A) + p(B) - 1\right] \leq p(A \cap B) \leq \min\left[p(A), p(B)\right] \tag{3.62}$$

Proof. The lower and upper bounds can be calculated in two steps

$$
\begin{aligned}
p(A \cap B) &= p(A) + p(B) - p(A \cup B); \; p(A \cup B) \leq 1 \\
p(A \cap B) &\geq p(A) + p(B) - 1 \\
p(A \cap B) &\geq 0 \\
p(A \cap B) &\geq \max\left[0, p(A) + p(B) - 1\right]
\end{aligned}
$$

$$
\begin{aligned}
p(A \cap B) &= p(B|A)p(A); \; p(A|B) \leq 1 \\
p(A \cap B) &\leq p(A) \\
p(A \cap B) &= p(A|B)p(B); \; p(A|B) \leq 1 \\
p(A \cap B) &\leq p(B) \\
p(A \cap B) &\leq \min\left[p(A), p(B)\right]
\end{aligned}
$$

□

Theorem 5. *The probability $p(A \cup B)$ for the occurrence of at least one of the two events A or B is bounded by*

$$\max[p(A), p(B)] \leq p(A \cup B) \leq \min[1, p(A) + p(B)] \tag{3.63}$$

Proof. The lower and upper bounds can be calculated in two steps

$$
\begin{aligned}
p(A \cup B) &= p(A) + p(B) - p(A \cap B) \\
p(A \cup B) &= p(A) + p(B) - p(A|B)p(B); \; p(A|B) \leq 1 \\
p(A \cup B) &\geq p(A) + p(B) - p(B) \\
p(A \cup B) &\geq p(A) \\
p(A \cup B) &= p(A) + p(B) - p(B|A)p(A); \; p(B|A) \leq 1 \\
p(A \cup B) &\geq p(A) + p(B) - p(A) \\
p(A \cup B) &\geq p(B) \\
p(A \cup B) &\geq \max[p(A), p(B)] \\
\\
p(A \cup B) &= p(A) + p(B) - p(A \cap B); \; p(A \cap B) \geq 0 \\
p(A \cup B) &\leq p(A) + p(B) \\
p(A \cup B) &\leq 1 \\
p(A \cup B) &\leq \min[1, p(A) + p(B)]
\end{aligned}
$$

□

The above inequalities were generalized by Maurice René Fréchet (1878–1973) for multiple events [195]

$$\max\left[0, \sum_{i=1}^{k} p(A_i) - (k-1)\right] \leq p\left(\bigcap_{i=1}^{k} A_i\right) \leq \min[p(A_1), p(A_2), \ldots, p(A_k)] \tag{3.64}$$

$$\max[p(A_1), p(A_2), \ldots, p(A_k)] \leq p\left(\bigcup_{i=1}^{k} A_i\right) \leq \min\left[1, \sum_{i=1}^{k} p(A_i)\right] \tag{3.65}$$

Classical probabilities could be used to define the dependence or independence of physical events.

Definition 3.20. *(Independent events) Two events A and B are said to be independent if the probability for the joint occurrence of events A and B is equal to the product of the individual probabilities $p(A)$ and $p(B)$*

$$p(A \cap B) = p(A)p(B) \tag{3.66}$$

Physical events that do not satisfy the above criterion for independence are said to be *dependent events*.

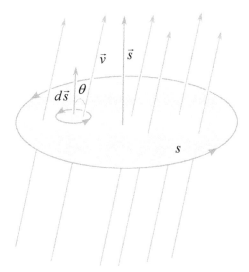

Figure 3.13   The particle flux Φ is given by the dot product of the particle velocity $\vec{v}$ through the surface s and the surface normal vector $\vec{s}$. Through every small surface element $d\vec{s}$ there is a differential flux $d\Phi = \vec{v} \cdot d\vec{s}$.

3.11   Particle and field fluxes

The particle flux Φ is a scalar physical quantity defined by the expression

$$\Phi = \lim_{\Delta t \to 0} \frac{\Delta V}{\Delta t} = \frac{dV}{dt} \tag{3.67}$$

where

$$dV = \vec{s} \cdot d\vec{\ell} = s d\ell \cos \theta \tag{3.68}$$

denotes a *volume element* with length $d\vec{\ell}$ that is filled with *fluid* that for time dt passes with velocity

$$\vec{v} = \frac{d\vec{\ell}}{dt} \tag{3.69}$$

through any cross section s of dV, and θ is the angle between the two vectors $\vec{s}$ and $\vec{v}$. It is worth recollecting that $\vec{s}$ has the direction of the positive normal $+n$ of surface area s, and its magnitude is proportional to the surface area s (Fig. 3.9). Substitution of Eq. (3.68) into Eq. (3.67) gives

$$\Phi = \vec{v} \cdot \vec{s} = vs \cos \theta \tag{3.70}$$

Thus, we have obtained that the particle flux is a scalar product of two vectors: the particle velocity vector $\vec{v}$ and the surface vector $\vec{s}$ (Fig. 3.13).

We could define an analogous scalar quantity when we investigate physical fields. The field flux $\Phi_A = \vec{A} \cdot \vec{s}$ is a scalar product of the field intensity $\vec{A}$ through surface $\vec{s}$.

3.12   Axioms of classical mechanics

Classically, the physical world is composed of a finite number n of massive particles. Each massive particle has a *position* and a *momentum* in the physical 3-dimensional space. The momentum $\vec{p}$ of a massive particle is given by the product of the particle *mass* m and the particle *velocity* $\vec{v}$ as follows

$$\vec{p} = m\vec{v} \tag{3.71}$$

Definition 3.21. *(Phase space) A multidimensional space in which every degree of freedom of a physical system is represented as an axis is called a* phase space.

The classical state of the physical system of n particles could be represented by a point in a *phase space*. In the phase space, each of the x, y and z components of the position vector of any particle would correspond to one of $3n$ *canonical position* coordinates q_i, and each of the x, y and z components of the momentum vector of any particle would correspond to one of $3n$ *canonical momentum* coordinates p_i [326, pp. 7–9]. The index i in the phase space is assumed to run from 1 to $3n$ and list the three components of position or momentum for each particle in a consecutive fashion (for example, $i = 1, 2, 3$ refers to the x, y, z components of the first particle, $i = 4, 5, 6$ refers to the x, y, z components of the second particle, etc.). The canonical positions q_i together with their corresponding canonical momenta p_i give rise to a $6n$-dimensional phase space $\mathbb{R}^{6n}$ that uniquely parametrizes solutions of the equations of motion of the classical physical system.

The world of classical mechanics can be summarized in the following three axioms [525].

Axiom 3.12.1. *(States) The state of a classical mechanical system composed of n particles is completely specified by a point in the phase space $\mathbb{R}^{6n}$, with position coordinates q_i and momentum coordinates p_i, for $i = 1, 2, \ldots, 3n$.*

Axiom 3.12.2. *(Observables) The observables of a classical mechanical system composed of n particles are the functions $f(q_i, p_i, t)$ on the phase space $\mathbb{R}^{6n}$.*

Axiom 3.12.3. *(Dynamics) There is a distinguished observable corresponding to the total energy of the system, called the Hamiltonian function $H(q_i, p_i, t)$, that uniquely determines the time evolution of the system according to Hamilton's equations*

$$\frac{dq_i}{dt} = \frac{\partial H}{\partial p_i} \tag{3.72}$$

$$\frac{dp_i}{dt} = -\frac{\partial H}{\partial q_i} \tag{3.73}$$

The time evolution of a classical *observable* $f(q_i, p_i, t)$ can be calculated with the use of the total derivative in time as

$$\frac{df}{dt} = \frac{\partial f}{\partial t} + \sum_{i=1}^{n} \left(\frac{\partial f}{\partial q_i} \frac{dq_i}{dt} + \frac{\partial f}{\partial p_i} \frac{dp_i}{dt} \right) \tag{3.74}$$

Further substitution using each of Hamilton's equations gives

$$\frac{df}{dt} = \frac{\partial f}{\partial t} + \sum_{i=1}^{n}\left(\frac{\partial f}{\partial q_i}\frac{\partial H}{\partial p_i} - \frac{\partial f}{\partial p_i}\frac{\partial H}{\partial q_i}\right) \tag{3.75}$$

The above result can be written in even more concise form if we introduce Poisson brackets.

Definition 3.22. *(Poisson bracket) The* Poisson bracket *of two functions $f_1(p_i, q_i, t)$ and $f_2(p_i, q_i, t)$ on the phase space is defined as*

$$\{f_1, f_2\} = \sum_{i=1}^{n}\left(\frac{\partial f_1}{\partial q_i}\frac{\partial f_2}{\partial p_i} - \frac{\partial f_1}{\partial p_i}\frac{\partial f_2}{\partial q_i}\right) \tag{3.76}$$

Thus, the time evolution of an observable $f(q_i, p_i, t)$ can be concisely written with the use of a Poisson bracket with the Hamiltonian H as

$$\frac{df}{dt} = \{f, H\} + \frac{\partial f}{\partial t} \tag{3.77}$$

With the latter result for the time evolution of the Hamiltonian we obtain

$$\frac{dH}{dt} = \{H, H\} + \frac{\partial H}{\partial t} = 0 + \frac{\partial H}{\partial t} = \frac{\partial H}{\partial t} \tag{3.78}$$

Here, the total derivative $\frac{dH}{dt}$ is the actual rate of change of H as the motion proceeds, with all of the coordinates q_i and p_i changing as t advances, whereas the partial derivative $\frac{\partial H}{\partial t}$ is the rate of change if we vary t, holding the coordinates q_i and p_i fixed [473, pp. 530–531]. If the Hamiltonian H does not depend explicitly on t, the partial derivative will be *zero*, $\frac{\partial H}{\partial t} = 0$. For a closed system the Hamiltonian does not vary with time, and the total energy H of the system is conserved.

3.13   Solving Hamilton's equations

The dynamics of a 1-dimensional harmonic oscillator (Fig. 3.14) could be used to illustrate Hamilton's approach to classical mechanics. The sum of the kinetic energy $\frac{1}{2}mv^2$ and the potential energy $\frac{1}{2}kx^2$ of the oscillator gives the total energy of the system

$$H = \frac{1}{2}mv^2 + \frac{1}{2}kx^2 \tag{3.79}$$

Expressed in terms of $p = mv$ and $q = x$, the Hamiltonian becomes

$$H = \frac{1}{2}\frac{p^2}{m} + \frac{1}{2}kq^2 \tag{3.80}$$

After substitution in Hamilton's equations, we find

$$\frac{dq}{dt} = \frac{\partial}{\partial p}\left(\frac{1}{2}\frac{p^2}{m} + \frac{1}{2}kq^2\right) = \frac{p}{m} \tag{3.81}$$

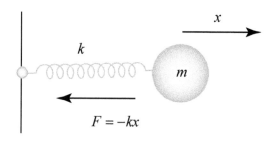

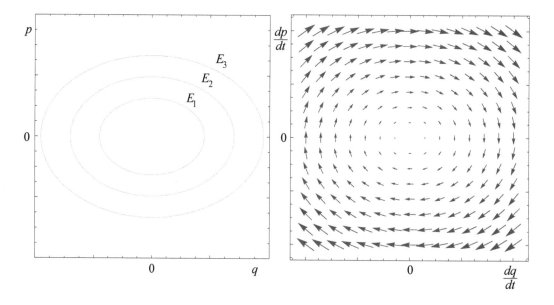

Figure 3.14   A 1-dimensional harmonic oscillator (top panel) represented by a mass m attached to a spring with spring constant k. The curves of constant energy (bottom left panel) are ellipses in phase space. The corresponding phase space flow (bottom right panel) defines motion in the clockwise direction along the ellipses of constant energy.

$$\frac{dp}{dt} = -\frac{\partial}{\partial q}\left(\frac{1}{2}\frac{p^2}{m} + \frac{1}{2}kq^2\right) = -kq \tag{3.82}$$

Since the velocity is $v = \frac{dq}{dt}$, it is easy to recognize that Eq. (3.81) is just $p = mv$. Moreover, since we have set $q = x$ and the force is $F = \frac{dp}{dt}$, it can be seen that Eq. (3.82) is just Hooke's law for the spring, namely, $F = -kx$. In the phase space, Eq. (3.80) determines ellipses with constant energies E_1, E_2, E_3, ... The time evolution of the system proceeds as a clockwise rotation along an ellipse of constant energy (Fig. 3.14). At this point, the picture of the classical world as a deterministic ticking clockwork mechanism is essentially complete.

3.14   Classical electrodynamics

Electrically charged physical objects generate electromagnetic fields that affect other charged objects in the vicinity of the field through electromagnetic forces.

Definition 3.23. *(Electric field) The electric field $\vec{E}$ is a vector field generated by electric charges or time-varying magnetic fields. The direction of the field is taken to be the direction of the force it would exert on a positive test charge. The electric field emanates radially outward from positive charges and penetrates radially in toward negative charges (Fig. 3.15). The electric intensity $\vec{E}$ is defined as the ratio of the electric force $\vec{F}_E$ acting upon a charged body and the charge q of the body*

$$\vec{E} = \lim_{\Delta q \to 0} \frac{\Delta \vec{F}_E}{\Delta q} = \frac{d\vec{F}_E}{dq} \tag{3.83}$$

The electric field of a point charge Q can be obtained from Coulomb's law as

$$\vec{E} = \frac{\vec{F}_E}{q} = \frac{1}{q} \frac{1}{4\pi\varepsilon_o} \frac{Qq}{r^2} \vec{r} = \frac{1}{4\pi\varepsilon_o} \frac{Q}{r^2} \vec{r} \tag{3.84}$$

where ε_0 is the electric permittivity of the vacuum and $\vec{r}$ is the unit vector that points from the source charge to the point of interest.

It should be noted that the electric field is a *potential field*, namely, the work W along any closed contour ℓ is *zero*

$$W = \oint_\ell \vec{F}_E \cdot d\vec{\ell} = 0 \tag{3.85}$$

Every point in the electric field has an *electric potential V* defined with the specific (for unit charge) work needed to carry the unit charge from this point to *infinity* along the path ℓ. A point x inside a given electric field has potential $V(x)$ defined by

$$V(x) = \int_x^\infty \vec{E} \cdot d\vec{\ell} + V_\infty \tag{3.86}$$

where $V_\infty = 0$ is the potential at a point infinitely far away. The electric potential difference between two points x_1 and x_2 defines a *voltage ΔV* (also known as *electromotive force* or *potential drop*)

$$\Delta V = \int_{V_1}^{V_2} dV = \int_{x_1}^{x_2} \vec{E} \cdot d\vec{\ell} \tag{3.87}$$

The link between the electric intensity and the gradient of the voltage is $\vec{E} = -\nabla V$.

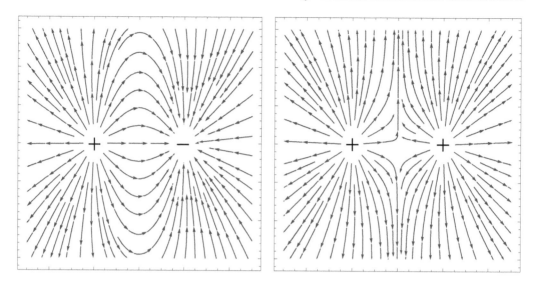

Figure 3.15   The lines of the electric field $\vec{E}$ emanate from positive charges and penetrate into negative charges. Charged particles whose charges have different signs attract one another, whereas particles whose charges have the same sign repel.

Another vector that could be used to describe the electric field is the vector of *electric induction* $\vec{D}$. For an isotropic dielectric with *electric permittivity* ε, the electric induction is defined as

$$\vec{D} = \varepsilon\vec{E} \tag{3.88}$$

The flux Φ_D of electric induction $\vec{D}$ through any closed surface s is equal to the charge q located inside the space region enclosed by s that excites the electric field

$$\Phi_D = \oiint_s \vec{D} \cdot d\vec{s} = q \tag{3.89}$$

If the normal $+n$ of the surface s and the vector $\vec{D}$ form an angle θ, the flux Φ_D through the surface s is given by

$$\Phi_D = \iint_s \vec{D} \cdot d\vec{s} = \vec{D} \cdot \vec{s} = Ds\cos\theta \tag{3.90}$$

whereas the differential flux through a small element of the surface is

$$d\Phi_D = \vec{D} \cdot d\vec{s} = Dds\cos\theta \tag{3.91}$$

From Eqs. (3.88) and (3.89) follows the *Gauss' law of electricity*

$$\Phi_E = \oiint_s \vec{E} \cdot d\vec{s} = \frac{q}{\varepsilon} \tag{3.92}$$

where Φ_E is the flux of electric intensity $\vec{E}$ through the closed surface s.

Definition 3.24. *(Electric current) The electric current i is the flux of physical charges that could be defined with the use of both scalar and vector quantities as*

$$i = \lim_{\Delta t \to 0} \frac{\Delta q}{\Delta t} = \frac{dq}{dt} = \Phi_J = \iint_S \vec{J} \cdot d\vec{s} \tag{3.93}$$

where $\vec{J}$ is the density of the electric current. As a scalar quantity the current density J is defined by the formula

$$J = \lim_{\Delta s \to 0} \frac{\Delta i}{\Delta s} = \frac{di}{ds} \tag{3.94}$$

where s is the cross section of the current flux Φ_J. It is useful to note that i denotes the flow of positive charges. The flow of negative charges could be described by a positive current with equal magnitude but opposite direction.

According to *Ohm's law*, the current i flowing through a cable depends on the voltage V, conductance G and resistance R as follows

$$i = VG = \frac{V}{R} \tag{3.95}$$

Definition 3.25. *(Magnetic field) The magnetic field $\vec{B}$ is a vector field generated by moving electric charges or time-varying electric fields. The vector of* magnetic induction $\vec{B}$ *(also known as* magnetic field strength *or* magnetic flux density*) is perpendicular to the vector of the electric intensity $\vec{E}$. The magnetic field acts only on moving charges. It manifests itself via the magnetic force $\vec{F}_M$ acting upon flowing currents inside the region where the magnetic field is distributed. According to* Laplace's law *the magnetic force $\vec{F}_M$, which acts upon an electric current-conveying cable immersed in a magnetic field with magnetic induction $\vec{B}$, is equal to the vector product*

$$\vec{F}_M = i \cdot \vec{\ell} \times \vec{B} \tag{3.96}$$

$$d\vec{F}_M = i \cdot d\vec{\ell} \times \vec{B} \tag{3.97}$$

If we have a magnetic dipole, the direction of the vector of magnetic induction is from the south pole (S) to the north pole (N) inside the dipole, and from N to S outside it.

The magnetic field could be excited either by a flowing electric current i or by changes in an existing electric field $\vec{E}$.

In the first case, if we have a cable with current i, it will generate a magnetic field with magnetic induction $\vec{B}$ given by *Ampère's circuital law*

$$\oint_\ell \vec{B} \cdot d\vec{\ell} = \mu_0 \iint_S \vec{J} \cdot d\vec{s} = \mu_0 i \tag{3.98}$$

where μ_0 is the *magnetic permeability of vacuum* and $\vec{B}$ points in the direction of rotation of a right-handed screw piercing in the direction of the current i (Fig. 3.16).

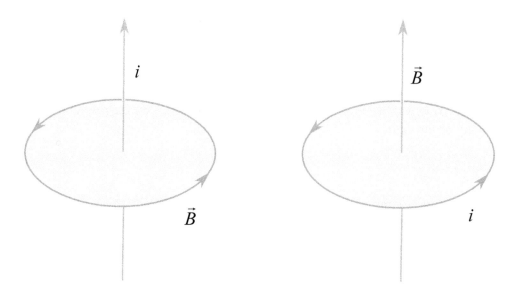

Figure 3.16   Direction of the vector of magnetic induction $\vec{B}$ around the axis with current i on the left panel and along the axis of contour with current i on the right panel. The current i by convention denotes the flux of positive charges.

In the second case, for $J = 0$, the magnetic induction by a changing electric field is defined by the *Maxwell–Ampère law* as

$$\oint_{\ell} \vec{B} \cdot d\vec{\ell} = \frac{1}{c^2} \frac{d}{dt} \Phi_E = \frac{1}{c^2} \frac{d}{dt} \iint_{s} \vec{E} \cdot d\vec{s} \qquad (3.99)$$

The total electromagnetic field exerts an electromagnetic force $\vec{F}_{EM}$ upon charged physical objects defined by the Lorentz formula

$$\vec{F}_{EM} = q\left(\vec{E} + \vec{v} \times \vec{B}\right) \qquad (3.100)$$

where $\vec{v}$ is the velocity and q is the charge of the object that is being acted upon.

If we have magnetically isotropic media with *magnetic permeability* μ, we could define the vector of *magnetic intensity* $\vec{H}$ as

$$\vec{H} = \frac{\vec{B}}{\mu} \qquad (3.101)$$

The circulation of the vector of magnetic intensity $\vec{H}$ along the closed contour ℓ_1 that interweaves in its core another contour ℓ_2 with current i flowing through ℓ_2 is defined by

$$\oint_{\ell_1} \vec{H} \cdot d\vec{\ell} = i \qquad (3.102)$$

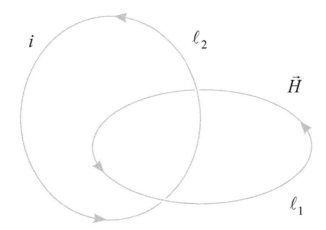

Figure 3.17 The circulation of the vector of magnetic intensity $\vec{H}$ along the closed contour ℓ_1 equals the current i flowing through the interwoven contour ℓ_2. The vector $\vec{H}$ at a point x of ℓ_1 has its direction aligned along the tangent passing through x.

It can be seen that the magnetic field is a *nonpotential field*, since the lines of field intensity $\vec{H}$ are closed and do always interweave the contour with the excitatory current i (Fig. 3.17). The circulation of the vector $\vec{H}$ will be *zero* only along closed contours which do not interweave in their cores any current i.

The flux Φ_B of magnetic induction $\vec{B}$ is given by

$$\Phi_B = \iint\limits_{s} \vec{B} \cdot d\vec{s} \tag{3.103}$$

Changing magnetic flux Φ_B generates *induced voltage V* according to *Faraday's law*

$$V = \oint\limits_{\ell} \vec{E} \cdot d\vec{\ell} = -\frac{d}{dt}\Phi_B \tag{3.104}$$

Thus, there will be induced voltage and electric current if there is a static cable inside a changing magnetic field

$$V = -\iint\limits_{s} \frac{d}{dt}\vec{B} \cdot d\vec{s} \tag{3.105}$$

or if the cable is moving inside a static magnetic field

$$V = -\left(\vec{v} \times \vec{B}\right)\vec{\ell} \tag{3.106}$$

3.15   Vector operators

The *gradient* ∇ acts on a scalar field φ and returns a vector

$$\nabla \varphi = \frac{\partial}{\partial x} \varphi \, \vec{e}_x + \frac{\partial}{\partial y} \varphi \, \vec{e}_y + \frac{\partial}{\partial z} \varphi \, \vec{e}_z = \begin{pmatrix} \frac{\partial}{\partial x} \varphi \\ \frac{\partial}{\partial y} \varphi \\ \frac{\partial}{\partial z} \varphi \end{pmatrix} \tag{3.107}$$

where we have written the unit vectors $\vec{e}_x$, $\vec{e}_y$ and $\vec{e}_z$ as column vectors

$$\vec{e}_x = \begin{pmatrix} 1 \\ 0 \\ 0 \end{pmatrix}, \vec{e}_y = \begin{pmatrix} 0 \\ 1 \\ 0 \end{pmatrix}, \vec{e}_z = \begin{pmatrix} 0 \\ 0 \\ 1 \end{pmatrix} \tag{3.108}$$

The *divergence* $\nabla \cdot$ acts on a vector field $\vec{A}$ and returns a scalar

$$\nabla \cdot \vec{A} = \nabla \cdot \begin{pmatrix} A_x \\ A_y \\ A_z \end{pmatrix} = \frac{\partial}{\partial x} A_x + \frac{\partial}{\partial y} A_y + \frac{\partial}{\partial z} A_z \tag{3.109}$$

The *curl* $\nabla \times$ acts on a vector field $\vec{A}$ and returns a vector

$$\nabla \times \vec{A} = \nabla \times \begin{pmatrix} A_x \\ A_y \\ A_z \end{pmatrix} = \begin{pmatrix} \frac{\partial}{\partial y} A_z - \frac{\partial}{\partial z} A_y \\ \frac{\partial}{\partial z} A_x - \frac{\partial}{\partial x} A_z \\ \frac{\partial}{\partial x} A_y - \frac{\partial}{\partial y} A_x \end{pmatrix} \tag{3.110}$$

The *Laplacian* ∇^2 acts on a scalar field φ and returns a scalar. It is defined as the divergence of the gradient

$$\nabla^2 \varphi = \nabla \cdot (\nabla \varphi) = \frac{\partial^2}{\partial x^2} \varphi + \frac{\partial^2}{\partial y^2} \varphi + \frac{\partial^2}{\partial z^2} \varphi \tag{3.111}$$

The *vector Laplacian* ∇^2 acts on a vector field $\vec{A}$ and returns a vector

$$\nabla^2 \vec{A} = \nabla^2 \begin{pmatrix} A_x \\ A_y \\ A_z \end{pmatrix} = \begin{pmatrix} \nabla^2 A_x \\ \nabla^2 A_y \\ \nabla^2 A_z \end{pmatrix} \tag{3.112}$$

$$\nabla^2 \vec{A} = \nabla \left(\nabla \cdot \vec{A} \right) - \nabla \times \left(\nabla \times \vec{A} \right) \tag{3.113}$$

Further rewriting gives the *curl of the curl identity*

$$\nabla \times \left(\nabla \times \vec{A} \right) = \nabla \left(\nabla \cdot \vec{A} \right) - \nabla^2 \vec{A} \tag{3.114}$$

3.16 Maxwell's equations

Maxwell's equations axiomatize neatly classical electrodynamics. If we denote with ρ the charge density, ε_0 the electric permittivity of vacuum, μ_0 the magnetic permeability of vacuum, and c the velocity of light in vacuum, we can write Maxwell's equations in integral form as

$$\oiint_s \vec{E} \cdot d\vec{s} = \frac{q}{\varepsilon_0} \tag{3.115}$$

$$\oiint_s \vec{B} \cdot d\vec{s} = 0 \tag{3.116}$$

$$\oint_\ell \vec{E} \cdot d\vec{\ell} = -\frac{d}{dt} \iint_s \vec{B} \cdot d\vec{s} \tag{3.117}$$

$$\oint_\ell \vec{B} \cdot d\vec{\ell} = \mu_0 \iint_s \vec{J} \cdot d\vec{s} + \frac{1}{c^2} \frac{d}{dt} \iint_s \vec{E} \cdot d\vec{s} \tag{3.118}$$

and in differential form as

$$\nabla \cdot \vec{E} = \frac{\rho}{\varepsilon_0} \tag{3.119}$$

$$\nabla \cdot \vec{B} = 0 \tag{3.120}$$

$$\nabla \times \vec{E} = -\frac{\partial \vec{B}}{\partial t} \tag{3.121}$$

$$\nabla \times \vec{B} = \mu_0 \vec{J} + \frac{1}{c^2} \frac{\partial \vec{E}}{\partial t} \tag{3.122}$$

Remarkably, the three basic physical constants in electrodynamics are linked by

$$\varepsilon_0 \mu_0 c^2 = 1 \tag{3.123}$$

The latter equation exposes a crucial fact, namely, that in electrodynamics the minimal number of physical units is four: length, time, mass and electric charge.

Maxwell's equations in vacuum, in the absence of charges $\rho = 0$ and currents $J = 0$, reduce to

$$\nabla \cdot \vec{E} = 0 \tag{3.124}$$

$$\nabla \cdot \vec{B} = 0 \tag{3.125}$$

$$\nabla \times \vec{E} = -\frac{\partial \vec{B}}{\partial t} \tag{3.126}$$

$$\nabla \times \vec{B} = \frac{1}{c^2} \frac{\partial \vec{E}}{\partial t} \tag{3.127}$$

With the use of the curl of the curl identity (3.114), from Eqs. (3.126) and (3.127) we can obtain the wave equations

$$\frac{1}{c^2}\frac{\partial^2 \vec{E}}{\partial t^2} - \nabla^2 \vec{E} = 0 \tag{3.128}$$

$$\frac{1}{c^2}\frac{\partial^2 \vec{B}}{\partial t^2} - \nabla^2 \vec{B} = 0 \tag{3.129}$$

Proof. The derivatives with respect to different variables commute with each other. We multiply Eq. (3.126) on both sides by $\nabla\times$ and use the fact that the curl is a spatial derivative so it commutes with the time derivative

$$\nabla \times \nabla \times \vec{E} = -\frac{\partial}{\partial t}\nabla \times \vec{B}$$

Using Eqs. (3.114) and (3.124) on the left-hand side, and Eq. (3.127) on the right-hand side results in

$$-\nabla^2 \vec{E} = -\frac{1}{c^2}\frac{\partial^2 \vec{E}}{\partial t^2}$$

Similarly, we multiply Eq. (3.127) on both sides by $\nabla\times$ to get

$$\nabla \times \nabla \times \vec{B} = \frac{1}{c^2}\frac{\partial}{\partial t}\nabla \times \vec{E}$$

Using Eqs. (3.114) and (3.125) on the left-hand side, and Eq. (3.126) on the right-hand side results in

$$-\nabla^2 \vec{B} = -\frac{1}{c^2}\frac{\partial^2 \vec{B}}{\partial t^2}$$

$\square$

After introduction of the *D'Alembertian* operator

$$\square = \frac{1}{c^2}\frac{\partial^2}{\partial t^2} - \nabla^2 \tag{3.130}$$

the wave equations (3.128) and (3.129) can be further simplified to

$$\square \vec{E} = 0 \tag{3.131}$$

$$\square \vec{B} = 0 \tag{3.132}$$

The solutions of Maxwell's equations can be written in the form

$$\vec{E} = \vec{E}_0(\vec{r}, t)\cos\left(\omega t - \vec{k}\cdot\vec{r} + \phi_0\right) \tag{3.133}$$

$$\vec{B} = \vec{B}_0(\vec{r}, t)\cos\left(\omega t - \vec{k}\cdot\vec{r} + \phi_0\right) \tag{3.134}$$

where ω is the wave angular frequency, $\vec{k} = (k_x, k_y, k_z)$ is the wave vector, $\vec{r} = (x, y, z)$ is the position vector and ϕ_0 is an arbitrary phase angle. The two vectors $\vec{E}$ and $\vec{B}$ are mutually perpendicular and located in a plane transverse to the direction of wave propagation $\vec{v}$ (Fig. 3.18).

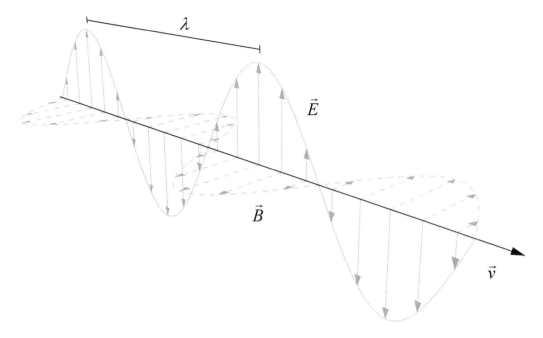

Figure 3.18 The electromagnetic waves that comprise electromagnetic radiation are self-propagating transverse oscillating waves of electric $\vec{E}$ and magnetic $\vec{B}$ fields. The electric and magnetic fields are always in phase and perpendicular to each other. Electromagnetic waves carry energy, momentum and angular momentum away from their source and can impart those quantities to material particles with which they interact.

3.17 Light

Light is made of electromagnetic waves, which are synchronized oscillations of electric and magnetic fields (Fig. 3.18). Maxwell's equations correctly predict that electromagnetic waves are radiated (emitted) by moving charges. All electromagnetic radiation occurs in the form of waves that could be characterized by the *frequency* ν or by the *wavelength* λ of their oscillation. The frequency and the wavelength are inversely related through the *speed of light in vacuum c*

$$\lambda = \frac{c}{\nu} \tag{3.135}$$

Electromagnetic interactions between charged physical systems are due to emission and absorption of electromagnetic radiation. The range of all possible frequencies of electromagnetic radiation is called the *electromagnetic spectrum*. Whereas in physics the term *light* stands for electromagnetic radiation, the term *visible light* refers to only a narrow part of the electromagnetic spectrum that can be detected by our eyes (Fig. 3.19). Ultraviolet light, X-rays and γ-rays have shorter wavelengths than visible light, whereas infrared light, microwaves and radio waves have longer wavelengths. In Section 7.3, when we discuss quantum physics, we shall see that light quanta with shorter wavelengths contain more energy according to the Planck–Einstein relation (7.2).

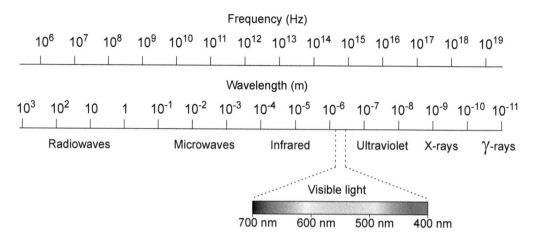

Figure 3.19 The electromagnetic spectrum. Electromagnetic waves are completely characterized by their frequency ν or corresponding wavelength λ.

Biologically, the most important effect of electromagnetic waves is the *transfer of energy* that can be used for building complex biomolecules. Plants use the energy of the sun light in the process of photosynthesis to produce sugar molecules such as glucose from carbon dioxide and water. Because the plants are subsequently used for food by animals, it is the energy delivered by the sun in the form of electromagnetic radiation that sustains the life and biodiversity on our planet. In addition, electromagnetic waves can be used for *transfer of information*. For example, we see the world around us through images generated by visible light that is scattered from physical objects.

The speed of light in vacuum c is a universal physical constant that sets the maximal speed at which all matter and electromagnetic radiation can travel in the universe. The existence of such a maximal speed, however, leads to a clash with our everyday intuition for summation of velocities. The Italian polymath Galileo Galilei (1564–1642) observed that if a person is walking on the deck of a ship with velocity $v_1 > 0$ toward the front of the ship, and if the velocity of the ship relative to the shore is $v_2 > 0$, then the velocity of the person as seen from the shore will appear to be $v = v_1 + v_2$. If the speed of light in vacuum c is the maximal possible speed, however, the Galilean addition of velocities cannot hold true for a beam of light emitted from a candle located on the ship, since in such a case we will obtain $c + v_2 > c$. In 1905, Albert Einstein (1879–1955) showed that the correct way to sum velocities is

$$v = \frac{v_1 + v_2}{1 + \frac{v_1 v_2}{c^2}} \tag{3.136}$$

Einstein's formula gives results that correspond well with our intuition only when the velocities v_1 and v_2 are much smaller than the speed of light in vacuum c. For velocities comparable with the speed of light c, relativistic effects such as length contraction and time dilation have to be taken into account [155, 157].

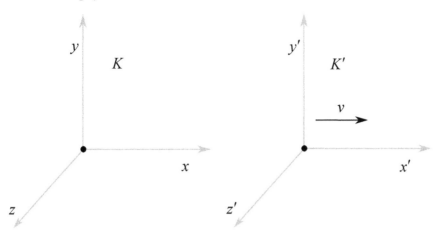

Figure 3.20 Two inertial reference systems K and K'. The system K' is moving relative to the system K with a constant velocity v along the x-direction.

3.18 Special relativity

Einstein's principle of relativity asserts that all physical processes take place in accordance with the same laws in all inertial systems. This means that no mechanical, electromagnetic or optical experiments can determine whether a given inertial system is at rest or in motion [155, 157]. Hence, being at rest and being in motion are not absolute but relative concepts.

Special relativity can be axiomatized with the following three axioms:

Axiom 3.18.1. *There is a set of physical systems that are related by linear transformations. All inertial physical systems belong to this set.*

Axiom 3.18.2. *Physical laws are the same in all inertial systems.*

Axiom 3.18.3. *There is a maximal speed at which matter and electromagnetic radiation can travel in the universe. This maximal speed is the speed of light in vacuum c.*

Einstein's axioms imply that space and time are no longer independent but are interwoven into a single continuum known as *spacetime*. To see that, consider two inertial reference systems K and K' that have parallel coordinate axes: $x \parallel x'$, $y \parallel y'$ and $z \parallel z'$. If the system K' is moving relative to the system K with a constant velocity v along the x-direction (Fig. 3.20) and if at time $t = t' = 0$, the origins of both coordinate systems coincide, $(x, y, z) = (x', y', z') = (0, 0, 0)$, the coordinate transformations between K and K' are given by *Lorentz formulas*

$$x = \frac{1}{\sqrt{1 - \frac{v^2}{c^2}}}(x' + vt') \qquad x' = \frac{1}{\sqrt{1 - \frac{v^2}{c^2}}}(x - vt) \tag{3.137}$$

$$y = y' \qquad y' = y \tag{3.138}$$

$$z = z' \qquad z' = z \tag{3.139}$$

$$t = \frac{1}{\sqrt{1 - \frac{v^2}{c^2}}}\left(t' + \frac{v}{c^2}x'\right) \qquad t' = \frac{1}{\sqrt{1 - \frac{v^2}{c^2}}}\left(t - \frac{v}{c^2}x\right) \tag{3.140}$$

Spatial and temporal dimensions from the inertial system K are linked with both spatial and temporal dimensions from the inertial system K'. Thus, spatial and temporal dimensions get interwoven. Each inertial system has its own space and its own time. Lorentz transformations imply that for moving objects the time appears to slow down and space appears to be contracted in the direction of motion. Noteworthy, the *simultaneity of events* is not absolute, so that simultaneous events from the viewpoint of one inertial system K are not simultaneous from the viewpoint of another inertial system K' moving with velocity v relative to K.

3.19   Classical information

Classical physics is built upon the concept of *classical information*. The *classical bits* of information can be read, copied, multiplied, processed, stored or erased. At first glance, it seems impossible to imagine a kind of information that cannot be read, copied, multiplied, processed, stored or erased, because intuitively all these properties are defining what information is. In Section 4.20, however, we shall see that there is another type of physical information that does not conform to the intuitions obtained from our everyday use of spoken and written language, books, personal computers, tablets, phones, other electronic devices, etc.

Example 3.1. *(Classical information) The string of bits, 0s and 1s, that encodes a digital movie recorded onto a DVD is a form of classical information. One can read the DVD and watch the movie using appropriate software and hardware. The movie file can also be copied, multiplied, processed, converted into another video format, or deleted, if needed. In the case when the DVD cannot be read, we say that the information has been corrupted or lost, so we no longer have the information.*

Each classical bit of information represents an answer to a single yes-or-no question (Section 2.2.2). For quantifying classical information, the semantic content of the yes-or-no question does not really matter provided that the possible answers are only two. Furthermore, the labeling of the two answers, yes or no, 0 or 1, is irrelevant; it suffices that the two labels are distinguishable. In essence, classical information is what allows us to distinguish between objects, entities, phenomena or thoughts. Next, with the use of the classical physics laws, we will prove several important theorems pertinent to classical information.

Theorem 6. *Classical information contained in an unknown classical physical state can be read.*

Proof. The string of bits, 0s and 1s, that represents any message needs to be encoded in at least two different physical states. Encoding that uses only a single physical state will invariably return 0000...0, which is not an answer to a yes-or-no question, hence contains no information. But if 0s and 1s are encoded in different physical states, they can always be read, since all physical states in classical physics are *observable* (Axiom 3.12.2), and the string of bits can be precisely determined. $\square$

Theorem 7. *Classical information contained in an unknown classical physical state can be cloned.*

Proof. Cloning requires copying and multiplying the original information. The string of bits, 0s and 1s, that represents a message encoded in a physical system can be read. Once we read and know what the message is, we can encode it in as many classical physical systems as we want. If we choose the number of physical system to be $n \geq 2$, we have produced at least one extra copy of the original message. □

Theorem 8. *Classical information contained in an unknown classical physical state can be broadcast.*

Proof. Broadcasting requires that the same string of bits, 0s and 1s, is conveyed to two or more recipients. Since the same classical information can be copied multiple times into different physical carriers, it can always be conveyed to more than two recipients by moving the physical carriers of information in space. □

Theorem 9. *Classical information cannot be transferred between physical systems faster than the speed of light in vacuum c.*

Proof. Transfer of classical information requires that the physical objects encoding the string of bits, 0s and 1s, are either transported themselves or the message is encoded in some other physical carrier that is then transported from the sender to the receiver. From Einstein's theory of relativity (Axiom 3.18.3), however, it follows that no physical carrier that is composed of material particles or light is able to travel through space faster than the speed of light in vacuum c. Therefore, the exchange of classical information cannot overcome that speed limit too. □

Theorem 10. *Classical information contained in an unknown classical physical state can be deleted.*

We will not attempt proving the latter theorem here, because it would require Einstein's theory of general relativity [157] and a detailed consideration of black hole dynamics [375]. In a nutshell, the argument involves throwing physical objects encoding a string of bits into a classical black hole and then using known *no-hair theorems* [27, 273] to show that the resulting black hole can be completely characterized by only three externally observable classical parameters: mass, electric charge and angular momentum. The deletion of classical information follows from the fact that one can throw many different physical strings of bits in a given initial black hole in order to produce the same final black hole state.

CHAPTER 4

The world of quantum physics

Quantum physics is the most successful description of the physical world we live in. Quantum theoretical predictions have been experimentally tested and confirmed up to the astonishing 14 digits of precision in terms of the *anomalous magnetic moment of the electron* [9, 235, 243, 427]. Regardless of this high level of experimental precision, however, quantum theory has been a source of great discomfort for physicists ever since the birth of quantum mechanics in the 1920s. The main reason for that discomfort is the fact that the quantum world at the fundamental level is not composed of material entities that are perfectly observable (measurable) and whose behavior is completely predictable (deterministic) at all times. Instead, the fundamental constituents of all quantum physical systems are complex-valued *probability amplitudes* that give rise to a set of probabilities for possible future states [456]. When a quantum system is measured, it undergoes indeterministic transition that actualizes only one of the states from the set of all physically possible future states of the system. Thus, the quantum description of the world is substantially different from a ticking clockwork mechanism.

The behavior of the physical world according to quantum mechanics could be concisely formalized in six fundamental axioms (Section 4.15). Central among those axioms is the *Schrödinger equation*, which describes the time evolution of closed quantum physical systems [422]. The solution of the Schrödinger equation is the *quantum wave function* ψ of the quantum system. At each point (x, y, z) in space at a time t, the value of the quantum wave function is a complex number $\psi(x, y, z, t)$, referred to as a *quantum probability amplitude*. Since the quantum wave function ψ provides a complete description of the quantum state of a physical system, it follows that the fabric of the quantum world is made of quantum probability amplitudes $\psi(x, y, z, t)$. The squared modulus $|\psi(x, y, z, t)|^2$ of each quantum probability amplitude gives a corresponding *quantum probability* for a physical event to occur at the given point in space and time [55, 174]. Here, it should be noted that the quantum probabilities do not arise due to our ignorance of what the state of the quantum system is, but rather the quantum probabilities represent inherent *propensities* of the quantum systems to produce certain outcomes under experimental measurement [394]. Thus, the quantum probability amplitudes are qualitatively different from all mathematical entities that appear in classical theories of physics, and as a result, quantum theory cannot be derived from a deeper physical theory based on classical principles alone [467, 468].

In order to be able to formulate precisely the axioms of quantum mechanics, several mathematical preliminaries will be needed.

4.1   Complex numbers

One of the amazing things in quantum theory is the appearance of *complex numbers*. Despite their name, the complex numbers are not complicated and have a lot of nice features that are especially helpful when one is working with objects that undergo periodic oscillations.

Real numbers $\mathbb{R}$ are inadequate for solving equations of the type $z^2 + 1 = 0$. In order to be able to solve the latter equation, one needs to introduce an *imaginary unit* $\imath$ with the property $\imath^2 = -1$. Then, the set of *complex numbers* $\mathbb{C}$ is defined as the collection of all numbers z that can be written in the binomial form $z = x + \imath y$, where x and y are real numbers, $\{x, y\} \in \mathbb{R}$ [202]. Thus, the complex numbers $\mathbb{C}$ contain the ordinary real numbers while extending them in order to solve problems that are unsolvable with real numbers only.

Geometrically, the complex numbers extend the idea of the 1-dimensional *real line* to the 2-dimensional *complex plane* by using the horizontal number line for the *real part* and adding a vertical axis to plot the *imaginary part*. Every complex number z can be represented as a point (x, y) in the *complex plane* (Fig. 4.1). In the complex plane, the *real numbers* correspond to points on the x-axis, whereas the *imaginary numbers* correspond to points on the y-axis. Hence, the x-axis is called the *real axis*, whereas the y-axis is called the *imaginary axis*.

Each complex number z is an inherently 2-dimensional mathematical entity that can also be thought of as an *ordered pair* of real numbers (x, y). Two ordered pairs (x_1, y_1) and (x_2, y_2) are equal only if $x_1 = x_2$ and $y_1 = y_2$. The first number x in the ordered pair (x, y) is the real part of the complex number z. We write $x = \text{Re}(z)$. The set of complex numbers $\mathbb{C}$ includes the real numbers $\mathbb{R}$ because the ordered pair $(x, 0)$ corresponds to the real number x. The second number y in the ordered pair (x, y) is the imaginary part of the complex number z. We write $y = \text{Im}(z)$. The ordered pair $(0, 1)$ is denoted with the symbol $\imath$ and is called *imaginary unit*.

Complex numbers form a mathematical *field* (Definition 3.2), and as such satisfy all of the field axioms, including *commutativity* and *associativity* of addition and multiplication, and *distributivity* of multiplication over addition.

The sum of two complex numbers $z_1 = (x_1, y_1)$ and $z_2 = (x_2, y_2)$ is given by

$$z_1 + z_2 = (x_1 + x_2, y_1 + y_2) \tag{4.1}$$

The product of two complex numbers $z_1 = (x_1, y_1)$ and $z_2 = (x_2, y_2)$ is given by

$$z_1 z_2 = (x_1 x_2 - y_1 y_2, x_1 y_2 + x_2 y_1) \tag{4.2}$$

That each complex number written as an ordered pair $z = (x, y)$ can also be written in the binomial form $z = x + \imath y$ can be shown by using Eqs. (4.1) and (4.2) as follows

$$z = x + \imath y = (x, 0) + (0, 1)(y, 0) = (x, 0) + (0, y) = (x, y)$$

From Eq. (4.2) it also follows that

$$\ldots, \imath^{-2} = -1, \imath^{-1} = -\imath, \imath^0 = 1, \imath^1 = \imath, \imath^2 = -1, \imath^3 = -\imath, \imath^4 = 1, \imath^5 = \imath, \imath^6 = -1, \ldots \tag{4.3}$$

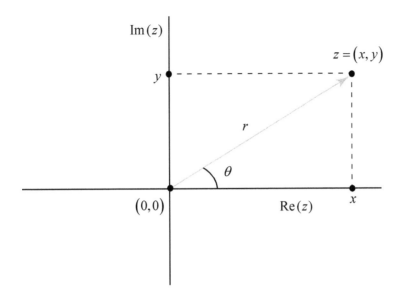

Figure 4.1   Representation of a complex number z in the complex plane.

Definition 4.1. *(Complex conjugation) Complex numbers that have equal real parts but opposite imaginary parts such as $z = x + \imath y$ and $z^* = x - \imath y$ are referred to as* complex conjugated numbers. *The operation of flipping the sign of the imaginary part of a given complex number is called* complex conjugation *and is denoted with the symbol* *.

The sum and the product of two complex conjugated numbers are given by

$$z + z^* = 2\text{Re}(z) = 2x \tag{4.4}$$

$$zz^* = [\text{Re}(z)]^2 + [\text{Im}(z)]^2 = x^2 + y^2 \tag{4.5}$$

The difference of two complex numbers $z_1 - z_2$ is a number z that satisfies the condition $z + z_2 = z_1$.

$$z_1 - z_2 = (x_1 - x_2, y_1 - y_2) \tag{4.6}$$

The fraction of two complex numbers $\frac{z_1}{z_2}$ is a number z that satisfies the condition $zz_2 = z_1$.

$$\frac{z_1}{z_2} = \left(\frac{x_1 x_2 + y_1 y_2}{x_2^2 + y_2^2}, \frac{x_2 y_1 - x_1 y_2}{x_2^2 + y_2^2} \right) \tag{4.7}$$

The division of two complex numbers can be easily performed using multiplication of both the numerator and denominator by the complex conjugated denominator followed by separation of the real and imaginary parts

$$\frac{z_1}{z_2} = \frac{z_1 z_2^*}{z_2 z_2^*} = \frac{(x_1 + \imath y_1)(x_2 - \imath y_2)}{(x_2 + \imath y_2)(x_2 - \imath y_2)} = \frac{x_1 x_2 + y_1 y_2}{x_2^2 + y_2^2} + \imath \frac{x_2 y_1 - x_1 y_2}{x_2^2 + y_2^2}$$

So far we have seen that each complex number z can be represented as an ordered pair (x,y) or as a binomial $x + \imath y$. There is a third way to represent the complex number, known as the trigonometric form of the complex number.

The complex number z described by the *Cartesian coordinates* (x,y) can be written also in *polar coordinates* (r,θ), where r is the length of the *radius vector* $\vec{r}$ that joins the origin of the coordinate system $(0,0)$ and the point z, and θ is the angle between the radius vector $\vec{r}$ and the positive direction of the x-axis (Fig. 4.1). The length r of the radius vector $\vec{r}$ is called the *modulus of the complex number z*, whereas the angle θ is called the *argument of the complex number*

$$r = |z| = \sqrt{zz^*} \tag{4.8}$$

$$\theta = \arg(z) \tag{4.9}$$

The relationship between Cartesian coordinates and polar coordinates can be summarized in the following *system of equations*

$$\begin{cases} x = r\cos\theta \\ y = r\sin\theta \end{cases} \tag{4.10}$$

$$\begin{cases} r = |z| = \sqrt{zz^*} = \sqrt{x^2 + y^2} \\ \theta = \arg(z) = \text{atan2}(x,y) = \begin{cases} \arctan(\frac{y}{x}) & \text{if } x > 0 \\ \arctan(\frac{y}{x}) + \pi & \text{if } x < 0 \text{ and } y \geq 0 \\ \arctan(\frac{y}{x}) - \pi & \text{if } x < 0 \text{ and } y < 0 \\ \frac{\pi}{2} & \text{if } x = 0 \text{ and } y > 0 \\ -\frac{\pi}{2} & \text{if } x = 0 \text{ and } y < 0 \\ \text{indeterminate} & \text{if } x = 0 \text{ and } y = 0 \end{cases} \end{cases} \tag{4.11}$$

Thus, the modulus is a single-valued function, whereas the argument is a *multivalued function*. The argument can change by any multiple of 2π and still give the same complex number: $\arg(z) = \theta + 2\pi k$, where $k = 0, \pm 1, \pm 2, \ldots$. The *principal value* of $\arg(z)$ is obtained when $k = 0$. Conversion from Cartesian coordinates to polar coordinates is done with the use of the two-argument function $\text{atan2}(x,y)$ defined on the interval $(-\pi, \pi]$, instead of the one-argument function $\arctan(\frac{y}{x})$ defined on the interval $(-\frac{\pi}{2}, \frac{\pi}{2}]$ (Fig. 4.2). The disadvantage of the arctan function is that it has a fraction as an input, but fractions do not keep memory of negative signs, $-\frac{y}{x} = \frac{-y}{x} = \frac{y}{-x}$ and $\frac{y}{x} = \frac{-y}{-x}$.

The relationship between Cartesian coordinates and polar coordinates allows one to express the complex number $z = x + \imath y$ in *trigonometric form* as

$$z = re^{\imath\theta} = r(\cos\theta + \imath\sin\theta) \tag{4.12}$$

where we have used the famous Euler formula

$$e^{\imath\theta} = (\cos\theta + \imath\sin\theta) \tag{4.13}$$

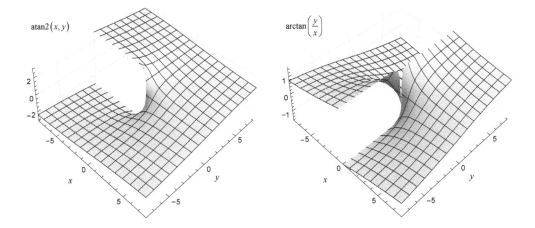

Figure 4.2   Plots of the two-argument function atan2(x, y) and the one-argument function arctan$(\frac{y}{x})$.

Trigonometrically, the multiplication of two complex numbers is easier

$$z_1 z_2 = r_1 r_2 e^{i(\theta_1 + \theta_2)} = r_1 r_2 \left[\cos(\theta_1 + \theta_2) + i \sin(\theta_1 + \theta_2) \right] \tag{4.14}$$

The modulus of the product is the product of the individual moduli

$$|z_1 z_2| = r_1 r_2 \tag{4.15}$$

whereas the argument of the product is the sum of the individual arguments

$$\arg(z_1 z_2) = \theta_1 + \theta_2 \tag{4.16}$$

Thus, multiplication of complex numbers leads to scaling and rotation in the complex plane. Here, a caveat should be made. The arguments can change by any multiple of 2π and still give the same complex numbers: $\arg(z_1) = \theta_1 + 2\pi k_1$, $k_1 = 0, \pm 1, \pm 2, \ldots$ and $\arg(z_2) = \theta_2 + 2\pi k_2$, $k_2 = 0, \pm 1, \pm 2, \ldots$ Because addition of these multiples of 2π produces again a multiple of 2π, one can work only with θ_1 and θ_2 without explicitly tracking the multiples of 2π. We shall see, however, that the multiples of 2π cannot be omitted in the formula for rooting (Eq. 4.22).

Division of two complex numbers can be easily derived using the result (4.14). We have $z_1 = \frac{z_1}{z_2} z_2$, from which it follows that

$$\left| \frac{z_1}{z_2} \right| = \frac{r_1}{r_2}; \qquad \arg\left(\frac{z_1}{z_2} \right) = \theta_1 - \theta_2 \tag{4.17}$$

Therefore

$$\frac{z_1}{z_2} = \frac{r_1}{r_2} e^{i(\theta_1 - \theta_2)} = \frac{r_1}{r_2} \left[\cos(\theta_1 - \theta_2) + i \sin(\theta_1 - \theta_2) \right] \tag{4.18}$$

Exponentiation of a complex number z to any power given by natural number n is derived from repeated application of the multiplication rule

$$z^n = \left(re^{i\theta}\right)^n = r^n e^{in\theta} = r^n \left[\cos\left(n\theta\right) + i\sin\left(n\theta\right)\right] \tag{4.19}$$

For the modulus and argument of the complex number z^n we have

$$|z^n| = r^n \tag{4.20}$$

$$\arg\left(z^n\right) = n\theta \tag{4.21}$$

Rooting can be derived using the result (4.14), but we need to be cautious about the $2\pi k$ terms as follows

$$\sqrt[n]{z} = r^{\frac{1}{n}} e^{i\left(\frac{\theta + 2\pi k}{n}\right)} = r^{\frac{1}{n}}\left[\cos\left(\frac{\theta + 2\pi k}{n}\right) + i\sin\left(\frac{\theta + 2\pi k}{n}\right)\right] \tag{4.22}$$

$$\left|\sqrt[n]{z}\right| = \left|z^{\frac{1}{n}}\right| = r^{\frac{1}{n}} \tag{4.23}$$

$$\arg\left(\sqrt[n]{z}\right) = \arg\left(z^{\frac{1}{n}}\right) = \frac{\theta + 2\pi k}{n} \tag{4.24}$$

Here the introduction of the $2\pi k$ term is essential because division of $2\pi k$ by n does not necessarily produce a multiple of 2π. Indeed, for $k = 0, 1, 2, \ldots, n-1$ we obtain exactly n distinct roots for $\sqrt[n]{z}$. For k that is different from the aforementioned values $0, 1, 2, \ldots, n-1$ we will obtain roots that are among those, which we already know.

4.2   Wave functions

In quantum mechanics, the solutions of the Schrödinger equation (4.157) are given by *quantum wave functions* ψ. If a quantum wave function is represented in position basis as $\psi(x)$, it gives the value of the quantum amplitude at each point x. The marvelous thing about quantum amplitudes is that they are *complex numbers*. Since the positions x are always real, it follows that the quantum wave function is a function with the domain being the real numbers $\mathbb{R}$ and the codomain being the complex numbers $\mathbb{C}$, or symbolically $\psi(x) : \mathbb{R} \to \mathbb{C}$. Because each complex dimension is inherently 2-dimensional, in order to plot a quantum wave function in one spatial dimension, we need 3-dimensional plots. To illustrate this, let us have a quantum wave function whose analytical expression is a Morlet wavelet [17]

$$\psi(x) = \left(e^{-i\omega x} - e^{-\frac{1}{2}\omega^2}\right)e^{-\frac{1}{2}x^2} \tag{4.25}$$

where the parameter ω is connected with both the energy and the frequency of the wavelet. Visualization of a Morlet wavelet with $\omega = 13$ is shown in Figure 4.3. For most points in space x the wave function $\psi(x)$ has both non-zero real and non-zero imaginary parts. The real and imaginary parts can be plotted separately as functions of x with real domains and codomains, $\text{Re}\left[\psi(x)\right] : \mathbb{R} \to \mathbb{R}$ and

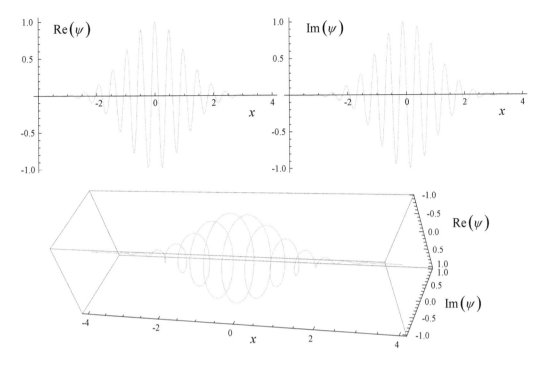

Figure 4.3   Plots of the real $\mathrm{Re}\,[\psi(x)]$, imaginary $\mathrm{Im}\,[\psi(x)]$, and both real and imaginary parts of the quantum wave function $\psi(x)$ of a Morlet wavelet with $\omega = 13$.

$\mathrm{Im}\,[\psi(x)] : \mathbb{R} \to \mathbb{R}$, using two coupled 2-dimensional plots (Fig. 4.3, top). Better comprehension of the geometry of the wavelet, however, can be achieved by a single 3-dimensional plot of the quantum wave function $\psi(x) : \mathbb{R} \to \mathbb{C}$ with real domain and complex codomain (Fig. 4.3, bottom).

Quantum systems in three real spatial dimensions have quantum wave functions $\psi(x, y, z)$ that have complex values at each point in space (x, y, z). Complete visualization of such wave functions $\psi(x, y, z)$ is impossible. One way to develop an intuition for the geometry and behavior of quantum systems in higher numbers of dimensions is to study lower dimensional plots in which there is only one non-fixed variable, with the rest of the variables being fixed to particular values.

According to the Born rule, the complex quantum probability amplitudes of the quantum wave function $\psi(x, y, z, t)$ determine the probability for finding the quantum particle at point (x, y, z) at time t with the use of the squared modulus of the wave function, which is real and non-negative [55]

$$\mathrm{prob}(x, y, z, t) = |\psi(x, y, z, t)|^2 \geq 0 \tag{4.26}$$

Because the particle has to be located somewhere in space with absolute certainty, we require that the quantum wave function $\psi(x, y, z, t)$ is a square-integrable function normalized to one

$$\int_{-\infty}^{\infty} \int_{-\infty}^{\infty} \int_{-\infty}^{\infty} |\psi(x, y, z, t)|^2 dx\,dy\,dz = 1 \tag{4.27}$$

4.3   Vector spaces

Quantum wave functions participate in interference phenomenona in which two waves superpose to form a resultant wave of greater, lower, or the same amplitude. Constructive interference occurs when the phase difference between the waves is an even multiple of π, whereas destructive interference occurs when the difference is an odd multiple of π. For intermediate phase differences between these two extremes, the amplitude of the resultant wave lies between the minimal and maximal interference values. Because quantum wave functions are able to sum with each other, they behave like vectors, and as solutions of the Schrödinger equation they form a vector space.

Definition 4.2. *(Vector space) A set of objects V is a* vector space *over a* field F, *if for all vector elements $\vec{x}, \vec{y}, \vec{z} \in V$ and any scalars $a, b \in F$, the following eight axioms hold:*

Axiom 4.3.1. *Commutativity of vector addition*

$$\vec{x} + \vec{y} = \vec{y} + \vec{x} \tag{4.28}$$

Axiom 4.3.2. *Associativity of vector addition*

$$(\vec{x} + \vec{y}) + \vec{z} = \vec{x} + (\vec{y} + \vec{z}) \tag{4.29}$$

Axiom 4.3.3. *For all vectors there exists an additive identity*

$$\vec{x} + \vec{0} = \vec{x} \tag{4.30}$$

Axiom 4.3.4. *For all vectors there exists an additive inverse*

$$\vec{x} + (-\vec{x}) = \vec{0} \tag{4.31}$$

Axiom 4.3.5. *Associativity of scalar multiplication*

$$a(b\vec{x}) = (a \cdot b)\vec{x} \tag{4.32}$$

Axiom 4.3.6. *Distributivity of scalar sums*

$$(a + b)\vec{x} = a\vec{x} + b\vec{x} \tag{4.33}$$

Axiom 4.3.7. *Distributivity of vector sums*

$$a(\vec{x} + \vec{y}) = a\vec{x} + a\vec{y} \tag{4.34}$$

Axiom 4.3.8. *For all vectors there exists a scalar multiplication identity*

$$1 \cdot \vec{x} = \vec{x} \tag{4.35}$$

If the vector space is over the *field* of real numbers $\mathbb{R}$, it is called a *real vector space*. Alternatively, if the vector space is over the *field* of complex numbers $\mathbb{C}$, it is called a *complex vector space*.

4.4   Inner product spaces

The eight axioms that define a vector space (Section 4.3) do not have a notion of *vector multiplication*. In a vector space, the *inner product* provides a way to multiply vectors together, resulting in a *scalar* output. A vector space for which an *inner product* is defined is called an *inner product space*. Since in quantum mechanics we are primarily interested in complex vector spaces, below we will provide a generalized definition of an inner product that is applicable to complex vector spaces.

Definition 4.3. *(Inner product space) The* inner product · *on a complex vector space is defined for all vector elements* $\vec{x}, \vec{y}, \vec{z} \in V$ *and any scalar number* $a \in \mathbb{C}$, *with the following four axioms:*

Axiom 4.4.1. *The inner product is complex conjugate symmetric*

$$\vec{x} \cdot \vec{y} = (\vec{y} \cdot \vec{x})^* \tag{4.36}$$

Axiom 4.4.2. *The inner product is distributive over vector addition*

$$(\vec{x} + \vec{y}) \cdot \vec{z} = \vec{x} \cdot \vec{z} + \vec{y} \cdot \vec{z} \tag{4.37}$$

Axiom 4.4.3. *The inner product is linear in the first vector variable and complex conjugate symmetric in the second vector variable*

$$a\vec{x} \cdot \vec{y} = a(\vec{x} \cdot \vec{y}) \tag{4.38}$$

$$\vec{x} \cdot a\vec{y} = a^*(\vec{x} \cdot \vec{y}) \tag{4.39}$$

Axiom 4.4.4. *The inner product is positive definite, which means that the inner product of a vector* $\vec{x}$ *with itself is non-negative and is zero if and only if* $\vec{x}$ *is the zero vector*

$$\vec{x} \cdot \vec{x} \geq 0 \tag{4.40}$$

$$\vec{x} \cdot \vec{x} = 0 \leftrightarrow \vec{x} = \vec{0} \tag{4.41}$$

The inner product introduces the *norm* $|\cdot|$ that gives the *length* of any vector $\vec{x}$

$$|\vec{x}| = \sqrt{\vec{x} \cdot \vec{x}} \tag{4.42}$$

From the properties of the inner product it follows that the vector norm is positive definite, positive scalable and satisfies the triangle inequality

$$|\vec{x}| \geq 0 \tag{4.43}$$

$$|a\vec{x}| = |a||\vec{x}| \tag{4.44}$$

$$|\vec{x} + \vec{y}| \leq |\vec{x}| + |\vec{y}| \tag{4.45}$$

4.5   Metric

The eight axioms that define a vector space (Section 4.3) do not have a notion of *distance* between the elements of the space. The notion of distance needs to be explicitly defined with the use of a *distance function* called a *metric*.

Definition 4.4. *(Metric) A metric of a set M is a non-negative function $g(x,y) \geq 0$ describing the distance between every two elements $x,y \in M$. The metric has to satisfy the following three axioms:*

Axiom 4.5.1. *The metric is symmetric*

$$g(x,y) = g(y,x) \tag{4.46}$$

Axiom 4.5.2. *The metric between two elements is zero if and only if the two elements coincide*

$$g(x,y) = 0 \leftrightarrow x = y \tag{4.47}$$

Axiom 4.5.3. *The metric satisfies the triangle inequality*

$$g(x,y) + g(y,z) \geq g(x,z) \tag{4.48}$$

If to the eight axioms that define a vector space (Section 4.3) we further add the four axioms that define an inner product (Section 4.4), we obtain an inner product space. Inner product spaces are always metric spaces, because we can use the *norm* of a vector given by Eq. (4.42) to define a metric on the vector space V as

$$g(\vec{x},\vec{y}) = |\vec{x} - \vec{y}| \tag{4.49}$$

In such a case, we say that the metric g is *induced by* the norm $|\cdot|$.

Definition 4.5. *(Metric space) A metric space M is a set for which distances between any two members of the set $x,y \in M$ are defined with a metric $g(x,y)$.*

Definition 4.6. *(Complete metric space) A metric space M is complete if every infinite Cauchy sequence $\{x_n\}_{n=1}^{\infty}$ of elements in M has a limit that is also in M. Here under Cauchy sequence is understood an infinite sequence $\{x_n\}_{n=1}^{\infty} = x_1, x_2, x_3, \ldots$ whose elements become arbitrarily close to each other as the sequence progresses (Fig. 4.4), or formally, if for every positive real number $\varepsilon > 0$, there is a positive integer N such that the distance given by the metric $g(x_m, x_n) < \varepsilon$ for all natural numbers $m, n > N$.*

4.6   Hilbert space

A powerful way to study quantum wave functions ψ_i is to represent them as vectors $\vec{\psi}_i$ in a complex *Hilbert space* $\mathcal{H}$. Such a representation is possible because the set of all quantum wave functions ψ_i that are solutions of the Schrödinger equation satisfy the axioms of a complex Hilbert space [140]. Satisfying these axioms is primarily due to the linearity of the Schrödinger equation, which implies that any linear sum of quantum wave function solutions $\sum_i a_i \psi_i$, where $a_i \in \mathbb{C}$ are arbitrary complex coefficients, is also a solution of the Schrödinger equation.

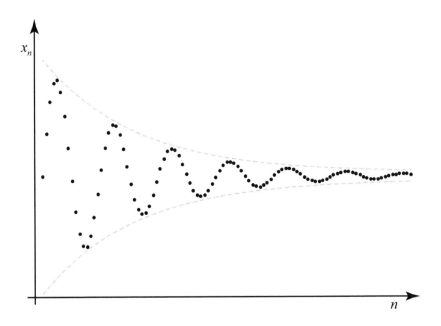

Figure 4.4   Example of an infinite Cauchy sequence $\{x_n\}_{n=1}^{\infty} = x_1, x_2, x_3, \ldots$ of elements in a metric space M. The elements of the sequence become arbitrarily close to each other as the sequence progresses.

Definition 4.7. *(Hilbert space) A Hilbert space $\mathcal{H}$ is a vector space possessing an inner product $\vec{x} \cdot \vec{y}$ such that the norm defined by $|\vec{x}| = \sqrt{\vec{x} \cdot \vec{x}}$ turns the space $\mathcal{H}$ into a complete metric space. Completeness of the metric space requires that the limit of every infinite Cauchy sequence of vectors $\{\vec{x}_n\}_{n=1}^{\infty}$ in $\mathcal{H}$ is a vector that also belongs to $\mathcal{H}$.*

If the Hilbert space $\mathcal{H}$ is over the *field* of real numbers $\mathbb{R}$, it is called a *real Hilbert space*. Alternatively, if the Hilbert space is over the *field* of complex numbers $\mathbb{C}$, it is called a *complex Hilbert space*.

4.7   Bra-ket notation

The bra-ket notation, composed of angle brackets and vertical bars, enhances the clarity of mathematical operations and improves the comprehension of quantum mechanical expressions. Motivated by the apparently different roles played by the first and the second vector variables in the axioms defining the inner product of two vectors (Section 4.4), Paul Dirac (1902–1984), who shared the 1933 Nobel Prize in Physics for his work on quantum mechanics, decided to introduce a better notation [139] for denoting vectors in complex Hilbert space $\mathcal{H}$. Instead of simply denoting the quantum wave function as a vector with the symbol $\vec{\psi}$, Dirac proposed to call it a *ket* $|\psi\rangle$ in $\mathcal{H}$. If the complex Hilbert space is n-dimensional, and if an orthogonal representation basis $|\psi_i\rangle$ is chosen, the ket can be written as

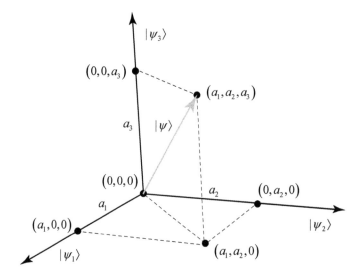

Figure 4.5   Illustration of a ket vector $|\psi\rangle$ whose coordinates a_1, a_2, a_3 are represented in respect to a set of orthogonal basis vectors $|\psi_1\rangle, |\psi_2\rangle, |\psi_3\rangle$ in 3-dimensional Hilbert space.

an $n \times 1$ matrix that is a column vector

$$|\psi\rangle = \begin{pmatrix} a_1 \\ a_2 \\ \vdots \\ a_n \end{pmatrix} \tag{4.50}$$

The representation basis $|\psi_i\rangle$ consists of orthogonal unit vectors $|\psi_1\rangle, |\psi_2\rangle, \ldots, |\psi_n\rangle$ that are perpendicular to each other and provide a coordinate system in which any ket vector $|\psi\rangle$ can be specified with a set of coordinates $a_1, a_2, \ldots, a_n$ produced by respective perpendicular (orthogonal) projections of the ket vector $|\psi\rangle$ to each of the basis vectors $|\psi_1\rangle, |\psi_2\rangle, \ldots, |\psi_n\rangle$ (Fig. 4.5). It is important to note that the representation of the ket $|\psi\rangle$ as a column vector is basis dependent because different choices of basis vectors for the coordinate system will produce different coordinates a_i for the same ket $|\psi\rangle$. The column representation of a ket vector is equivalent to the following expression in which the basis vectors are given explicitly

$$|\psi\rangle = a_1|\psi_1\rangle + a_2|\psi_2\rangle + \ldots + a_n|\psi_n\rangle = \sum_{i=1}^{n} a_i|\psi_i\rangle \tag{4.51}$$

In order to be able to write inner products, to each ket $|\psi\rangle$ Dirac also defined a *bra* $\langle\psi|$ that is the complex conjugate transpose of the ket

$$\langle\psi| = \begin{pmatrix} a_1^* & a_2^* & \cdots & a_n^* \end{pmatrix} \tag{4.52}$$

It is convenient to think of the bra $\langle\psi|$ as a $1 \times n$ matrix that is a row vector in a *dual Hilbert space* $\mathcal{H}^*$.

The operation composed of complex conjugation and transposition of a matrix is denoted with the symbol †. Because the bra is the complex conjugate transpose of the ket, and vice versa, we can write

$$|\psi\rangle^\dagger = \langle\psi| \tag{4.53}$$

$$\langle\psi|^\dagger = |\psi\rangle \tag{4.54}$$

The inner product of two vectors $\vec{x}$ and $\vec{y}$ in $\mathcal{H}$ written in the bra-ket notation is

$$\vec{x} \cdot \vec{y} = \langle y|x\rangle \tag{4.55}$$

The inner product between $|x\rangle$ and $|y\rangle$ in $\mathcal{H}$ is the same as the scalar obtained by matrix multiplication of the row vector $\langle y|$ in $\mathcal{H}^*$ and the column vector $|x\rangle$ in $\mathcal{H}$.

With the use of Eq. (4.55) we can rewrite the four axioms for the inner product in bra-ket notation as follows:

Axiom 4.7.1. *The inner product is complex conjugate symmetric*

$$\langle y|x\rangle = \langle x|y\rangle^* \tag{4.56}$$

Axiom 4.7.2. *The inner product is distributive over vector addition*

$$\langle z|x + y\rangle = \langle z|x\rangle + \langle z|y\rangle \tag{4.57}$$

Axiom 4.7.3. *For any scalar $a \in \mathbb{C}$, the inner product is linear in the ket vector variable and complex conjugate symmetric in the bra vector variable*

$$\langle y|ax\rangle = a\langle y|x\rangle \tag{4.58}$$

$$\langle ay|x\rangle = a^*\langle y|x\rangle \tag{4.59}$$

Axiom 4.7.4. *The inner product is positive definite*

$$\langle x|x\rangle \geq 0 \tag{4.60}$$

Every vector space contains a *zero vector* $\vec{0}$. The zero vector in $\mathcal{H}$ is called the *zero ket* and is obtained when the scalar number *zero* is multiplied by any ket

$$0|\psi\rangle = \begin{pmatrix} 0 \\ 0 \\ \vdots \\ 0 \end{pmatrix} \tag{4.61}$$

Thus, the scalar zero is different from the zero ket. Nevertheless, in Dirac notation one writes the zero ket as 0. The reason for this abuse of mathematical notation is that in quantum information theory one wants to write the two basis states of a quantum bit as $|0\rangle$ and $|1\rangle$, where $|0\rangle$ is a vector with unit length such that $\langle 0|0\rangle = 1$.

In contrast, the inner product of the *zero ket* with itself is the scalar *zero*

$$\begin{pmatrix} 0 & 0 & \cdots & 0 \end{pmatrix} \begin{pmatrix} 0 \\ 0 \\ \vdots \\ 0 \end{pmatrix} = 0 \tag{4.62}$$

Since the symbol $|0\rangle$ in quantum information theory does not represent the zero vector, quantum physicists have decided to assign the symbol 0 to the zero ket and determine from the context whether the symbol 0 denotes a scalar or a vector. Such a determination from the context is always possible because scalars can never be summed with vectors [244, p. 17].

4.8   Matrix multiplication

The bra-ket notation provides a handy way for writing abstract inner products. However, to be able to actually multiply row vectors by column vectors in a given basis and calculate the value of inner products, one needs to know the rules for *matrix multiplication*. Matrix representation and computation of matrix products is also of great use for solving linear systems of equations that frequently arise in physical applications [163].

Definition 4.8. *(Matrix multiplication of vectors) Matrix multiplication between a row vector and a column vector is possible only for vectors with the same dimensionality that have the same number of entries. The result from matrix multiplication of an n-dimensional row vector*

$$\langle x| = \begin{pmatrix} b_1 & b_2 & \cdots & b_n \end{pmatrix} \tag{4.63}$$

with an n-dimensional column vector

$$|\psi\rangle = \begin{pmatrix} a_1 \\ a_2 \\ \vdots \\ a_n \end{pmatrix} \tag{4.64}$$

is a scalar sum of all individual products formed by multiplying the i row entry by the corresponding i column entry

$$\langle x|\psi\rangle = b_1 a_1 + b_2 a_2 + \ldots + b_n a_n = \sum_{i=1}^{n} b_i a_i \tag{4.65}$$

With the rule for multiplication of row vectors by column vectors, one is able to define the matrix multiplication for arbitrary matrices $\hat{A}$ and $\hat{B}$ such that the number of columns of $\hat{A}$ equals the number of rows of $\hat{B}$.

Definition 4.9. *(Matrix multiplication) Matrix multiplication, also known as the Cayley product of two matrices $\hat{A}$ and $\hat{B}$, is only possible if the matrix $\hat{A}$ has the same number of columns as the number of rows of the matrix $\hat{B}$ [163, pp. 19–20]. If we multiply an $n \times m$ matrix*

$$
\hat{A} = \begin{pmatrix}
A_{11} & A_{12} & \cdots & A_{1m} \\
A_{21} & A_{22} & \cdots & A_{2m} \\
\vdots & \vdots & \ddots & \vdots \\
A_{n1} & A_{n2} & \cdots & A_{nm}
\end{pmatrix}
\tag{4.66}
$$

by an $m \times p$ matrix

$$
\hat{B} = \begin{pmatrix}
B_{11} & B_{12} & \cdots & B_{1p} \\
B_{21} & B_{22} & \cdots & B_{2p} \\
\vdots & \vdots & \ddots & \vdots \\
B_{m1} & B_{m2} & \cdots & B_{mp}
\end{pmatrix}
\tag{4.67}
$$

the result is an $n \times p$ matrix

$$
\hat{A}\hat{B} = \begin{pmatrix}
(AB)_{11} & (AB)_{12} & \cdots & (AB)_{1p} \\
(AB)_{21} & (AB)_{22} & \cdots & (AB)_{2p} \\
\vdots & \vdots & \ddots & \vdots \\
(AB)_{n1} & (AB)_{n2} & \cdots & (AB)_{np}
\end{pmatrix}
\tag{4.68}
$$

such that the ij entry of the matrix $\hat{A}\hat{B}$ is produced by multiplication of the i row of the matrix $\hat{A}$ by the j column of the matrix $\hat{B}$

$$
(AB)_{ij} = A_{i1}B_{1j} + A_{i2}B_{2j} + \ldots + A_{im}B_{mj} = \sum_{k=1}^{m} A_{ik}B_{kj}
\tag{4.69}
$$

4.9   Operators

The matrices that appear in quantum mechanics are square $n \times n$ matrices, where n is the number of dimensions of the Hilbert space $\mathcal{H}$. Matrix multiplication of an $n \times n$ matrix by an $n \times 1$ column vector results in an $n \times 1$ column vector, whereas matrix multiplication of a $1 \times n$ row vector by an $n \times n$ matrix results in a $1 \times n$ row vector. Thus, every $n \times n$ matrix is an *operator* that is a mathematical *function* with input being a vector and output being a vector. Because the physical states in quantum theory are represented by vectors, it can be said that operators applied on physical states produce other physical states. Typically, operators are labeled with a hat symbol as $\hat{A}$, $\hat{B}$, etc. Matrix multiplication of an $n \times 1$ column vector by a $1 \times n$ row vector results in an $n \times n$ matrix. Thus, any expression of the form $|\psi\rangle\langle\chi|$ formed by matrix multiplication of a ket $|\psi\rangle$ and a bra $\langle\chi|$ is an *operator*. Applying the operator $|\psi\rangle\langle\chi|$ on the left of a ket $|\varphi\rangle$ produces

$$
|\psi\rangle\langle\chi|\varphi\rangle = (\langle\chi|\varphi\rangle)|\psi\rangle
\tag{4.70}
$$

Alternatively, applying the operator $|\psi\rangle\langle\chi|$ on the right of a bra $\langle\varphi|$ produces

$$\langle\varphi|\psi\rangle\langle\chi| = ((\langle\varphi|\psi\rangle))\langle\chi| \tag{4.71}$$

Because matrix multiplication is linear, it follows that quantum mechanical operators are linear too.

Definition 4.10. *(Linear operator) An operator $\hat{A}$ is linear if for every pair of ket vectors $|\psi\rangle$ and $|\chi\rangle$ and any scalar a, the following equations hold*

$$\hat{A}(|\psi\rangle + |\chi\rangle) = \hat{A}|\psi\rangle + \hat{A}|\chi\rangle \tag{4.72}$$

$$\hat{A}(a|\psi\rangle) = a\hat{A}|\psi\rangle \tag{4.73}$$

Definition 4.11. *(Eigenvectors and eigenvalues) Among all ket vectors $|\psi\rangle$ upon which an operator $\hat{A}$ can operate, there is a special set of vectors $\{|\psi_m\rangle\}_{m=1}^{n}$ such that*

$$\hat{A}|\psi_m\rangle = \lambda_m|\psi_m\rangle \tag{4.74}$$

namely, the action of the operator $\hat{A}$ returns back the original ket vector $|\psi_m\rangle$ multiplied by some scaling factor λ_m. Every such ket vector $|\psi_m\rangle$ is called an eigenvector *of the operator $\hat{A}$ with an* eigenvalue λ_m.

To every physical *observable* (measurable parameter) A there is an associated operator $\hat{A}$ in $\mathcal{H}$ such that when operating upon a wave function $|\psi_m\rangle$ associated with a definite value λ_m of the observable A will yield the value λ_m times the original wave function $|\psi_m\rangle$. Because physical measurements always return a result from the measurements, it is required that the set of eigenvectors of the operator $\hat{A}$ provides a complete basis of the complex Hilbert space $\mathcal{H}$. Furthermore, since the results from physical measurements are always real numbers, it is also required that every operator $\hat{A}$ associated with a physical observable is represented by a *Hermitian matrix*, namely, $\hat{A} = \hat{A}^{\dagger}$.

Definition 4.12. *(Spectral decomposition) Every Hermitian operator $\hat{A} = \hat{A}^{\dagger}$ on an n-dimensional Hilbert space $\mathcal{H}$ can be expressed in terms of its eigenvalues λ_m and eigenvectors $|\psi_m\rangle$ as*

$$\hat{A} = \lambda_1|\psi_1\rangle\langle\psi_1| + \lambda_2|\psi_2\rangle\langle\psi_2| + \ldots + \lambda_n|\psi_n\rangle\langle\psi_n| = \sum_{m=1}^{n}\lambda_m|\psi_m\rangle\langle\psi_m| \tag{4.75}$$

where each projection operator $|\psi_m\rangle\langle\psi_m|$ projects input ket vectors onto the ray defined by the ket $|\psi_m\rangle$. The linear combination of pairwise orthogonal projections $|\psi_m\rangle\langle\psi_m|$ given by Eq. (4.75) is called the spectral decomposition *of the operator $\hat{A}$.*

Definition 4.13. *(Eigenstates) Eigenvectors of Hermitian matrices $\hat{A} = \hat{A}^{\dagger}$ representing physical observables are referred to as* eigenstates. *Each eigenstate represents a quantum physical state in which the given observable possesses a definite eigenvalue.*

In general, matrix multiplication of two operators is not commutative and the order in which the two operators appear is important, $\hat{A}\hat{B} \neq \hat{B}\hat{A}$. Still, there are some pairs of operators that commute with each other and $\hat{A}\hat{B} = \hat{B}\hat{A}$ holds.

Definition 4.14. *(Commutator) The* commutator $\left[\hat{A},\hat{B}\right]$ *of two operators* $\hat{A}$ *and* $\hat{B}$ *is given by*

$$\left[\hat{A},\hat{B}\right] = \hat{A}\hat{B} - \hat{B}\hat{A} \tag{4.76}$$

Any two commuting operators $\hat{A}$ *and* $\hat{B}$ *have a* zero *commutator because if* $\hat{A}\hat{B} = \hat{B}\hat{A}$, *then* $\hat{A}\hat{B} - \hat{B}\hat{A} = \left[\hat{A},\hat{B}\right] = 0$. *Conversely, if two operators have a* zero *commutator then necessarily the operators are commuting.*

Theorem 11. *If two Hermitian operators commute, they possess a complete orthonormal set of common eigenvectors, and vice versa [341, p. 200].*

4.10   Orthonormal basis

With the use of inner products, we can express concisely the geometric properties of vectors in algebraic form.

Definition 4.15. *(Unit vector) A* unit vector *is any vector* $|\psi\rangle$ *with unit length* $|\psi| = 1$. *From Eq. (4.42) it follows that the inner product of the unit vector* $|\psi\rangle$ *with itself is equal to one*

$$\langle\psi|\psi\rangle = 1 \tag{4.77}$$

Definition 4.16. *(Orthogonal vectors)* Orthogonal vectors *(vectors perpendicular to each other) are any two vectors* $|\psi_1\rangle$ *and* $|\psi_2\rangle$ *that have* zero *inner product*

$$\langle\psi_1|\psi_2\rangle = 0 \tag{4.78}$$

Definition 4.17. *(Orthonormal vectors) Orthogonal vectors with unit length are called* orthonormal vectors.

Definition 4.18. *(Linear independence) A set of vectors* $\{|\psi_1\rangle,|\psi_2\rangle,|\psi_3\rangle,\dots,|\psi_n\rangle\}$ *is* linearly independent *if none of the vectors in the set can be expressed as a linear sum of the other vectors. A set of mutually orthogonal non-zero vectors is always linearly independent.*

Definition 4.19. *(Orthogonal basis) For a vector space* V *with n dimensions, the maximal number of mutually orthogonal non-zero vectors that can be found is n. Having a set containing the maximal number of mutually orthogonal non-zero vectors* $\{|\psi_i\rangle\}_{i=1}^{n} = \{|\psi_1\rangle,|\psi_2\rangle,\dots,|\psi_n\rangle\}$, *one can express any vector in the vector space V as a linear combination of the vectors in the set* $\{|\psi_i\rangle\}_{i=1}^{n}$. *Consequently, we say that the vector set* $\{|\psi_i\rangle\}_{i=1}^{n}$ *forms an* orthogonal basis *for the vector space V.*

Definition 4.20. *(Orthonormal basis) An orthogonal basis* $\{|\psi_i\rangle\}_{i=1}^n$ *of a vector space* V *composed of unit vectors is called an* orthonormal basis. *For any two vectors in the orthonormal basis the inner product is given by*

$$\langle \psi_i | \psi_j \rangle = \delta_{ij} \tag{4.79}$$

where δ_{ij} *is the* Kronecker delta

$$\delta_{ij} = \begin{cases} 0 & \text{if } i \neq j \\ 1 & \text{if } i = j \end{cases} \tag{4.80}$$

Any ket vector $|\psi\rangle$ in a complex Hilbert space $\mathcal{H}$ can be expressed using inner products with the unit vectors that form an orthonormal basis $\{|\psi_i\rangle\}_{i=1}^n$ of $\mathcal{H}$. For that purpose we decompose the *unit operator* $\hat{I}$ (also known as the *identity operator*) in terms of the basis vectors $\{|\psi_i\rangle\}_{i=1}^n$ as

$$\hat{I} = \sum_{i=1}^n |\psi_i\rangle\langle\psi_i| \tag{4.81}$$

Because the action of the unit operator $\hat{I}$ on any ket $|\psi\rangle$ is to leave the ket unchanged, we have

$$|\psi\rangle = \sum_{i=1}^n |\psi_i\rangle\langle\psi_i|\psi\rangle = \sum_{i=1}^n \langle\psi_i|\psi\rangle|\psi_i\rangle \tag{4.82}$$

Recognizing that the orthogonal projection a_i of the ket vector $|\psi\rangle$ onto a basis vector $|\psi_i\rangle$ (Fig. 4.5) is given by the corresponding inner product, namely, $a_i = \langle\psi_i|\psi\rangle$, we obtain the known result from Eq. (4.51)

$$|\psi\rangle = \sum_{i=1}^n a_i|\psi_i\rangle = a_1|\psi_1\rangle + a_2|\psi_2\rangle + \ldots + a_n|\psi_n\rangle \tag{4.83}$$

4.11   Quantum wave function representations

The *quantum state vector* $|\psi\rangle$ and the *quantum wave function* ψ in a Hilbert space $\mathcal{H}$ are abstract mathematical concepts that are basis independent. For explicit matrix calculations, however, the quantum state vector $|\psi\rangle$ and the quantum wave function ψ need to be represented in a certain basis. The domain of the quantum wave function ψ is not unique, but is fixed when a representation basis is chosen. Particular representations can be obtained with the use of orthogonal decompositions of the unit operator

$$\hat{I} = \int_{-\infty}^{\infty} |x\rangle\langle x|dx = \int_{-\infty}^{\infty} |p\rangle\langle p|dp \tag{4.84}$$

Most common are the position $\psi(x)$ and the momentum $\psi(p)$ representations

$$|\psi\rangle = \int_{-\infty}^{\infty} |x\rangle\langle x|\psi\rangle dx = \int_{-\infty}^{\infty} \psi(x,t)|x\rangle dx \tag{4.85}$$

$$|\psi\rangle = \int_{-\infty}^{\infty} |p\rangle\langle p|\psi\rangle dp = \int_{-\infty}^{\infty} \psi(p,t)|p\rangle dp \tag{4.86}$$

The complex numbers $\langle x|\psi\rangle$ and $\langle p|\psi\rangle$ are the *quantum probability amplitudes* to find the particle in the state $|\psi\rangle$ at position x or with momentum p, correspondingly. In general, the quantum probability amplitudes will vary with position x or momentum p. Such varying is described by a mathematical function, $\psi(x,t)$ or $\psi(p,t)$, which is called the quantum wave function

$$\langle x|\psi\rangle = \psi(x,t) \tag{4.87}$$
$$\langle p|\psi\rangle = \psi(p,t) \tag{4.88}$$

Strictly speaking, one should distinguish between the function ψ and its value $\psi(x)$ at a point x. Nevertheless, mathematical notation is often abused and expressions such as $\psi(x)$ or $\psi(p)$ are used to highlight the wave function domain and the representation basis. To avoid possible confusion arising from such abuse of notation, we will refer to $\psi(x)$ as a *quantum wave function* when we treat it as a mathematical function varying with x, whereas we will refer to $\psi(x)$ as a *quantum probability amplitude* when we treat it as the value of the quantum wave function ψ at the point x.

The position and the momentum representations of the wave function are related through Fourier transform [453, p. 54]. For one dimension, we have

$$\psi(x,t) = \frac{1}{\sqrt{2\pi\hbar}} \int_{-\infty}^{\infty} e^{\frac{i}{\hbar}px} \psi(p,t)dp \tag{4.89}$$

$$\psi(p,t) = \frac{1}{\sqrt{2\pi\hbar}} \int_{-\infty}^{\infty} e^{-\frac{i}{\hbar}px} \psi(x,t)dx \tag{4.90}$$

Proof. Consider the inner product $\langle x|p\rangle = (\langle p|x\rangle)^*$ that is explicitly given by

$$\langle x|p\rangle = \frac{1}{\sqrt{2\pi\hbar}} e^{\frac{i}{\hbar}px} \tag{4.91}$$

$$\langle p|x\rangle = \frac{1}{\sqrt{2\pi\hbar}} e^{-\frac{i}{\hbar}px} \tag{4.92}$$

Introducing orthogonal decompositions of the unit operator we have [490, p. 160]

$$\langle x|\psi\rangle = \int_{-\infty}^{\infty} \langle x|p\rangle\langle p|\psi\rangle dp = \frac{1}{\sqrt{2\pi\hbar}} \int_{-\infty}^{\infty} e^{\frac{i}{\hbar}px}\langle p|\psi\rangle dp \tag{4.93}$$

$$\langle p|\psi\rangle = \int_{-\infty}^{\infty} \langle p|x\rangle\langle x|\psi\rangle dx = \frac{1}{\sqrt{2\pi\hbar}} \int_{-\infty}^{\infty} e^{-\frac{i}{\hbar}px}\langle x|\psi\rangle dx \tag{4.94}$$

Substituting $\langle x|\psi\rangle = \psi(x,t)$ and $\langle p|\psi\rangle = \psi(p,t)$ gives Eqs. (4.89) and (4.90). $\square$

   With the use of the Dirac delta function, we can show that if the wave function is normalized in the position basis, it is also normalized in the momentum basis.

Definition 4.21. *(Delta function) The* Dirac delta function $\delta(x)$ *is the limit of a normal distribution centered at $\mu = 0$ with vanishing variance $\sigma^2 \to 0$*

$$\delta(x) = \lim_{\sigma^2 \to 0} \frac{1}{\sqrt{2\pi\sigma^2}} e^{-\frac{(x-\mu)^2}{2\sigma^2}} \tag{4.95}$$

Thus, $\delta(x)$ is zero everywhere except at $x = 0$, where it is infinite, and has an integral of one over the entire real line [140, pp. 58–60]

$$\delta(x) = \begin{cases} +\infty, & x = 0 \\ 0, & x \neq 0 \end{cases} \tag{4.96}$$

$$\int_{-\infty}^{\infty} \delta(x)dx = \int_{-\infty}^{\infty} \delta(x - x')dx = 1 \tag{4.97}$$

Some useful identities involving the Dirac delta function are

$$\delta(x) = \delta(-x) \tag{4.98}$$

$$\delta(ax) = \frac{1}{|a|}\delta(x), \qquad a \neq 0 \tag{4.99}$$

$$f(x)\delta(x - x') = f(x')\delta(x - x') \tag{4.100}$$

$$f(x') = \int_{-\infty}^{\infty} f(x)\delta(x - x')dx \tag{4.101}$$

$$\delta(x - x') = \frac{1}{2\pi} \int_{-\infty}^{\infty} e^{ik(x-x')}dk \tag{4.102}$$

In the position basis, we have

$$|x'\rangle = \int_{-\infty}^{\infty} |x\rangle\langle x|x'\rangle dx = \int_{-\infty}^{\infty} |x\rangle\delta(x - x')dx \tag{4.103}$$

where $\langle x|x'\rangle = \delta(x - x')$. Together with the normalization condition $\langle\psi|\psi\rangle = 1$ this further gives

$$\begin{aligned} \langle\psi|\psi\rangle &= \int_{-\infty}^{\infty}\int_{-\infty}^{\infty} \langle\psi|x\rangle\langle x|x'\rangle\langle x'|\psi\rangle dxdx' \\ &= \int_{-\infty}^{\infty}\int_{-\infty}^{\infty} \langle\psi|x\rangle\delta(x - x')\langle x'|\psi\rangle dxdx' \\ &= \int_{-\infty}^{\infty} \langle\psi|x\rangle\langle x|\psi\rangle dx \\ &= \int_{-\infty}^{\infty} |\langle x|\psi\rangle|^2 dx \\ &= \int_{-\infty}^{\infty} |\psi(x)|^2 dx = 1 \end{aligned} \tag{4.104}$$

Similar calculation shows that

$$
\int_{-\infty}^{\infty} |\psi(p)|^2 dp = \int_{-\infty}^{\infty} \langle \psi | p \rangle \langle p | \psi \rangle dp
$$

$$
= \int_{-\infty}^{\infty} \int_{-\infty}^{\infty} \int_{-\infty}^{\infty} \langle \psi | x \rangle \langle x | p \rangle \langle p | x' \rangle \langle x' | \psi \rangle dx dx' dp
$$

and after substituting the inner products given by Eqs. (4.91) and (4.92)

$$
\int_{-\infty}^{\infty} |\psi(p)|^2 dp = \frac{1}{2\pi\hbar} \int_{-\infty}^{\infty} \int_{-\infty}^{\infty} \int_{-\infty}^{\infty} \langle \psi | x \rangle e^{\frac{i}{\hbar} p(x-x')} \langle x' | \psi \rangle dx dx' dp
$$

$$
= \frac{1}{2\pi} \int_{-\infty}^{\infty} \int_{-\infty}^{\infty} \int_{-\infty}^{\infty} \langle \psi | x \rangle e^{ik(x-x')} \langle x' | \psi \rangle dx dx' dk
$$

$$
= \int_{-\infty}^{\infty} \int_{-\infty}^{\infty} \langle \psi | x \rangle \delta(x-x') \langle x' | \psi \rangle dx dx'
$$

$$
= \int_{-\infty}^{\infty} \langle \psi | x \rangle \langle x | \psi \rangle dx = \int_{-\infty}^{\infty} |\psi(x)|^2 dx = 1 \qquad (4.105)
$$

Thus, if the wave function is normalized in the position basis $\int_{-\infty}^{\infty} |\psi(x)|^2 dx = 1$, it is also normalized in the momentum basis $\int_{-\infty}^{\infty} |\psi(p)|^2 dp = 1$. This could be viewed as a consequence of the fact that *plane waves* e^{ikx} integrate to Dirac delta functions

$$
\int_{-\infty}^{\infty} e^{ikx} dk = 2\pi\delta(x) \qquad (4.106)
$$

$$
\int_{-\infty}^{\infty} e^{ikx} dx = 2\pi\delta(k) \qquad (4.107)
$$

4.12   Two-level quantum systems

The Pauli spin matrices $\hat{\sigma}_x$, $\hat{\sigma}_y$ and $\hat{\sigma}_z$ are operators commonly encountered in quantum mechanical problems related to two-level quantum systems, or qubits, such as spin-$\frac{1}{2}$ particles. Each Pauli spin matrix represents an observable, namely, the direction of the spin, up or down, along the corresponding x-, y- or z-axis in the real 3-dimensional Euclidean space $\mathbb{R}^3$. If we choose $| \uparrow_z \rangle, | \downarrow_z \rangle$ as a representation basis in the Hilbert space $\mathcal{H} = \mathbb{C}^2$, the Pauli spin matrices can be written as

$$
\hat{\sigma}_x = \begin{pmatrix} 0 & 1 \\ 1 & 0 \end{pmatrix} \qquad (4.108)
$$

$$
\hat{\sigma}_y = \begin{pmatrix} 0 & -i \\ i & 0 \end{pmatrix} \qquad (4.109)
$$

$$
\hat{\sigma}_z = \begin{pmatrix} 1 & 0 \\ 0 & -1 \end{pmatrix} \qquad (4.110)
$$

The corresponding eigenvectors and eigenvalues are

$$|\uparrow_x\rangle = \frac{1}{\sqrt{2}}\begin{pmatrix} 1 \\ 1 \end{pmatrix}, \qquad \lambda_+ = +1 \tag{4.111}$$

$$|\downarrow_x\rangle = \frac{1}{\sqrt{2}}\begin{pmatrix} 1 \\ -1 \end{pmatrix}, \qquad \lambda_- = -1 \tag{4.112}$$

$$|\uparrow_y\rangle = \frac{1}{\sqrt{2}}\begin{pmatrix} 1 \\ \imath \end{pmatrix}, \qquad \lambda_+ = +1 \tag{4.113}$$

$$|\downarrow_y\rangle = \frac{1}{\sqrt{2}}\begin{pmatrix} 1 \\ -\imath \end{pmatrix}, \qquad \lambda_- = -1 \tag{4.114}$$

$$|\uparrow_z\rangle = \begin{pmatrix} 1 \\ 0 \end{pmatrix}, \qquad \lambda_+ = +1 \tag{4.115}$$

$$|\downarrow_z\rangle = \begin{pmatrix} 0 \\ 1 \end{pmatrix}, \qquad \lambda_- = -1 \tag{4.116}$$

Expressing the eigenstates of spin oriented along the x-axis or y-axis through the eigenstates of spin oriented along the z-axis, and vice versa, could be done as follows

$$|\uparrow_x\rangle = \frac{1}{\sqrt{2}}(|\uparrow_z\rangle + |\downarrow_z\rangle); \quad |\downarrow_x\rangle = \frac{1}{\sqrt{2}}(|\uparrow_z\rangle - |\downarrow_z\rangle) \tag{4.117}$$

$$|\uparrow_z\rangle = \frac{1}{\sqrt{2}}(|\uparrow_x\rangle + |\downarrow_x\rangle); \quad |\downarrow_z\rangle = \frac{1}{\sqrt{2}}(|\uparrow_x\rangle - |\downarrow_x\rangle) \tag{4.118}$$

$$|\uparrow_y\rangle = \frac{1}{\sqrt{2}}(|\uparrow_z\rangle + \imath|\downarrow_z\rangle); \quad |\downarrow_y\rangle = \frac{1}{\sqrt{2}}(|\uparrow_z\rangle - \imath|\downarrow_z\rangle) \tag{4.119}$$

$$|\uparrow_z\rangle = \frac{1}{\sqrt{2}}\Big(|\uparrow_y\rangle + |\downarrow_y\rangle\Big); \quad |\downarrow_z\rangle = -\frac{1}{\sqrt{2}}\imath\Big(|\uparrow_y\rangle - |\downarrow_y\rangle\Big) \tag{4.120}$$

The Pauli spin matrices for spin-$\frac{1}{2}$ obey the identities

$$\hat{\sigma}_x\hat{\sigma}_x = \hat{\sigma}_y\hat{\sigma}_y = \hat{\sigma}_z\hat{\sigma}_z = -\imath\hat{\sigma}_x\hat{\sigma}_y\hat{\sigma}_z = \hat{I} \tag{4.121}$$

and the commutation relations

$$[\hat{\sigma}_x, \hat{\sigma}_y] = 2\imath\hat{\sigma}_z; \qquad [\hat{\sigma}_y, \hat{\sigma}_z] = 2\imath\hat{\sigma}_x; \qquad [\hat{\sigma}_z, \hat{\sigma}_x] = 2\imath\hat{\sigma}_y \tag{4.122}$$

The corresponding spin operators $\hat{S}_x$, $\hat{S}_y$ and $\hat{S}_z$ for spin-$\frac{1}{2}$ are given by

$$\hat{S}_x = \frac{\hbar}{2}\hat{\sigma}_x; \qquad \hat{S}_y = \frac{\hbar}{2}\hat{\sigma}_y; \qquad \hat{S}_z = \frac{\hbar}{2}\hat{\sigma}_z \tag{4.123}$$

4.13 Three-level quantum systems

Spin-1 particles are an example of three-level quantum mechanical systems, or qutrits. Generalized Pauli spin matrices $\hat{\sigma}_x$, $\hat{\sigma}_y$ and $\hat{\sigma}_z$ represent three-level spin observables that have spin eigenvalues of +1, 0 or −1, along the corresponding x-, y- or z-axes in the real 3-dimensional Euclidean space $\mathbb{R}^3$. If we choose $|\uparrow_z\rangle, |0_z\rangle, |\downarrow_z\rangle$ as a representation basis in the Hilbert space $\mathcal{H} = \mathbb{C}^3$, the generalized Pauli spin matrices for spin-1 can be written as

$$\hat{\sigma}_x = \frac{1}{\sqrt{2}} \begin{pmatrix} 0 & 1 & 0 \\ 1 & 0 & 1 \\ 0 & 1 & 0 \end{pmatrix} \tag{4.124}$$

$$\hat{\sigma}_y = \frac{1}{\sqrt{2}} \begin{pmatrix} 0 & -\imath & 0 \\ \imath & 0 & -\imath \\ 0 & \imath & 0 \end{pmatrix} \tag{4.125}$$

$$\hat{\sigma}_z = \begin{pmatrix} 1 & 0 & 0 \\ 0 & 0 & 0 \\ 0 & 0 & -1 \end{pmatrix} \tag{4.126}$$

The corresponding eigenvectors and eigenvalues are

$$|\uparrow_x\rangle = \frac{1}{2} \begin{pmatrix} 1 \\ \sqrt{2} \\ 1 \end{pmatrix}, \qquad \lambda_+ = +1 \tag{4.127}$$

$$|0_x\rangle = \frac{1}{\sqrt{2}} \begin{pmatrix} -1 \\ 0 \\ 1 \end{pmatrix}, \qquad \lambda_0 = 0 \tag{4.128}$$

$$|\downarrow_x\rangle = \frac{1}{2} \begin{pmatrix} 1 \\ -\sqrt{2} \\ 1 \end{pmatrix}, \qquad \lambda_- = -1 \tag{4.129}$$

$$|\uparrow_y\rangle = \frac{1}{2} \begin{pmatrix} 1 \\ \imath\sqrt{2} \\ -1 \end{pmatrix}, \qquad \lambda_+ = +1 \tag{4.130}$$

$$|0_y\rangle = \frac{1}{\sqrt{2}} \begin{pmatrix} 1 \\ 0 \\ 1 \end{pmatrix}, \qquad \lambda_0 = 0 \tag{4.131}$$

$$|\downarrow_y\rangle = \frac{1}{2} \begin{pmatrix} -1 \\ \imath\sqrt{2} \\ 1 \end{pmatrix}, \qquad \lambda_- = -1 \tag{4.132}$$

$$|\uparrow_z\rangle = \begin{pmatrix} 1 \\ 0 \\ 0 \end{pmatrix}, \qquad \lambda_+ = +1 \tag{4.133}$$

$$|0_z\rangle = \begin{pmatrix} 0 \\ 1 \\ 0 \end{pmatrix}, \qquad \lambda_0 = 0 \tag{4.134}$$

$$|\downarrow_z\rangle = \begin{pmatrix} 0 \\ 0 \\ 1 \end{pmatrix}, \qquad \lambda_- = -1 \tag{4.135}$$

The generalized Pauli spin matrices for spin-1 obey the commutation relations

$$[\hat{\sigma}_x, \hat{\sigma}_y] = i\hat{\sigma}_z; \qquad [\hat{\sigma}_y, \hat{\sigma}_z] = i\hat{\sigma}_x; \qquad [\hat{\sigma}_z, \hat{\sigma}_x] = i\hat{\sigma}_y \tag{4.136}$$

The corresponding spin operators $\hat{S}_x$, $\hat{S}_y$ and $\hat{S}_z$ for spin-1 are given by

$$\hat{S}_x = \hbar\hat{\sigma}_x; \qquad \hat{S}_y = \hbar\hat{\sigma}_y; \qquad \hat{S}_z = \hbar\hat{\sigma}_z \tag{4.137}$$

An important property that we will need for proving the Kochen–Specker theorem in Section 4.20, is that the squared spin-1 operators along three perpendicular axes in the real 3-dimensional Euclidean space $\mathbb{R}^3$ commute with each other and can be measured simultaneously. From Theorem 11 it follows that $\hat{\sigma}_x^2$, $\hat{\sigma}_y^2$ and $\hat{\sigma}_z^2$ have a complete orthonormal set of common eigenvectors

$$\hat{\sigma}_x^2 = \begin{pmatrix} \frac{1}{2} & 0 & \frac{1}{2} \\ 0 & 1 & 0 \\ \frac{1}{2} & 0 & \frac{1}{2} \end{pmatrix}; \quad \hat{\sigma}_y^2 = \begin{pmatrix} \frac{1}{2} & 0 & -\frac{1}{2} \\ 0 & 1 & 0 \\ -\frac{1}{2} & 0 & \frac{1}{2} \end{pmatrix}; \quad \hat{\sigma}_z^2 = \begin{pmatrix} 1 & 0 & 0 \\ 0 & 0 & 0 \\ 0 & 0 & 1 \end{pmatrix} \tag{4.138}$$

$$|s_1\rangle = \begin{pmatrix} -1 \\ 0 \\ 1 \end{pmatrix}; \quad |s_2\rangle = \begin{pmatrix} 1 \\ 0 \\ 1 \end{pmatrix}; \quad |s_3\rangle = \begin{pmatrix} 0 \\ 1 \\ 0 \end{pmatrix} \tag{4.139}$$

Each of the simultaneous eigenvectors, however, has different eigenvalues for exactly two pairs of the squared spin-1 operators

$$\hat{\sigma}_x^2|s_1\rangle = 0; \qquad \hat{\sigma}_y^2|s_1\rangle = |s_1\rangle; \qquad \hat{\sigma}_z^2|s_1\rangle = |s_1\rangle \tag{4.140}$$

$$\hat{\sigma}_x^2|s_2\rangle = |s_2\rangle; \qquad \hat{\sigma}_y^2|s_2\rangle = 0; \qquad \hat{\sigma}_z^2|s_2\rangle = |s_2\rangle \tag{4.141}$$

$$\hat{\sigma}_x^2|s_3\rangle = |s_3\rangle; \qquad \hat{\sigma}_y^2|s_3\rangle = |s_3\rangle; \qquad \hat{\sigma}_z^2|s_3\rangle = 0 \tag{4.142}$$

Thus, the squared spin-1 operators along three perpendicular axes in the real 3-dimensional Euclidean space $\mathbb{R}^3$ always return two 1s and one 0. This result can be experimentally confirmed with high accuracy, and is known as the $1, 0, 1$ rule [99, 100, 101]. A direct corollary of the $1, 0, 1$ rule is that there can be no two 0s for the squared spin-1 operators along any two perpendicular axes in the real 3-dimensional Euclidean space $\mathbb{R}^3$.

4.14   Tensor products

Composite systems in quantum mechanics are described by *tensor products*. Forming a tensor product, denoted by the symbol $\otimes$, is always possible for any two matrices with arbitrary dimensions. Thus, the tensor product is an entirely different operation compared to ordinary matrix multiplication (Section 4.8).

Definition 4.22. *(Tensor product) The tensor product, also known as the* Kronecker product, *of an $n \times m$ matrix*

$$
\hat{A} = \begin{pmatrix}
A_{11} & A_{12} & \cdots & A_{1m} \\
A_{21} & A_{22} & \cdots & A_{2m} \\
\vdots & \vdots & \ddots & \vdots \\
A_{n1} & A_{n2} & \cdots & A_{nm}
\end{pmatrix}
\tag{4.143}
$$

and a $p \times q$ matrix

$$
\hat{B} = \begin{pmatrix}
B_{11} & B_{12} & \cdots & B_{1q} \\
B_{21} & B_{22} & \cdots & B_{2q} \\
\vdots & \vdots & \ddots & \vdots \\
B_{p1} & B_{p2} & \cdots & B_{pq}
\end{pmatrix}
\tag{4.144}
$$

is another $np \times mq$ matrix

$$
\hat{A} \otimes \hat{B} = \begin{pmatrix}
A_{11}\hat{B} & A_{12}\hat{B} & \cdots & A_{1m}\hat{B} \\
A_{21}\hat{B} & A_{22}\hat{B} & \cdots & A_{2m}\hat{B} \\
\vdots & \vdots & \ddots & \vdots \\
A_{n1}\hat{B} & A_{n2}\hat{B} & \cdots & A_{nm}\hat{B}
\end{pmatrix}
\tag{4.145}
$$

or written explicitly in blocks

$$
\left(
\begin{bmatrix}
A_{11}B_{11} & A_{11}B_{12} & \cdots & A_{11}B_{1q} \\
A_{11}B_{21} & A_{11}B_{22} & \cdots & A_{11}B_{2q} \\
\vdots & \vdots & \ddots & \vdots \\
A_{11}B_{p1} & A_{11}B_{p2} & \cdots & A_{11}B_{pq}
\end{bmatrix}
\cdots
\begin{bmatrix}
A_{1m}B_{11} & A_{1m}B_{12} & \cdots & A_{1m}B_{1q} \\
A_{1m}B_{21} & A_{1m}B_{22} & \cdots & A_{1m}B_{2q} \\
\vdots & \vdots & \ddots & \vdots \\
A_{1m}B_{p1} & A_{1m}B_{p2} & \cdots & A_{1m}B_{pq}
\end{bmatrix}
\right.
$$
$$
\vdots \qquad \ddots \qquad \vdots
$$
$$
\left.
\begin{bmatrix}
A_{n1}B_{11} & A_{n1}B_{12} & \cdots & A_{n1}B_{1q} \\
A_{n1}B_{21} & A_{n1}B_{22} & \cdots & A_{n1}B_{2q} \\
\vdots & \vdots & \ddots & \vdots \\
A_{n1}B_{p1} & A_{n1}B_{p2} & \cdots & A_{n1}B_{pq}
\end{bmatrix}
\cdots
\begin{bmatrix}
A_{nm}B_{11} & A_{nm}B_{12} & \cdots & A_{nm}B_{1q} \\
A_{nm}B_{21} & A_{nm}B_{22} & \cdots & A_{nm}B_{2q} \\
\vdots & \vdots & \ddots & \vdots \\
A_{nm}B_{p1} & A_{nm}B_{p2} & \cdots & A_{nm}B_{pq}
\end{bmatrix}
\right)
$$

The blocks in the tensor product matrix do not need to be separated with brackets. Here the brackets were inserted in order to highlight the structure of the tensor product. It can be seen that, in general, the tensor products are not commutative.

The tensor product is linear and associative

$$\left(\hat{A} + \hat{B}\right) \otimes \hat{C} = \hat{A} \otimes \hat{C} + \hat{B} \otimes \hat{C} \tag{4.146}$$

$$\hat{A} \otimes \left(\hat{B} + \hat{C}\right) = \hat{A} \otimes \hat{B} + \hat{A} \otimes \hat{C} \tag{4.147}$$

$$k\left(\hat{A} \otimes \hat{B}\right) = \left(k\hat{A}\right) \otimes \hat{B} = \hat{A} \otimes \left(k\hat{B}\right) \tag{4.148}$$

$$\left(\hat{A} \otimes \hat{B}\right) \otimes \hat{C} = \hat{A} \otimes \left(\hat{B} \otimes \hat{C}\right) \tag{4.149}$$

where $\hat{A}$, $\hat{B}$ and $\hat{C}$ are matrices and k is a scalar.

The tensor product and ordinary matrix multiplication can be mixed as follows

$$\left(\hat{A} \otimes \hat{B}\right)\left(\hat{C} \otimes \hat{D}\right) = \left(\hat{A}\hat{C}\right) \otimes \left(\hat{B}\hat{D}\right) \tag{4.150}$$

provided that the dimensions of the matrices are such that all matrix multiplications are possible.

Tensor products of kets $|\psi_A\rangle \otimes |\psi_B\rangle$ are used to describe composite quantum mechanical states in which each component subsystem A and B has a definite ket vector $|\psi_A\rangle$ or $|\psi_B\rangle$. If the Hilbert space of subsystem A is $\mathcal{H}_A$ and the Hilbert space of subsystem B is $\mathcal{H}_B$, then the state of the composite system will be in the tensor product Hilbert space $\mathcal{H}_A \otimes \mathcal{H}_B$.

Because ordinary matrix multiplication is not defined for two ket vectors, it is a common practice to omit the tensor product symbol $\otimes$ when writing ket tensor products

$$|\psi_A\rangle \otimes |\psi_B\rangle = |\psi_A\rangle|\psi_B\rangle \tag{4.151}$$

For multiparticle states, the tensor product notation could be compressed even further

$$|\psi_1\rangle \otimes |\psi_2\rangle \otimes |\psi_3\rangle \otimes \ldots \otimes |\psi_n\rangle = |\psi_1 \psi_2 \psi_3 \ldots \psi_n\rangle \tag{4.152}$$

The complex conjugate transpose † distributes over tensor products

$$\left(\hat{A} \otimes \hat{B}\right)^\dagger = \hat{A}^\dagger \otimes \hat{B}^\dagger \tag{4.153}$$

Thus, for tensor product quantum states we have

$$\left(|\psi_A\rangle \otimes |\psi_B\rangle\right)^\dagger = \left(|\psi_A\rangle|\psi_B\rangle\right)^\dagger = \langle\psi_A| \otimes \langle\psi_B| = \langle\psi_A|\langle\psi_B| \tag{4.154}$$

With the use of the mixed-product property given by Eq. (4.150), one can easily calculate *inner products*

$$\left(\langle\psi_A| \otimes \langle\psi_B|\right)\left(|\psi_A'\rangle \otimes |\psi_B'\rangle\right) = \langle\psi_A|\psi_A'\rangle\langle\psi_B|\psi_B'\rangle \tag{4.155}$$

or produce *tensor product operators*

$$\left(|\psi_A'\rangle \otimes |\psi_B'\rangle\right)\left(\langle\psi_A| \otimes \langle\psi_B|\right) = \left(|\psi_A'\rangle\langle\psi_A|\right) \otimes \left(|\psi_B'\rangle\langle\psi_B|\right) \tag{4.156}$$

4.15 Axioms of quantum mechanics

With the use of the mathematical preliminaries discussed in the preceding sections, we can summarize the quantum description of the world in six axioms, whose empirical adequacy has been tested and confirmed with an astounding precision in numerous experiments [18, 66, 132, 133, 134, 235].

Axiom 4.15.1. *(State) To every closed physical system that does not interact with the rest of the world is associated an n-dimensional complex Hilbert space $\mathcal{H}$ called the state space of the system (the number of dimensions n does not need to be finite). The physical state of the closed system is completely described by a unit state vector $|\psi\rangle$ in $\mathcal{H}$. The unit length of the state vector $|\psi\rangle$ reflects the fact that the probability for the quantum system to be in the given state $|\psi\rangle$ is 1.*

Axiom 4.15.2. *(Composition) If the physical system is composite, then the state space $\mathcal{H}$ is the tensor product of the state spaces of the component physical subsystems. Thus, if we have k component subsystems with corresponding complex Hilbert spaces $\mathcal{H}_1, \mathcal{H}_2, \ldots, \mathcal{H}_k$, the state space of the total composite system is $\mathcal{H} = \mathcal{H}_1 \otimes \mathcal{H}_2 \otimes \ldots \otimes \mathcal{H}_k$.*

Axiom 4.15.3. *(Born rule) If $|\psi\rangle$ is the vector representing the state of a system and if $|\chi\rangle$ represents another physical state, there exists a probability $|\langle\psi|\chi\rangle|^2$ of finding the system in the state $|\chi\rangle$ upon measurement.*

Axiom 4.15.4. *(Observables) To every observable physical property A there exists an associated Hermitian operator $\hat{A} = \hat{A}^\dagger$, which acts in the Hilbert space of states $\mathcal{H}$. The eigenvalues of the operator are the possible values of the physical properties. The Hermitian operator $\hat{A}$ is to be used in conjunction with the wave function $|\psi\rangle$ of the physical system. In particular, the expectation value of the operator is $\langle\psi|\hat{A}|\psi\rangle$.*

Axiom 4.15.5. *(Dynamics) The time evolution of a closed physical system is given by the Schrödinger equation*

$$i\hbar\frac{\partial}{\partial t}|\psi\rangle = \hat{H}|\psi\rangle \tag{4.157}$$

where the Hamiltonian $\hat{H} = \hat{H}^\dagger$ is a Hermitian operator corresponding to the total energy of the system.

Axiom 4.15.6. *(Wave function collapse) If $\hat{A}$ is an observable with eigenvalues $\{\lambda_m\}_{m=1}^n$ and orthonormal eigenvectors $\{|\psi_m\rangle\}_{m=1}^n$, given a system in the state $|\psi\rangle$, the probability of obtaining λ_m as the measurement outcome of $\hat{A}$ is equal to $|\langle\psi_m|\psi\rangle|^2$. After the measurement the system jumps abruptly to the state projected on the subspace of the eigenvalue λ_m corresponding to the measurement outcome, $|\psi\rangle \rightarrow |\psi_m\rangle$.*

The wave function collapse described by the abrupt jump $|\psi\rangle \rightarrow |\psi_m\rangle$ may appear to be incompatible with the time evolution by the Schrödinger equation [214]. In no collapse models of quantum mechanics, this incompatibility is claimed to be only apparent because when the system is measured it becomes an open system, hence the time evolution of a system that is not closed does not have to obey the

Schrödinger equation [531, 532]. In collapse models of quantum mechanics, however, this incompatibility is accounted for by the introduction of a novel physical process that generates discontinuous quantum jumps whenever sufficiently large quantum systems reach a certain energy threshold $\mathcal{E}$ (see Sections 6.1 and 6.5.1).

In essence, the world of quantum mechanics is fundamentally different from the ticking clockwork classical world. According to the quantum axioms, the fabric of the quantum physical states is made of *complex-valued probability amplitudes*. Consequently, different physical events could occur only with certain *probabilities* and the behavior of quantum systems is inherently *indeterministic*.

4.16   Quantum superpositions

The Hilbert vector space structure of the space of quantum states follows from the linearity of the Schrödinger equation (4.157).

Theorem 12. *If $|\psi_1\rangle$ and $|\psi_2\rangle$ are two different solutions of the Schrödinger equation, then any linear combination of solutions $|\psi\rangle = a|\psi_1\rangle + b|\psi_2\rangle$ is also a solution of the Schrödinger equation.*

Proof. Suppose that $|\psi_1\rangle$ and $|\psi_2\rangle$ are two different solutions of the Schrödinger equation (4.157). Using two arbitrary complex scalars $a, b \in \mathbb{C}$ we can write

$$a\imath\hbar\frac{\partial}{\partial t}|\psi_1\rangle = a\hat{H}|\psi_1\rangle \tag{4.158}$$

$$b\imath\hbar\frac{\partial}{\partial t}|\psi_2\rangle = b\hat{H}|\psi_2\rangle \tag{4.159}$$

Summing both expressions gives

$$a\imath\hbar\frac{\partial}{\partial t}|\psi_1\rangle + b\imath\hbar\frac{\partial}{\partial t}|\psi_2\rangle = a\hat{H}|\psi_1\rangle + b\hat{H}|\psi_2\rangle \tag{4.160}$$

Using the linearity of the differential operator $\frac{\partial}{\partial t}$ and the Hamiltonian $\hat{H}$, we have

$$\imath\hbar\frac{\partial}{\partial t}\left(a|\psi_1\rangle + b|\psi_2\rangle\right) \;=\; \hat{H}\left(a|\psi_1\rangle + b|\psi_2\rangle\right) \tag{4.161}$$

$$\imath\hbar\frac{\partial}{\partial t}|\psi\rangle \;=\; \hat{H}|\psi\rangle \tag{4.162}$$

Thus, any linear combination of the form $|\psi\rangle = a|\psi_1\rangle + b|\psi_2\rangle$ also solves the Schrödinger equation.   □

Definition 4.23. *(Quantum superposition) A linear combination of two or more distinct quantum states $|\psi_1\rangle$, $|\psi_2\rangle$, ..., $|\psi_k\rangle$ given by*

$$|\psi\rangle = \sum_{i=1}^{k} a_i|\psi_i\rangle = a_1|\psi_1\rangle + a_2|\psi_2\rangle + \ldots a_k|\psi_k\rangle \tag{4.163}$$

where $a_1, a_2, \ldots, a_k$ are non-zero, is called a quantum superposition *of these states.*

4.17 Quantum entanglement

Consider a closed composite quantum system in which the component subsystems interact with each other. The composite system as a whole will evolve in time by the Schrödinger equation, but it is possible that none of the component subsystems evolves by the Schrödinger equation. As a result, the composite system will have a state vector $|\psi\rangle$, whereas none of the component subsystems will have its own state vector [158]. Mathematically, even though the state space of the composite system is given by a tensor product $\mathcal{H} = \mathcal{H}_1 \otimes \mathcal{H}_2 \otimes \mathcal{H}_3 \otimes \ldots \otimes \mathcal{H}_k$, it is not true that every vector $|\psi\rangle$ in the Hilbert space $\mathcal{H}$ is also factorizable in the form of a tensor product $|\psi\rangle = |\psi_1\rangle \otimes |\psi_2\rangle \otimes |\psi_3\rangle \otimes \ldots \otimes |\psi_k\rangle$. Indeed, in the tensor product space there are non-factorizable *quantum entangled states* for which none of the component subsystems would have its own state vector, and consequently, there would be no component state vectors to evolve by the Schrödinger equation.

Definition 4.24. *(Quantum entangled state) A quantum state $|\psi\rangle$ of a composite system that cannot be expressed as a tensor product of individual states of the component subsystems, namely, $|\psi\rangle \neq |\psi_1\rangle \otimes |\psi_2\rangle \otimes \ldots \otimes |\psi_k\rangle$, is called a* quantum entangled state.

The *singlet state* of two spin-$\frac{1}{2}$ particles is a quantum entangled state

$$|\psi_{AB}\rangle = \frac{1}{\sqrt{2}}(|\uparrow_A\rangle|\downarrow_B\rangle - |\downarrow_A\rangle|\uparrow_B\rangle) \qquad (4.164)$$

Because the singlet state cannot be written as a composite tensor product state of particles A and B, neither particle A nor particle B has its own state vector.

Proof. Every tensor product state can be written as $|\psi_{AB}\rangle = |\psi_A\rangle \otimes |\psi_B\rangle$, where both particles A and B have their own unit state vectors

$$|\psi_A\rangle = a_1|\uparrow_A\rangle + a_2|\downarrow_A\rangle \qquad (4.165)$$

$$|\psi_B\rangle = b_1|\uparrow_B\rangle + b_2|\downarrow_B\rangle \qquad (4.166)$$

that satisfy the normalization conditions

$$\langle\psi_A|\psi_A\rangle = a_1^* a_1 + a_2^* a_2 = |a_1|^2 + |a_2|^2 = 1 \qquad (4.167)$$

$$\langle\psi_B|\psi_B\rangle = b_1^* b_1 + b_2^* b_2 = |b_1|^2 + |b_2|^2 = 1 \qquad (4.168)$$

Expanding the tensor product state using Eqs. (4.165) and (4.166) leads to

$$
\begin{aligned}
|\psi_{AB}\rangle &= (a_1|\uparrow_A\rangle + a_2|\downarrow_A\rangle) \otimes (b_1|\uparrow_B\rangle + b_2|\downarrow_B\rangle) \\
&= a_1 b_1|\uparrow_A\rangle|\uparrow_B\rangle + a_1 b_2|\uparrow_A\rangle|\downarrow_B\rangle + a_2 b_1|\downarrow_A\rangle|\uparrow_B\rangle + a_2 b_2|\downarrow_A\rangle|\downarrow_B\rangle
\end{aligned}
$$

In the singlet state, the coefficients in front of the basis states are

$$a_1 b_1 = a_2 b_2 = 0 \qquad (4.169)$$

$$a_1 b_2 = a_2 b_1 = \frac{1}{\sqrt{2}} \qquad (4.170)$$

The latter two equations are incompatible with the normalization conditions given by Eqs. (4.167) and (4.168). If $a_1 = 0$ to satisfy Eq. (4.169), then $|a_2| = 1$ from Eq. (4.167), and necessarily $b_2 = 0$ again from Eq. (4.169). This leads to a contradiction, because $a_1 b_2 = 0 \neq \frac{1}{\sqrt{2}}$. Similarly, if $b_1 = 0$ to satisfy Eq. (4.169), then $|b_2| = 1$ from Eq. (4.167), and necessarily $a_2 = 0$ again from Eq. (4.169). This also leads to a contradiction, because $a_2 b_1 = 0 \neq \frac{1}{\sqrt{2}}$. The contradictions in both cases and the lack of other alternatives imply that the assumption that the singlet state can be written as a tensor product is false. $\square$

4.18   Density matrices

Definition 4.25. *(Density matrix) The density matrix* $\hat{\rho} = \hat{\rho}^\dagger$ *of a quantum system Q is a Hermitian statistical operator that can be used to predict the expectation values for* all *local physical observables* $\hat{A}$ *that involve only the system Q. The expectation value* $\langle \hat{A} \rangle$ *of the local observable* $\hat{A}$ *is given by*

$$\langle \hat{A} \rangle = Tr\left(\hat{\rho}\hat{A}\right) \tag{4.171}$$

where the trace *of a matrix is defined by the sum of all main diagonal entries*

$$Tr(\hat{A}) = a_{11} + a_{22} + \ldots + a_{nn} = \sum_{i=1}^{n} a_{ii} \tag{4.172}$$

Every physically valid density matrix has a unit trace $Tr(\hat{\rho}) = 1$ *due to normalization of probabilities. For pure quantum systems that have their own state vector* $|\psi\rangle$, *the* pure *density matrix is given by*

$$\hat{\rho} = |\psi\rangle\langle\psi| \tag{4.173}$$

For mixed quantum systems that are quantum entangled with other quantum systems and do not have their own state vector, the mixed *density matrix is given by*

$$\hat{\rho} = \sum_{i} p_i |\psi_i\rangle\langle\psi_i| \tag{4.174}$$

where $\{|\psi_i\rangle\}$ *is some set of pure states, not necessarily orthogonal, and the probabilities* $0 \leq p_i < 1$ *sum up to unity* $\sum_i p_i = 1$. *It is incorrect to interpret each* $|\psi_i\rangle$ *as a putative state vector in which the system is with probability* p_i, *because the mixed system is in none of the states* $\{|\psi_i\rangle\}$ *as a consequence of the fact that, orthogonal or not, the set of states* $\{|\psi_i\rangle\}$ *is not unique.*

In order to be able to calculate explicitly the reduced density matrix $\hat{\rho}_A$ of a quantum system A that is quantum entangled with another system B, we will need to define the *partial trace*, which is an operator-valued function on operators.

Definition 4.26. (*Reduced density matrix*) *Consider a composite quantum system consisting of two component subsystems A and B with corresponding m-dimensional Hilbert space $\mathcal{H}_A$ and n-dimensional Hilbert space $\mathcal{H}_B$. Let $\{|a_i\rangle\}_{i=1}^m$ be an orthonormal basis of $\mathcal{H}_A$ and $\{|b_j\rangle\}_{j=1}^n$ be an orthonormal basis of $\mathcal{H}_B$. The density matrix $\hat{\rho}_{AB}$ on $\mathcal{H}_A \otimes \mathcal{H}_B$ can be written as*

$$\hat{\rho}_{AB} = \sum_{i=1}^m \sum_{i'=1}^m \sum_{j=1}^n \sum_{j'=1}^n \rho_{ii'jj'} |a_i\rangle |b_j\rangle \langle a_{i'}| \langle b_{j'}| \tag{4.175}$$

$$= \sum_{i=1}^m \sum_{i'=1}^m \sum_{j=1}^n \sum_{j'=1}^n \rho_{ii'jj'} |a_i\rangle \langle a_{i'}| \otimes |b_j\rangle \langle b_{j'}| \tag{4.176}$$

The partial trace Tr_B traces out the component subsystem B, leaving a reduced density matrix $\hat{\rho}_A$ on $\mathcal{H}_A$ given by

$$Tr_B(\hat{\rho}_{AB}) = \sum_{j=1}^n \langle b_j | \hat{\rho}_{AB} | b_j \rangle \tag{4.177}$$

Taking into account that

$$\langle b_{j'} | b_j \rangle = \delta_{j'j} = \begin{cases} 0, & j' \neq j \\ 1, & j' = j \end{cases} \tag{4.178}$$

we obtain

$$Tr_B(\hat{\rho}_{AB}) = \sum_{i=1}^m \sum_{i'=1}^m \sum_{j=1}^n \sum_{j'=1}^n \rho_{ii'jj'} |a_i\rangle \langle a_{i'}| \langle b_j | b_j \rangle \langle b_{j'} | b_j \rangle \tag{4.179}$$

$$= \sum_{i=1}^m \sum_{i'=1}^m \sum_{j=1}^n \rho_{ii'jj} |a_i\rangle \langle a_{i'}| \tag{4.180}$$

$$= \hat{\rho}_A \tag{4.181}$$

Similarly, the partial trace Tr_A traces out the component subsystem A, leaving a reduced density matrix $\hat{\rho}_B$ on $\mathcal{H}_B$.

Example 4.1. (*Tracing out component subsystems*) *Consider a composite quantum system that has a 2-level component subsystem A and a 3-level component subsystem B. The composite density matrix $\hat{\rho}_{AB}$ on $\mathcal{H}_A \otimes \mathcal{H}_B$ can be explicitly written with its $\rho_{ii'jj'}$ entries grouped by the ii' indices as*

$$\hat{\rho}_{AB} = \left(\begin{array}{cc} \begin{bmatrix} \rho_{1111} & \rho_{1112} & \rho_{1113} \\ \rho_{1121} & \rho_{1122} & \rho_{1123} \\ \rho_{1131} & \rho_{1132} & \rho_{1133} \end{bmatrix} & \begin{bmatrix} \rho_{1211} & \rho_{1212} & \rho_{1213} \\ \rho_{1221} & \rho_{1222} & \rho_{1223} \\ \rho_{1231} & \rho_{1232} & \rho_{1233} \end{bmatrix} \\ \begin{bmatrix} \rho_{2111} & \rho_{2112} & \rho_{2113} \\ \rho_{2121} & \rho_{2122} & \rho_{2123} \\ \rho_{2131} & \rho_{2132} & \rho_{2133} \end{bmatrix} & \begin{bmatrix} \rho_{2211} & \rho_{2212} & \rho_{2213} \\ \rho_{2221} & \rho_{2222} & \rho_{2223} \\ \rho_{2231} & \rho_{2232} & \rho_{2233} \end{bmatrix} \end{array} \right) \tag{4.182}$$

Tracing out the second component B is done using a multi-diagonal pattern

$$
Tr_B(\hat{\rho}_{AB}) =
\begin{pmatrix}
Tr\begin{bmatrix} \rho_{1111} & \rho_{1112} & \rho_{1113} \\ \rho_{1121} & \rho_{1122} & \rho_{1123} \\ \rho_{1131} & \rho_{1132} & \rho_{1133} \end{bmatrix} & Tr\begin{bmatrix} \rho_{1211} & \rho_{1212} & \rho_{1213} \\ \rho_{1221} & \rho_{1222} & \rho_{1223} \\ \rho_{1231} & \rho_{1232} & \rho_{1233} \end{bmatrix} \\[6pt]
Tr\begin{bmatrix} \rho_{2111} & \rho_{2112} & \rho_{2113} \\ \rho_{2121} & \rho_{2122} & \rho_{2123} \\ \rho_{2131} & \rho_{2132} & \rho_{2133} \end{bmatrix} & Tr\begin{bmatrix} \rho_{2211} & \rho_{2212} & \rho_{2213} \\ \rho_{2221} & \rho_{2222} & \rho_{2223} \\ \rho_{2231} & \rho_{2232} & \rho_{2233} \end{bmatrix}
\end{pmatrix}
\tag{4.183}
$$

$$
Tr_B(\hat{\rho}_{AB}) = \hat{\rho}_A =
\begin{pmatrix}
[\rho_{1111}+\rho_{1122}+\rho_{1133}] & [\rho_{1211}+\rho_{1222}+\rho_{1233}] \\
[\rho_{2111}+\rho_{2122}+\rho_{2133}] & [\rho_{2211}+\rho_{2222}+\rho_{2233}]
\end{pmatrix}
\tag{4.184}
$$

Tracing out the first component A is done using a long-diagonal pattern

$$
Tr_A(\hat{\rho}_{AB}) = Tr_A
\begin{pmatrix}
\begin{bmatrix} \rho_{1111} & \rho_{1112} & \rho_{1113} \\ \rho_{1121} & \rho_{1122} & \rho_{1123} \\ \rho_{1131} & \rho_{1132} & \rho_{1133} \end{bmatrix} & \begin{bmatrix} \rho_{1211} & \rho_{1212} & \rho_{1213} \\ \rho_{1221} & \rho_{1222} & \rho_{1223} \\ \rho_{1231} & \rho_{1232} & \rho_{1233} \end{bmatrix} \\[6pt]
\begin{bmatrix} \rho_{2111} & \rho_{2112} & \rho_{2113} \\ \rho_{2121} & \rho_{2122} & \rho_{2123} \\ \rho_{2131} & \rho_{2132} & \rho_{2133} \end{bmatrix} & \begin{bmatrix} \rho_{2211} & \rho_{2212} & \rho_{2213} \\ \rho_{2221} & \rho_{2222} & \rho_{2223} \\ \rho_{2231} & \rho_{2232} & \rho_{2233} \end{bmatrix}
\end{pmatrix}
\tag{4.185}
$$

$$
Tr_A(\hat{\rho}_{AB}) =
\begin{bmatrix} \rho_{1111} & \rho_{1112} & \rho_{1113} \\ \rho_{1121} & \rho_{1122} & \rho_{1123} \\ \rho_{1131} & \rho_{1132} & \rho_{1133} \end{bmatrix} +
\begin{bmatrix} \rho_{2211} & \rho_{2212} & \rho_{2213} \\ \rho_{2221} & \rho_{2222} & \rho_{2223} \\ \rho_{2231} & \rho_{2232} & \rho_{2233} \end{bmatrix}
\tag{4.186}
$$

$$
Tr_A(\hat{\rho}_{AB}) = \hat{\rho}_B =
\begin{pmatrix}
[\rho_{1111}+\rho_{2211}] & [\rho_{1112}+\rho_{2212}] & [\rho_{1113}+\rho_{2213}] \\
[\rho_{1121}+\rho_{2221}] & [\rho_{1122}+\rho_{2222}] & [\rho_{1123}+\rho_{2223}] \\
[\rho_{1131}+\rho_{2231}] & [\rho_{1132}+\rho_{2232}] & [\rho_{1133}+\rho_{2233}]
\end{pmatrix}
\tag{4.187}
$$

The composite density matrix $\hat{\rho}_{AB}$ on $\mathcal{H}_A \otimes \mathcal{H}_B$ can also be written with its $\rho_{ii'jj'}$ entries grouped by the jj' indices as

$$
\hat{\rho}_{AB} =
\begin{pmatrix}
\begin{bmatrix} \rho_{1111} & \rho_{1211} \\ \rho_{2111} & \rho_{2211} \end{bmatrix} & \begin{bmatrix} \rho_{1112} & \rho_{1212} \\ \rho_{2112} & \rho_{2212} \end{bmatrix} & \begin{bmatrix} \rho_{1113} & \rho_{1213} \\ \rho_{2113} & \rho_{2213} \end{bmatrix} \\[6pt]
\begin{bmatrix} \rho_{1121} & \rho_{1221} \\ \rho_{2121} & \rho_{2221} \end{bmatrix} & \begin{bmatrix} \rho_{1122} & \rho_{1222} \\ \rho_{2122} & \rho_{2222} \end{bmatrix} & \begin{bmatrix} \rho_{1123} & \rho_{1223} \\ \rho_{2123} & \rho_{2223} \end{bmatrix} \\[6pt]
\begin{bmatrix} \rho_{1131} & \rho_{1231} \\ \rho_{2131} & \rho_{2231} \end{bmatrix} & \begin{bmatrix} \rho_{1132} & \rho_{1232} \\ \rho_{2132} & \rho_{2232} \end{bmatrix} & \begin{bmatrix} \rho_{1133} & \rho_{1233} \\ \rho_{2133} & \rho_{2233} \end{bmatrix}
\end{pmatrix}
\tag{4.188}
$$

Tracing out the second component B is now done using the long-diagonal pattern

$$
Tr_B(\hat{\rho}_{AB}) = Tr_B
\begin{pmatrix}
\begin{bmatrix} \rho_{1111} & \rho_{1211} \\ \rho_{2111} & \rho_{2211} \end{bmatrix} & \begin{bmatrix} \rho_{1112} & \rho_{1212} \\ \rho_{2112} & \rho_{2212} \end{bmatrix} & \begin{bmatrix} \rho_{1113} & \rho_{1213} \\ \rho_{2113} & \rho_{2213} \end{bmatrix} \\[6pt]
\begin{bmatrix} \rho_{1121} & \rho_{1221} \\ \rho_{2121} & \rho_{2221} \end{bmatrix} & \begin{bmatrix} \rho_{1122} & \rho_{1222} \\ \rho_{2122} & \rho_{2222} \end{bmatrix} & \begin{bmatrix} \rho_{1123} & \rho_{1223} \\ \rho_{2123} & \rho_{2223} \end{bmatrix} \\[6pt]
\begin{bmatrix} \rho_{1131} & \rho_{1231} \\ \rho_{2131} & \rho_{2231} \end{bmatrix} & \begin{bmatrix} \rho_{1132} & \rho_{1232} \\ \rho_{2132} & \rho_{2232} \end{bmatrix} & \begin{bmatrix} \rho_{1133} & \rho_{1233} \\ \rho_{2133} & \rho_{2233} \end{bmatrix}
\end{pmatrix}
\tag{4.189}
$$

$$Tr_B(\hat{\rho}_{AB}) = \hat{\rho}_A = \begin{bmatrix} \rho_{1111} & \rho_{1211} \\ \rho_{2111} & \rho_{2211} \end{bmatrix} + \begin{bmatrix} \rho_{1122} & \rho_{1222} \\ \rho_{2122} & \rho_{2222} \end{bmatrix} + \begin{bmatrix} \rho_{1133} & \rho_{1233} \\ \rho_{2133} & \rho_{2233} \end{bmatrix} \quad (4.190)$$

The result is the same as in Eq. (4.184).

And tracing out the first component A is done using the multi-diagonal pattern

$$Tr_A(\hat{\rho}_{AB}) = \begin{pmatrix} Tr\begin{bmatrix} \rho_{1111} & \rho_{1211} \\ \rho_{2111} & \rho_{2211} \end{bmatrix} & Tr\begin{bmatrix} \rho_{1112} & \rho_{1212} \\ \rho_{2112} & \rho_{2212} \end{bmatrix} & Tr\begin{bmatrix} \rho_{1113} & \rho_{1213} \\ \rho_{2113} & \rho_{2213} \end{bmatrix} \\ Tr\begin{bmatrix} \rho_{1121} & \rho_{1221} \\ \rho_{2121} & \rho_{2221} \end{bmatrix} & Tr\begin{bmatrix} \rho_{1122} & \rho_{1222} \\ \rho_{2122} & \rho_{2222} \end{bmatrix} & Tr\begin{bmatrix} \rho_{1123} & \rho_{1223} \\ \rho_{2123} & \rho_{2223} \end{bmatrix} \\ Tr\begin{bmatrix} \rho_{1131} & \rho_{1231} \\ \rho_{2131} & \rho_{2231} \end{bmatrix} & Tr\begin{bmatrix} \rho_{1132} & \rho_{1232} \\ \rho_{2132} & \rho_{2232} \end{bmatrix} & Tr\begin{bmatrix} \rho_{1133} & \rho_{1233} \\ \rho_{2133} & \rho_{2233} \end{bmatrix} \end{pmatrix}$$

$$(4.191)$$

The result is the same as in Eq. (4.187). The visual differences in the tracing patterns for the first or the second component subsystems are artifacts created by the grouping of the $\rho_{ii'jj'}$ entries either by the ii' or by the jj' indices. The physics also remains the same if the composite density matrix $\hat{\rho}_{AB}$ on $\mathcal{H}_A \otimes \mathcal{H}_B$ is written as $\hat{\rho}_{BA}$ on $\mathcal{H}_B \otimes \mathcal{H}_A$.

Quantum physicists often claim that the density matrices contain all the useful information about an arbitrary quantum state. This is incorrect, however, because mixed density matrices can only predict the outcomes of local measurements and are clueless about the existence of possible quantum correlations between the measured quantum system and other external quantum systems. Consider, for example, the quantum entangled singlet state $|\psi_{AB}\rangle$ of two qubits A and B given by

$$|\psi_{AB}\rangle = \frac{1}{\sqrt{2}}(|\uparrow_A\rangle|\downarrow_B\rangle - |\downarrow_A\rangle|\uparrow_B\rangle) \quad (4.192)$$

The density matrix of the composite system is

$$\hat{\rho}_{AB} = \begin{pmatrix} \begin{bmatrix} 0 & 0 \\ 0 & \frac{1}{2} \end{bmatrix} & \begin{bmatrix} 0 & 0 \\ -\frac{1}{2} & 0 \end{bmatrix} \\ \begin{bmatrix} 0 & -\frac{1}{2} \\ 0 & 0 \end{bmatrix} & \begin{bmatrix} \frac{1}{2} & 0 \\ 0 & 0 \end{bmatrix} \end{pmatrix} \quad (4.193)$$

and the reduced density matrices of each component qubit are

$$\hat{\rho}_A = \begin{pmatrix} \frac{1}{2} & 0 \\ 0 & \frac{1}{2} \end{pmatrix}; \qquad \hat{\rho}_B = \begin{pmatrix} \frac{1}{2} & 0 \\ 0 & \frac{1}{2} \end{pmatrix} \quad (4.194)$$

Local operators $\hat{M}_A$ on $\mathcal{H}_A$ that act only upon qubit A can be written as $\hat{M}_A \otimes \hat{I}_B$ on $\mathcal{H}_A \otimes \mathcal{H}_B$, and local operators $\hat{M}_B$ on $\mathcal{H}_B$ that act only upon qubit B can be written as $\hat{I}_A \otimes \hat{M}_B$ on $\mathcal{H}_A \otimes \mathcal{H}_B$. For local operators, we have

$$\text{Tr}(\hat{M}_A\hat{\rho}_A) = \text{Tr}\left[\hat{M}_A\text{Tr}_B(\hat{\rho}_{AB})\right] = \text{Tr}\left[\left(\hat{M}_A \otimes \hat{I}_B\right)\hat{\rho}_{AB}\right] \quad (4.195)$$

$$\text{Tr}(\hat{M}_B\hat{\rho}_B) = \text{Tr}\left[\hat{M}_B\text{Tr}_A(\hat{\rho}_{AB})\right] = \text{Tr}\left[\left(\hat{I}_A \otimes \hat{M}_B\right)\hat{\rho}_{AB}\right] \quad (4.196)$$

As an example, consider the operator $\hat{\sigma}_x$ acting only upon qubit A. The expectation value is $\text{Tr}(\hat{\sigma}_x \hat{\rho}_A) = 0$, which means that the eigenvalues ± 1 are obtained each with a probability of $\frac{1}{2}$. Explicit calculation using the density matrices shows

$$
\text{Tr}\left[\left(\hat{\sigma}_x \otimes \hat{I}_B\right)\hat{\rho}_{AB}\right] = \text{Tr}\left[\left(\begin{bmatrix} \begin{bmatrix} 0 & 0 \\ 0 & 0 \end{bmatrix} & \begin{bmatrix} 1 & 0 \\ 0 & 1 \end{bmatrix} \\ \begin{bmatrix} 1 & 0 \\ 0 & 1 \end{bmatrix} & \begin{bmatrix} 0 & 0 \\ 0 & 0 \end{bmatrix} \end{bmatrix}\right)\left(\begin{bmatrix} \begin{bmatrix} 0 & 0 \\ 0 & \frac{1}{2} \end{bmatrix} & \begin{bmatrix} 0 & 0 \\ -\frac{1}{2} & 0 \end{bmatrix} \\ \begin{bmatrix} 0 & -\frac{1}{2} \\ 0 & 0 \end{bmatrix} & \begin{bmatrix} \frac{1}{2} & 0 \\ 0 & 0 \end{bmatrix} \end{bmatrix}\right)\right]
$$

$$
= \text{Tr}\left[\text{Tr}_B\left(\begin{bmatrix} \begin{bmatrix} 0 & -\frac{1}{2} \\ 0 & 0 \end{bmatrix} & \begin{bmatrix} \frac{1}{2} & 0 \\ 0 & 0 \end{bmatrix} \\ \begin{bmatrix} 0 & 0 \\ 0 & \frac{1}{2} \end{bmatrix} & \begin{bmatrix} 0 & 0 \\ -\frac{1}{2} & 0 \end{bmatrix} \end{bmatrix}\right)\right] = \text{Tr}\begin{bmatrix} 0 & \frac{1}{2} \\ \frac{1}{2} & 0 \end{bmatrix}
$$

$$
= \text{Tr}\left[\begin{pmatrix} 0 & 1 \\ 1 & 0 \end{pmatrix}\begin{pmatrix} \frac{1}{2} & 0 \\ 0 & \frac{1}{2} \end{pmatrix}\right] = \text{Tr}\left[\hat{\sigma}_x \hat{\rho}_A\right] = 0 \qquad (4.197)
$$

The expectation value of $\hat{\sigma}_x \otimes \hat{I}_B$ is calculated using the state vector $|\psi_{AB}\rangle$ as

$$
\langle\psi_{AB}|\hat{\sigma}_x \otimes \hat{I}_B|\psi_{AB}\rangle = \begin{pmatrix} 0 \\ \frac{1}{\sqrt{2}} \\ -\frac{1}{\sqrt{2}} \\ 0 \end{pmatrix}^{\dagger} \begin{pmatrix} 0 & 0 & 1 & 0 \\ 0 & 0 & 0 & 1 \\ 1 & 0 & 0 & 0 \\ 0 & 1 & 0 & 0 \end{pmatrix} \begin{pmatrix} 0 \\ \frac{1}{\sqrt{2}} \\ -\frac{1}{\sqrt{2}} \\ 0 \end{pmatrix}
$$

$$
= \begin{pmatrix} 0 \\ \frac{1}{\sqrt{2}} \\ -\frac{1}{\sqrt{2}} \\ 0 \end{pmatrix}^{\dagger} \begin{pmatrix} -\frac{1}{\sqrt{2}} \\ 0 \\ 0 \\ \frac{1}{\sqrt{2}} \end{pmatrix} = 0 \qquad (4.198)
$$

The reduced density matrices $\hat{\rho}_A$ and $\hat{\rho}_B$ can be used only for local measurements of local observables, but not for nonlocal measurements of nonlocal observables that take into account the quantum correlations between the values of several quantum components. As an example, consider the product observables $\hat{\sigma}_x \otimes \hat{\sigma}_x$, $\hat{\sigma}_y \otimes \hat{\sigma}_y$ and $\hat{\sigma}_z \otimes \hat{\sigma}_z$ that measure the product of the corresponding spin values of qubit A and qubit B. Instead of the reduced density matrices $\hat{\rho}_A$ and $\hat{\rho}_B$, one needs to use either the composite density matrix $\hat{\rho}_{AB}$ or the quantum state vector $|\psi_{AB}\rangle$. Explicit calculation for the observable $\hat{\sigma}_x \otimes \hat{\sigma}_x$ using the density matrix $\hat{\rho}_{AB}$ shows

$$
\text{Tr}\left[\left(\hat{\sigma}_x \otimes \hat{\sigma}_x\right)\hat{\rho}_{AB}\right] = \text{Tr}\left[\left(\begin{bmatrix} \begin{bmatrix} 0 & 0 \\ 0 & 0 \end{bmatrix} & \begin{bmatrix} 0 & 1 \\ 1 & 0 \end{bmatrix} \\ \begin{bmatrix} 0 & 1 \\ 1 & 0 \end{bmatrix} & \begin{bmatrix} 0 & 0 \\ 0 & 0 \end{bmatrix} \end{bmatrix}\right)\left(\begin{bmatrix} \begin{bmatrix} 0 & 0 \\ 0 & \frac{1}{2} \end{bmatrix} & \begin{bmatrix} 0 & 0 \\ -\frac{1}{2} & 0 \end{bmatrix} \\ \begin{bmatrix} 0 & -\frac{1}{2} \\ 0 & 0 \end{bmatrix} & \begin{bmatrix} \frac{1}{2} & 0 \\ 0 & 0 \end{bmatrix} \end{bmatrix}\right)\right]
$$

$$
= \text{Tr}\left(\begin{bmatrix} \begin{bmatrix} 0 & 0 \\ 0 & -\frac{1}{2} \end{bmatrix} & \begin{bmatrix} 0 & 0 \\ \frac{1}{2} & 0 \end{bmatrix} \\ \begin{bmatrix} 0 & \frac{1}{2} \\ 0 & 0 \end{bmatrix} & \begin{bmatrix} -\frac{1}{2} & 0 \\ 0 & 0 \end{bmatrix} \end{bmatrix}\right) = -1 \qquad (4.199)
$$

The same result is also obtained using the quantum state vector $|\psi_{AB}\rangle$

$$\langle\psi_{AB}|\hat{\sigma}_x\otimes\hat{\sigma}_x|\psi_{AB}\rangle = \begin{pmatrix} 0 \\ \frac{1}{\sqrt{2}} \\ -\frac{1}{\sqrt{2}} \\ 0 \end{pmatrix}^\dagger \begin{pmatrix} 0 & 0 & 0 & 1 \\ 0 & 0 & 1 & 0 \\ 0 & 1 & 0 & 0 \\ 1 & 0 & 0 & 0 \end{pmatrix} \begin{pmatrix} 0 \\ \frac{1}{\sqrt{2}} \\ -\frac{1}{\sqrt{2}} \\ 0 \end{pmatrix}$$

$$= \begin{pmatrix} 0 \\ \frac{1}{\sqrt{2}} \\ -\frac{1}{\sqrt{2}} \\ 0 \end{pmatrix}^\dagger \begin{pmatrix} 0 \\ -\frac{1}{\sqrt{2}} \\ \frac{1}{\sqrt{2}} \\ 0 \end{pmatrix} = -1 \tag{4.200}$$

Here, it is important to note that the product of spin values along the x-axis of qubit A and qubit B is definitely -1 even though none of the qubits A or B has a definite eigenvalue ± 1 along the x-axis. This is because the singlet state $|\psi_{AB}\rangle$ is an eigenvector of $\hat{\sigma}_x\otimes\hat{\sigma}_x$ with an eigenvalue of -1. Similarly, the singlet state $|\psi_{AB}\rangle$ is an eigenvector of $\hat{\sigma}_y\otimes\hat{\sigma}_y$ and $\hat{\sigma}_z\otimes\hat{\sigma}_z$ with an eigenvalue of -1. Thus, we have

$$\hat{\sigma}_x\otimes\hat{\sigma}_x|\psi_{AB}\rangle = \hat{\sigma}_y\otimes\hat{\sigma}_y|\psi_{AB}\rangle = \hat{\sigma}_z\otimes\hat{\sigma}_z|\psi_{AB}\rangle = -|\psi_{AB}\rangle \tag{4.201}$$

The triple of operators $\hat{\sigma}_x\otimes\hat{\sigma}_x$, $\hat{\sigma}_y\otimes\hat{\sigma}_y$ and $\hat{\sigma}_z\otimes\hat{\sigma}_z$ can be measured simultaneously since these operators commute with each other. For the state $|\psi_{AB}\rangle$, the expectation values of the operators are $\langle\hat{\sigma}_x\otimes\hat{\sigma}_x\rangle = -1$, $\langle\hat{\sigma}_y\otimes\hat{\sigma}_y\rangle = -1$ and $\langle\hat{\sigma}_z\otimes\hat{\sigma}_z\rangle = -1$.

The triple of operators $\hat{\sigma}_x\otimes\hat{\sigma}_x$, $\hat{\sigma}_x\otimes\hat{I}_B$ and $\hat{I}_A\otimes\hat{\sigma}_x$ can also be measured simultaneously with expectation values $\langle\hat{\sigma}_x\otimes\hat{\sigma}_x\rangle = -1$, $\langle\hat{\sigma}_x\otimes\hat{I}_B\rangle = 0$ and $\langle\hat{I}_A\otimes\hat{\sigma}_x\rangle = 0$. Since the eigenvalues of $\hat{\sigma}_x$ are ± 1, it follows that each eigenvalue outcome for the spin along the x-axis of the individual qubits occurs with a probability of $\frac{1}{2}$ but in a correlated way such that whenever the obtained value for qubit A is -1 the value of qubit B is $+1$, and vice versa, hence the product of both values is always -1.

Noteworthy, the definite measurement outcome of the nonlocal operator $\hat{\sigma}_x\otimes\hat{\sigma}_x$ does not itself imply the existence of definite individual spin values along the x-axis. Indeed, suppose, to the contrary, that from the definite value of the composite product one could also infer definite spin values for the individual qubits. Then, by measuring together the commuting operators $\hat{\sigma}_x\otimes\hat{\sigma}_x$ and $\hat{\sigma}_y\otimes\hat{\sigma}_y$, one could derive that the spin of each qubit is parallel both to the x-axis and to the y-axis, which is a geometric contradiction. Therefore, the contrary premise must be wrong. Because $\hat{\sigma}_y\otimes\hat{\sigma}_y$ does not commute with $\hat{\sigma}_x\otimes\hat{I}_B$ and $\hat{I}_A\otimes\hat{\sigma}_x$, and similarly, $\hat{\sigma}_x\otimes\hat{\sigma}_x$ does not commute with $\hat{\sigma}_y\otimes\hat{I}_B$ and $\hat{I}_A\otimes\hat{\sigma}_y$, such triples of operators represent incompatible quantum measurements that cannot be performed simultaneously. In other words, once we measure simultaneously $\hat{\sigma}_x\otimes\hat{\sigma}_x$ and $\hat{\sigma}_y\otimes\hat{\sigma}_y$, none of the individual spin operators $\hat{\sigma}_x\otimes\hat{I}_B$, $\hat{I}_A\otimes\hat{\sigma}_x$, $\hat{\sigma}_y\otimes\hat{I}_B$ or $\hat{I}_A\otimes\hat{\sigma}_y$ can be said to have a definite value. The inability of noncommuting observables to be determined simultaneously is a manifestation of quantum complementarity [62]. Further examples of incompatible nonlocal quantum observables will be shown in Section 4.20 in the proofs of the Kochen–Specker theorem (for $n \geq 8$) and Bell's theorem without inequalities (for $n \geq 8$).

In essence, the quantum state vector $|\psi\rangle$ of a quantum system Q provides a complete physical description, because if the quantum system Q has its own quantum state vector, then it is not quantum entangled or correlated with any external physical system. On the other hand, the mixed state density matrix $\hat{\rho}$ of a quantum system Q' provides only an incomplete physical description, because one does not know with which external physical systems the system Q' is entangled and in what form that entanglement is. Since the mixed state quantum system Q' does not have its own state vector, in order to obtain a complete physical description one needs to consider the quantum state vector $|\psi''\rangle$ of an enlarged composite quantum system Q'' that has the system Q' as an entangled component. The density matrix description is as useful as the quantum state description only in rare cases when the density matrix is pure, the rationale for which is exactly the fact that pure state density matrices stand in one-to-one correspondence with quantum state vectors.

4.19   Solving the Schrödinger equation

Quantum tunneling of an electron inside a triple well potential could be used to illustrate the time dynamics of a quantum system resulting from the Schrödinger equation. Suppose that the normalized position states of the electron in each of the potential wells are $|A\rangle$, $|B\rangle$ and $|C\rangle$. The Hilbert space of this 3-level system will be 3-dimensional complex space $\mathcal{H} = \mathbb{C}^3$. Let the offset energies between different wells be Δ_1, Δ_2 and Δ_3, and the transitions between different wells be given by tunneling matrix elements such that κ_{12} is the quantum probability amplitude to transfer an electron through the barrier from well $|A\rangle$ to well $|B\rangle$, κ_{23} from well $|B\rangle$ to well $|C\rangle$, and κ_{13} from well $|A\rangle$ to well $|C\rangle$ (Fig. 4.6).

The Hamiltonian of the system in position basis [95] can be written as

$$\hat{H} = \begin{pmatrix} \Delta_1 & -\kappa_{12} & -\kappa_{13} \\ -\kappa_{12}^* & \Delta_2 & -\kappa_{23} \\ -\kappa_{13}^* & -\kappa_{23}^* & \Delta_3 \end{pmatrix} \tag{4.202}$$

For numerical calculation, we will assume no offset energy between wells and will set $\kappa_{12} = \kappa_{23} = \frac{1}{\sqrt{2}}$, $\kappa_{13} = 0$ [208]. The Hamiltonian becomes

$$\hat{H} = -\frac{1}{\sqrt{2}} \begin{pmatrix} 0 & 1 & 0 \\ 1 & 0 & 1 \\ 0 & 1 & 0 \end{pmatrix} \tag{4.203}$$

The energy eigenstates of the system are the eigenstates of the Hamiltonian. Finding the eigenvalues and eigenvectors of a matrix can be done using available eigen-

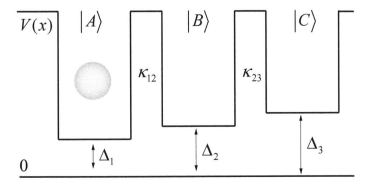

Figure 4.6   A quantum particle inside a triple well potential. The particle can move freely inside the wells $|A\rangle$, $|B\rangle$ and $|C\rangle$, however, it must tunnel through the walls in order to jump from one well to another. The quantum tunneling coefficients between adjacent well are κ_{12} and κ_{23}, and the offset energies for the wells are Δ_1, Δ_2 and Δ_3.

value algorithms. The normalized eigenvectors of the Hamiltonian are

$$|E_+\rangle = \frac{1}{2}|A\rangle - \frac{1}{\sqrt{2}}|B\rangle + \frac{1}{2}|C\rangle \tag{4.204}$$

$$|E_0\rangle = \frac{1}{\sqrt{2}}|A\rangle - \frac{1}{\sqrt{2}}|C\rangle \tag{4.205}$$

$$|E_-\rangle = \frac{1}{2}|A\rangle + \frac{1}{\sqrt{2}}|B\rangle + \frac{1}{2}|C\rangle \tag{4.206}$$

with eigenvalues

$$E_+ = +1, \quad E_0 = 0, \quad E_- = -1 \tag{4.207}$$

The normalization implies that

$$\langle E_+|E_+\rangle = \langle E_0|E_0\rangle = \langle E_-|E_-\rangle = 1 \tag{4.208}$$

If we express the state vector $|\psi\rangle$ in energy basis

$$|\psi\rangle = \sum_{j=1}^{n}|E_j\rangle\langle E_j|\psi\rangle = \sum_{j=1}^{n}\langle E_j|\psi\rangle|E_j\rangle = \sum_{j=1}^{n}\alpha_j|E_j\rangle \tag{4.209}$$

where $\alpha_j = \langle E_j|\psi\rangle$ are the quantum probability amplitudes for each energy eigenvector [467], we can turn the action of the Hamiltonian operator $\hat{H}$ on the state vector $|\psi\rangle$ in the Schrödinger equation into simple multiplication by the scalar eigenvalues E_j

$$i\hbar\frac{\partial}{\partial t}|\psi\rangle = \hat{H}|\psi\rangle = \hat{H}\sum_{j=1}^{n}\alpha_j|E_j\rangle = \sum_{j=1}^{n}\alpha_j E_j|E_j\rangle \tag{4.210}$$

Explicitly rewriting the quantum state vector $|\psi\rangle$ in the energy basis gives a system of n differential equations

$$
i\hbar\frac{\partial}{\partial t}\begin{pmatrix} \alpha_1 \\ \alpha_2 \\ \vdots \\ a_n \end{pmatrix} = \begin{pmatrix} \alpha_1 E_1 \\ \alpha_2 E_2 \\ \vdots \\ a_n E_n \end{pmatrix}
\tag{4.211}
$$

The solutions are of the form

$$
\alpha_j(t) = \alpha_j(0)e^{-\frac{i}{\hbar}E_j t}
\tag{4.212}
$$

where $\alpha_j(0)$ is the quantum probability amplitude at time $t = 0$ [468, pp. 119–124].

Since we already know the Hamiltonian $\hat{H}$ given by Eq. (4.203), the general solution can be applied to the triple well potential if we are given the initial state vector $|\psi(0)\rangle$. Suppose that the electron is initially in well A, as shown in Figure 4.6. The initial state $|\psi(0)\rangle = |A\rangle$ can be expressed in the energy basis as

$$
\begin{aligned}
|A\rangle &= \langle E_+|A\rangle|E_+\rangle + \langle E_0|A\rangle|E_0\rangle + \langle E_-|A\rangle|E_-\rangle \\
&= \frac{1}{2}|E_+\rangle + \frac{1}{\sqrt{2}}|E_0\rangle + \frac{1}{2}|E_-\rangle
\end{aligned}
\tag{4.213}
$$

The solution for $|\psi(t)\rangle$ in the energy basis after plugging in the eigenvalues E_j given by Eq. (4.207) is

$$
|\psi(t)\rangle = \frac{1}{2}e^{-\frac{i}{\hbar}t}|E_+\rangle + \frac{1}{\sqrt{2}}|E_0\rangle + \frac{1}{2}e^{\frac{i}{\hbar}t}|E_-\rangle
\tag{4.214}
$$

Having solved $|\psi(t)\rangle$ in the energy basis does not mean that we have to measure the state in the energy basis as well. For example, we can measure $|\psi(t)\rangle$ in the position basis and observe quantum interference phenomena. Using Eqs. (4.204), (4.205) and (4.206) we can rewrite $|\psi(t)\rangle$ in the position basis as

$$
\begin{aligned}
|\psi(t)\rangle &= \frac{1}{2}e^{-\frac{i}{\hbar}t}\left(\frac{1}{2}|A\rangle - \frac{1}{\sqrt{2}}|B\rangle + \frac{1}{2}|C\rangle\right) + \frac{1}{\sqrt{2}}\left(\frac{1}{\sqrt{2}}|A\rangle - \frac{1}{\sqrt{2}}|C\rangle\right) \\
&\quad + \frac{1}{2}e^{\frac{i}{\hbar}t}\left(\frac{1}{2}|A\rangle + \frac{1}{\sqrt{2}}|B\rangle + \frac{1}{2}|C\rangle\right)
\end{aligned}
\tag{4.215}
$$

This expression simplifies to

$$
|\psi(t)\rangle = \frac{1}{2}\left[1 + \cos\left(\frac{t}{\hbar}\right)\right]|A\rangle + \frac{1}{\sqrt{2}}i\sin\left(\frac{t}{\hbar}\right)|B\rangle + \frac{1}{2}\left[\cos\left(\frac{t}{\hbar}\right) - 1\right]|C\rangle
\tag{4.216}
$$

where we have used the relations

$$
\cos x = \mathrm{Re}\,(e^{ix}) = \frac{1}{2}(e^{ix} + e^{-ix})
\tag{4.217}
$$

$$
\sin x = \mathrm{Im}\,(e^{ix}) = \frac{1}{2i}(e^{ix} - e^{-ix})
\tag{4.218}
$$

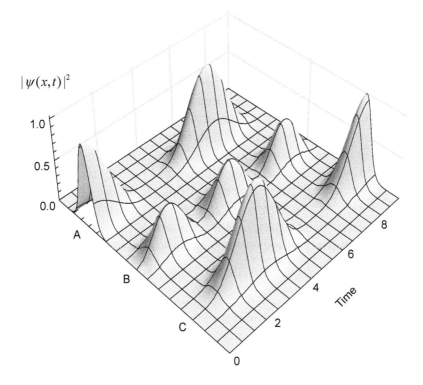

Figure 4.7   Quantum tunneling of an electron inside a triple well potential simulated for a period of time $t = 3\pi/\hbar$. At times $t = 2k\pi/\hbar$, $k = 0, 1, 2, \ldots$ the electron is localized in well A, whereas at times $t = (2k + 1)\pi/\hbar$, the electron is localized in well C. If unperturbed, the quantum system in a triple potential well tunnels from well A to wells B and C, returning periodically to its initial state. The probability $|\psi(x,t)|^2$ is normalized so that $\int_{-\infty}^{\infty} |\psi(x,t)|^2 dx = 1$.

Plotting the probability $|\psi(x,t)|^2$ as a function of time shows that the electron tunnels forth-and-back between the three wells as long as the system remains closed and unperturbed by external observation (Fig. 4.7). The presented quantum description of the world may look like a deterministic one, but there is an important difference. The solution of the Schrödinger equation together with the Born rule provides a deterministic evolution of quantum probabilities $|\psi(x,t)|^2$ in space and time (Fig. 4.7). However, these are probabilities for potential events to happen. For example, it can be seen that at certain times there is a non-zero probability for the electron to be in any of the three wells. If we indeed measure the electron at such a moment of time to see in which well it is, an indeterministic localization event will occur and we will always find the electron being in only one of the wells. At this point, we have just glimpsed the wondrous nature of the quantum world. To understand it better, we will need to study the properties of quantum information as revealed by known quantum no-go theorems.

4.20 Quantum information

The carriers of *quantum information* are referred to as *quantum bits*, or simply *qubits*. Peculiarly, the quantum information contained in the quantum state (wave function) of a qubit cannot be observed, read or deduced from experimental data as in the case of the classical bits stored on a DVD [7, 65]. If we have a quantum version of DVD storing a string of qubits, in general we cannot copy the string of qubits [131, 528], we cannot process the qubits using irreversible computational gates [502], and we can only swap the qubits but not erase them [367]. Furthermore, the Bell and Kochen–Specker no-go theorems [39, 295] imply that quantum information is non-local and quantum correlations are enforced with a speed that exceeds the speed of light in vacuum c. Next, we will show how these remarkable results can be proved as mathematical theorems using the standard Hilbert space formalism of quantum mechanics.

Theorem 13. *An unknown quantum state $|\Psi\rangle$ cannot be determined by a single measurement.*

Proof. Determination of the *unknown* state $|\Psi\rangle$ of a quantum system could be attempted by performing a measurement. In quantum theory, every measurement is represented by some *observable* that is a Hermitian operator $\hat{A} = \hat{A}^\dagger$ whose eigenvalues are exhibited as the measurement outcomes. The spectral decomposition of the operator $\hat{A}$ could be written as $\hat{A} = \sum_{i=1}^{n} \lambda_i |\psi_i\rangle\langle\psi_i|$. In order to be able to unambiguously decide from the outcomes whether or not the system was in a given state $|\psi_1\rangle$, there should be a unique (non-degenerate) outcome λ_1 which occurs with certainty if the state was $|\psi_1\rangle$, and which will certainly not occur if the state was some other arbitrary state $|\psi_2\rangle$ different from $|\psi_1\rangle$. Then, the expectation values of the projection operator $\hat{P}_{\lambda_1} = |\psi_1\rangle\langle\psi_1|$ that projects the measured unknown quantum state onto the eigenvector $|\psi_1\rangle$ of $\hat{A}$ for each of the two states $|\psi_1\rangle$ and $|\psi_2\rangle$ should be

$$\langle\psi_1|\hat{P}_{\lambda_1}|\psi_1\rangle = 1 \tag{4.219}$$

$$\langle\psi_2|\hat{P}_{\lambda_1}|\psi_2\rangle = 0 \tag{4.220}$$

These equations are equivalent to

$$\hat{P}_{\lambda_1}|\psi_1\rangle = 1|\psi_1\rangle = |\psi_1\rangle \tag{4.221}$$

$$\hat{P}_{\lambda_1}|\psi_2\rangle = 0|\psi_2\rangle = 0 \tag{4.222}$$

Therefore, $\langle\psi_2|\psi_1\rangle = \langle\psi_2|\hat{P}_{\lambda_1}|\psi_1\rangle = (\langle\psi_1|\hat{P}_{\lambda_1}|\psi_2\rangle)^* = 0$, which is to say that $|\psi_1\rangle$ and $|\psi_2\rangle$ are mutually orthogonal quantum states [65]. Conversely, if $|\psi_1\rangle$ and $|\psi_2\rangle$ are non-orthogonal quantum states, namely, $\langle\psi_2|\psi_1\rangle = \alpha \neq 0$, the expectation value for the state $|\psi_2\rangle$ will be greater than zero $\langle\psi_2|\hat{P}_{\lambda_1}|\psi_2\rangle = \langle\psi_2|\psi_1\rangle\langle\psi_1|\psi_2\rangle = \alpha\alpha^* > 0$, hence the states cannot be distinguished with absolute certainty. Because no measurement could distinguish unequivocally between any pair of non-orthogonal states, it follows that unambiguous determination of unknown quantum state $|\Psi\rangle$ is impossible by a single measurement of the quantum system [65, 366]. $\square$

Noteworthy, after the first measurement of the observable $\hat{A}$ performed upon the given quantum system, the unknown quantum state $|\Psi\rangle$ is transformed into the eigenstate $|\psi_i\rangle$ corresponding to the measured eigenvalue λ_i. This implies that, in general, we are unable to perform a second quantum measurement on the initial quantum state $|\Psi\rangle$. Thus, we can prove an even stronger result according to which unambiguous determination of unknown quantum state $|\Psi\rangle$ is impossible by either single or repeated measurement of the same quantum system. In particular, from Axiom 4.15.6 it follows that if the first measurement returns the eigenvalue λ_i, immediate repeated measurements of the observable $\hat{A}$ will return a string of results with the same eigenvalue $\lambda_i, \lambda_i, \lambda_i, \ldots$

Theorem 14. *(No-cloning theorem) Quantum information contained in an unknown quantum state $|\Psi\rangle$ cannot be cloned.*

Proof. Suppose we have a two-level quantum system (qubit) A, whose *unknown* quantum state $|\Psi_A\rangle$ we wish to copy. In general, the state can be written as

$$|\Psi_A\rangle = \alpha|0_A\rangle + \beta|1_A\rangle \tag{4.223}$$

where $|0_A\rangle$ and $|1_A\rangle$ are two orthogonal basis states in two-dimensional Hilbert space $\mathcal{H}_A = \mathbb{C}^2$, and the complex coefficients α and β are unknown. In order to make a copy, we take another qubit B with Hilbert space $\mathcal{H}_B = \mathbb{C}^2$ and initial state $|e_B\rangle$, which must be independent of $|\Psi_A\rangle$ of which we have no prior knowledge. The composite system is then described by the tensor product state $|\Psi_A\rangle \otimes |e_B\rangle$ in Hilbert space $\mathcal{H}_A \otimes \mathcal{H}_B$. There are only two ways to manipulate the composite system. One possibility is to perform an observation and measure the qubit A, which forces the system into some eigenstate of the observable and corrupts the information contained in the qubit A (Theorem 13). This precludes achieving a copy of qubit A. A second alternative is to control the Hamiltonian of the composite system, and thus the time evolution operator $\hat{U}$, which is *linear*. For any fixed time interval, the operator $\hat{U}$ would act as a copier provided that

$$\hat{U}|\Psi_A\rangle \otimes |e_B\rangle = |\Psi_A\rangle \otimes |\Psi_B\rangle \tag{4.224}$$

for all $|\Psi\rangle$. This must be true for the basis states as well, so

$$\hat{U}|0_A\rangle \otimes |e_B\rangle = |0_A\rangle \otimes |0_B\rangle \tag{4.225}$$

$$\hat{U}|1_A\rangle \otimes |e_B\rangle = |1_A\rangle \otimes |1_B\rangle \tag{4.226}$$

Then Eq. (4.223) and the linearity of $\hat{U}$ imply

$$\hat{U}|\Psi_A\rangle \otimes |e_B\rangle = \hat{U}(\alpha|0_A\rangle + \beta|1_A\rangle) \otimes |e_B\rangle = \alpha\hat{U}|0_A\rangle \otimes |e_B\rangle + \beta\hat{U}|1_A\rangle \otimes |e_B\rangle$$

$$= \alpha|0_A\rangle \otimes |0_B\rangle + \beta|1_A\rangle \otimes |1_B\rangle \tag{4.227}$$

In general, this is not equal to $|\Psi_A\rangle \otimes |\Psi_B\rangle$, as may be directly verified by plugging in $\alpha = \frac{3}{5}$ and $\beta = \frac{4}{5}$. Indeed, if one starts with $|\Psi_A\rangle$ being a superposition of the basis states $|0_A\rangle$ and $|1_A\rangle$, the time evolution operator $\hat{U}$ will create an *entangled state*, so $\hat{U}$ cannot act as a general copier. Thus, the unknown quantum state $|\Psi_A\rangle$ of a qubit A cannot be cloned to another qubit B [131, 528]. $\qquad\square$

The argument used in the proof of the no-cloning theorem illustrates something very interesting, namely, if $|\Psi_A\rangle$ is not in one of the two basis states for which our copying machine is designed, the putative copy will be *entangled* with the original qubit. If we measure the original qubit, we will corrupt the *entangled copy* as well. In general, the copy that we can achieve is a pseudo-copy, because it is entangled with the original qubit and will be corrupted when the original qubit is measured [207]. We should also note that quantum cloning by guessing the necessary copying machine is *improbable*: If $|\Psi_A\rangle$ were in one of the two basis states, the copy will be a true copy, yet we will not know this. Since there is an infinite number of possible basis states, the chance for production of a true copy is equal to the chance of guessing correctly the unknown quantum state, which is one out of an infinite number of possible states. In other words, there is an infinite number of copying machines that can copy only a pair of basis states, and since originally the quantum state is unknown to us, the probability of choosing at random the correct copying machine is *zero*.

Theorem 15. *Quantum information contained in an unknown quantum state $|\Psi\rangle$ cannot be read. The quantum state is unobservable.*

Proof. The ability to read quantum information requires a physical process by which an initially *unknown* quantum state becomes *known*. We have already proved that unambiguous determination of unknown quantum state $|\Psi\rangle$ of a qubit is impossible by a single measurement of the qubit (Theorem 13). Repeated measurement of the qubit does not help either, because after the first measurement of any observable $\hat{A}$, the initial quantum state gets corrupted by being projected into an eigenstate of $\hat{A}$. The only other alternative to consider is whether there is a way to copy the original qubit multiple times before we measure those copies as well in an attempt to reconstruct the unknown quantum state $|\Psi\rangle$ of the qubit. However, the no-cloning theorem eliminates all cloning scenarios from consideration. Thus, unambiguous determination of an unknown individual quantum state $|\Psi\rangle$ of a qubit is impossible [65]. The quantum state is not observable. The latter statement is also directly reflected in the mathematical formalism of quantum mechanics; the *quantum state* is represented by a ket vector $|\Psi\rangle$ that is an $n \times 1$ matrix, whereas every *observable* is represented by a Hermitian operator $\hat{A}$ that is an $n \times n$ matrix.   □

Noteworthy, quantum theory does not forbid the reading of a known quantum state, because if we know exactly what the state $|\Psi\rangle$ is, we could use the operator $|\Psi\rangle\langle\Psi|$ to measure the state $|\Psi\rangle$ with absolute certainty. Reading what we already know is not a generally useful form of reading, however, because at best we only verify what we already know.

Theorem 16. (*No-broadcasting theorem*) *Quantum information contained in an unknown quantum state* $|\Psi\rangle$ *cannot be broadcast.*

Proof. Broadcasting refers to the process of conveying an unknown quantum state $|\Psi\rangle$ to two or more recipients. Performing quantum measurement of any observable $\hat{A}$ invariably forces the measured qubit into some eigenstate $|a_i\rangle$ of the observable $\hat{A}$ with probability p_i. If we are interested in preparing the quantum state $|a_n\rangle$, we can measure the observable $\hat{A}$ of the qubit, and post-select only those measurement outcomes that return the eigenstate $|a_n\rangle$, discarding all other cases. The efficiency of the preparation procedure will be p_n, since we have to discard $(1 - p_n)$ of the cases that do not return $|a_n\rangle$. Even though we are in principle able to prepare any quantum state, the quantum information contained in an unknown quantum state $|\Psi\rangle$ cannot be broadcast, because we have to be able to read it first. Without reading (Theorem 13), there is no broadcasting. Generalization of the *no-broadcasting theorem* has been also proven for mixed quantum systems [29]. $\square$

Theorem 17. (*Bell's theorem*) *Quantum information is nonlocal and allows enforcement of quantum correlations faster than the speed of light in vacuum c.*

Proof. Suppose we have a local process that prepares two coins in two dark boxes and then sends them far away. Let the coins have three different two-valued properties: A: made of gold, 1; $\bar{A}$: made of silver, 0; B: with large diameter, 1; $\bar{B}$: with small diameter, 0; C: with a hole in the middle, 1; $\bar{C}$: without a hole in the middle, 0. Let also the two boxes be opened far away by two different observers, who have measuring devices that can measure only one of the properties A, B or C, but in the process the coin is destroyed [399]. The observers can then communicate by sending classical radio signals traveling with the speed of light in vacuum c and find out that the coins are always perfectly correlated so that whenever both observers have measured the same property A, B or C, they always get the same outcome of the measurements, 0 or 1. If we use indices 1 and 2 for denoting the first and the second coin, we can write

$$
\begin{aligned}
p(A_1|A_2) = p(A_2|A_1) = p(\bar{A}_1|\bar{A}_2) = p(\bar{A}_2|\bar{A}_1) = 1 \\
p(B_1|B_2) = p(B_2|B_1) = p(\bar{B}_1|\bar{B}_2) = p(\bar{B}_2|\bar{B}_1) = 1 \\
p(C_1|C_2) = p(C_2|C_1) = p(\bar{C}_1|\bar{C}_2) = p(\bar{C}_2|\bar{C}_1) = 1
\end{aligned} \tag{4.228}
$$

From Kolmogorov's probability, Eqs. (3.59) and (3.60), we can deduce that

$$
\begin{aligned}
A_1 \cap A_2 &= A_1 \cup A_2 = A_1 = A_2 = A \\
\bar{A}_1 \cap \bar{A}_2 &= \bar{A}_1 \cup \bar{A}_2 = \bar{A}_1 = \bar{A}_2 = \bar{A} \\
B_1 \cap B_2 &= B_1 \cup B_2 = B_1 = B_2 = B \\
\bar{B}_1 \cap \bar{B}_2 &= \bar{B}_1 \cup \bar{B}_2 = \bar{B}_1 = \bar{B}_2 = \bar{B} \\
C_1 \cap C_2 &= C_1 \cup C_2 = C_1 = C_2 = C \\
\bar{C}_1 \cap \bar{C}_2 &= \bar{C}_1 \cup \bar{C}_2 = \bar{C}_1 = \bar{C}_2 = \bar{C}
\end{aligned} \tag{4.229}
$$

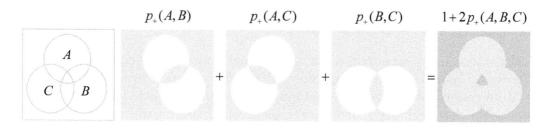

$$p_+(A,B) \qquad p_+(A,C) \qquad p_+(B,C) \qquad 1+2p_+(A,B,C)$$

Figure 4.8 Classical correlations obey Kolmogorov's axioms of probability and are thus bound by Bell inequality $p_+(A,B)+p_+(A,C)+p_+(B,C) \geq 1$, where $p_+(X,Y)$ is the probability that both properties X and Y have the same value, 0 or 1, and $p_+(X,Y,Z)$ is the probability that all three properties X, Y and Z have the same value, 0 or 1.

The existence of classical correlations is not surprising, because they can be set up by local processes at a point in the past when the two physical systems interacted with each other with signals bounded by the speed of light in vacuum c. An important feature of classical correlations, however, is that they obey Kolmogorov's axioms of probability and are thereby bound by a number of inequalities named after the quantum physicist John Stewart Bell (1928–1990) [39]. Here, we will prove a version of the Bell inequality derived by John Preskill [399] that bounds the probabilities $p_+(X,Y)$ for both properties X and Y to have the same value, 0 or 1.

$$p_+(A,B) + p_+(A,C) + p_+(B,C) \geq 1 \tag{4.230}$$

From Kolmogorov's axioms (Section 3.10) we have

$$p_+(A,B) = p(A \cap B \cap C) + p(A \cap B \cap \bar{C}) + p(\bar{A} \cap \bar{B} \cap C) + p(\bar{A} \cap \bar{B} \cap \bar{C})$$
$$p_+(A,C) = p(A \cap B \cap C) + p(A \cap \bar{B} \cap C) + p(\bar{A} \cap B \cap \bar{C}) + p(\bar{A} \cap \bar{B} \cap \bar{C})$$
$$p_+(B,C) = p(A \cap B \cap C) + p(\bar{A} \cap B \cap C) + p(A \cap \bar{B} \cap \bar{C}) + p(\bar{A} \cap \bar{B} \cap \bar{C}) \tag{4.231}$$

The probabilities in the whole sample space sum up to 1

$$\begin{aligned} 1 \;=\;& p(A \cap B \cap C) + p(A \cap \bar{B} \cap C) + p(A \cap B \cap \bar{C}) + p(A \cap \bar{B} \cap \bar{C}) \\ &+ p(\bar{A} \cap B \cap C) + p(\bar{A} \cap \bar{B} \cap C) + p(\bar{A} \cap B \cap \bar{C}) + p(\bar{A} \cap \bar{B} \cap \bar{C}) \end{aligned} \tag{4.232}$$

From Eqs. (4.231) and (4.232) it follows that

$$p_+(A,B) + p_+(A,C) + p_+(B,C) = 1 + 2p_+(A,B,C) \tag{4.233}$$

where

$$p_+(A,B,C) = p(A \cap B \cap C) + p(\bar{A} \cap \bar{B} \cap \bar{C}) \geq 0 \tag{4.234}$$

The Bell inequality (4.230) follows directly from Eqs. (4.233) and (4.234). Visually, the Bell inequality can be proved from Figure 4.8.

Now, using the axioms of quantum mechanics, we can show that quantum entangled systems violate the Bell inequality. Consider the maximally entangled state of two qubits

$$|\Psi\rangle = \frac{1}{\sqrt{2}}\left(|0_1\rangle|0_2\rangle + |1_1\rangle|1_2\rangle\right) \tag{4.235}$$

and a physical measurement of one of the following three quantum observables written in the $|0\rangle, |1\rangle$ basis as

$$\hat{A} = \begin{pmatrix} 0 & 0 \\ 0 & 1 \end{pmatrix} \tag{4.236}$$

$$\hat{B} = \begin{pmatrix} \frac{3}{4} & -\frac{\sqrt{3}}{4} \\ -\frac{\sqrt{3}}{4} & \frac{1}{4} \end{pmatrix} \tag{4.237}$$

$$\hat{C} = \begin{pmatrix} \frac{3}{4} & \frac{\sqrt{3}}{4} \\ \frac{\sqrt{3}}{4} & \frac{1}{4} \end{pmatrix} \tag{4.238}$$

The eigenstates and the corresponding eigenvalues of these observables are

$$A : \begin{cases} |\bar{A}\rangle = |0\rangle, & \lambda_0 = 0 \\ |A\rangle = |1\rangle, & \lambda_1 = 1 \end{cases} \tag{4.239}$$

$$B : \begin{cases} |\bar{B}\rangle = \frac{1}{2}|0\rangle + \frac{\sqrt{3}}{2}|1\rangle, & \lambda_0 = 0 \\ |B\rangle = \frac{\sqrt{3}}{2}|0\rangle - \frac{1}{2}|1\rangle, & \lambda_1 = 1 \end{cases} \tag{4.240}$$

$$C : \begin{cases} |\bar{C}\rangle = \frac{1}{2}|0\rangle - \frac{\sqrt{3}}{2}|1\rangle, & \lambda_0 = 0 \\ |C\rangle = \frac{\sqrt{3}}{2}|0\rangle + \frac{1}{2}|1\rangle, & \lambda_1 = 1 \end{cases} \tag{4.241}$$

The state $|\Psi\rangle$ can be rewritten as

$$|\Psi\rangle = \frac{1}{\sqrt{2}}\left(|A_1\rangle|A_2\rangle + |\bar{A}_1\rangle|\bar{A}_2\rangle\right) \tag{4.242}$$

$$= \frac{1}{\sqrt{2}}\left(|B_1\rangle|B_2\rangle + |\bar{B}_1\rangle|\bar{B}_2\rangle\right) \tag{4.243}$$

$$= \frac{1}{\sqrt{2}}\left(|C_1\rangle|C_2\rangle + |\bar{C}_1\rangle|\bar{C}_2\rangle\right) \tag{4.244}$$

Thus, whenever the same observable is measured for both qubits, 1 and 2, the resulting outcomes, 0 or 1, will always be the same. Therefore, the quantum state $|\Psi\rangle$ and the observables $\hat{A}$, $\hat{B}$ and $\hat{C}$ indeed satisfy the conditional probabilities required by Eqs. (4.228). Next, it is possible to measure one property on qubit 1 and another property on qubit 2 using projection operators. According to the Born rule, the expectation value of a quantum projection operator that measures different properties on each qubit $|X_1\rangle|Y_2\rangle\langle X_1|\langle Y_2| = |X_1\rangle\langle X_1| \otimes |Y_2\rangle\langle Y_2|$ is given by

$$\langle\Psi|X_1\rangle|Y_2\rangle\langle X_1|\langle Y_2|\Psi\rangle = |\langle X_1|\langle Y_2|\Psi\rangle|^2 \tag{4.245}$$

The probabilities of obtaining the same measurement outcomes on both qubits are

$$p_+(A_1, B_2) = |\langle A_1|\langle B_2\|\Psi\rangle|^2 + |\langle \bar{A}_1|\langle \bar{B}_2\|\Psi\rangle|^2$$

$$= \left|\left(\frac{\sqrt{3}}{2}\langle 1_1|\langle 0_2| - \frac{1}{2}\langle 1_1|\langle 1_2|\right)\frac{1}{\sqrt{2}}(|0_1\rangle|0_2\rangle + |1_1\rangle|1_2\rangle)\right|^2$$

$$+ \left|\left(\frac{1}{2}\langle 0_1|\langle 0_2| + \frac{\sqrt{3}}{2}\langle 0_1|\langle 1_2|\right)\frac{1}{\sqrt{2}}(|0_1\rangle|0_2\rangle + |1_1\rangle|1_2\rangle)\right|^2$$

$$= \left|-\frac{1}{2}\frac{1}{\sqrt{2}}\right|^2 + \left|\frac{1}{2}\frac{1}{\sqrt{2}}\right|^2 = \frac{1}{4}$$

$$p_+(A_1, C_2) = |\langle A_1|\langle C_2\|\Psi\rangle|^2 + |\langle \bar{A}_1|\langle \bar{C}_2\|\Psi\rangle|^2$$

$$= \left|\left(\frac{\sqrt{3}}{2}\langle 1_1|\langle 0_2| + \frac{1}{2}\langle 1_1|\langle 1_2|\right)\frac{1}{\sqrt{2}}(|0_1\rangle|0_2\rangle + |1_1\rangle|1_2\rangle)\right|^2$$

$$+ \left|\left(\frac{1}{2}\langle 0_1|\langle 0_2| - \frac{\sqrt{3}}{2}\langle 0_1|\langle 1_2|\right)\frac{1}{\sqrt{2}}(|0_1\rangle|0_2\rangle + |1_1\rangle|1_2\rangle)\right|^2$$

$$= \left|\frac{1}{2}\frac{1}{\sqrt{2}}\right|^2 + \left|\frac{1}{2}\frac{1}{\sqrt{2}}\right|^2 = \frac{1}{4}$$

$$p_+(B_1, C_2) = |\langle B_1|\langle C_2\|\Psi\rangle|^2 + |\langle \bar{B}_1|\langle \bar{C}_2\|\Psi\rangle|^2$$

$$= \left|\left(\frac{\sqrt{3}}{2}\langle 0_1| - \frac{1}{2}\langle 1_1|\right)\left(\frac{\sqrt{3}}{2}\langle 0_2| + \frac{1}{2}\langle 1_2|\right)\frac{1}{\sqrt{2}}(|0_1\rangle|0_2\rangle + |1_1\rangle|1_2\rangle)\right|^2$$

$$+ \left|\left(\frac{1}{2}\langle 0_1| + \frac{\sqrt{3}}{2}\langle 1_1|\right)\left(\frac{1}{2}\langle 0_2| - \frac{\sqrt{3}}{2}\langle 1_2|\right)\frac{1}{\sqrt{2}}(|0_1\rangle|0_2\rangle + |1_1\rangle|1_2\rangle)\right|^2$$

$$= \left|\left(\frac{3}{4} - \frac{1}{4}\right)\frac{1}{\sqrt{2}}\right|^2 + \left|\left(\frac{1}{4} - \frac{3}{4}\right)\frac{1}{\sqrt{2}}\right|^2 = \frac{1}{4}$$

Similarly to the classical case, the quantum probabilities are symmetric, meaning that it does not matter which of the two different properties is measured on qubit 1 and which on qubit 2

$$p_+(A_1, B_2) = p_+(A_2, B_1) = p_+(A, B) \tag{4.246}$$

$$p_+(A_1, C_2) = p_+(A_2, C_1) = p_+(A, C) \tag{4.247}$$

$$p_+(B_1, C_2) = p_+(B_2, C_1) = p_+(B, C) \tag{4.248}$$

In contrast to the classical case, however, the quantum probabilities manifestly violate the Bell inequality given by Eq. (4.230) since

$$p_+(A, B) + p_+(A, C) + p_+(B, C) = \frac{3}{4} < 1 \tag{4.249}$$

Thus, quantum correlations between systems that are far away from each other in space cannot be explained by some clever local arrangement of classical correlations between physical properties at some point in the past. Instead, quantum correlations are nonlocal and enforced instantaneously in accordance with the choice of measurements performed upon each particle, irrespective of the spatial distance between the quantum entangled particles [39]. □

Here, a caveat is needed. Even though the probabilities $p_+(A_1, B_2)$ and $p_+(A_2, B_1)$ are equal, the quantum physical outcomes $|A_1\rangle|B_2\rangle$ and $|A_2\rangle|B_1\rangle$ are not the same. For example, in the state $|A_1\rangle|B_2\rangle$, the qubit 1 is definitely in state $|A\rangle$ and the qubit 2 is definitely in state $|B\rangle$. If we check whether qubit 1 from the state $|A_1\rangle|B_2\rangle$ is in state $|B\rangle$, we will not find it in state $|B\rangle$ with certainty, but with probability $\frac{1}{4}$. In other words, if two non-commuting observables (such as $\hat{A}$ and $\hat{B}$) are measured on each qubit from an entangled pair, the result from the measurement on one of the qubits cannot be claimed to reveal what the state of the other qubit is. In the classical world, however, from the measurement performed on one of two perfectly correlated classical bits, the axioms of Kolmogorov's probability allow you to make inferences about the physical properties of the second bit using Eqs. (4.228) and (4.229).

It should also be noted that quantum mechanics does not have a meaningful interpretation of the expression $p_+(A, B, C)$ for the qubit pair, because two non-commuting observables such as $\hat{A}$ and $\hat{B}$ cannot be measured on the same qubit at the same time. It is only possible to simultaneously measure the observables $\hat{A}$ and $\hat{B}$ on two different quantum systems due to the appearance of tensor products in the calculation. Indeed, $\hat{A} \otimes \hat{I}$ and $\hat{I} \otimes \hat{B}$ commute with each other since

$$\left(\hat{A} \otimes \hat{I}\right)\left(\hat{I} \otimes \hat{B}\right) = \left(\hat{I} \otimes \hat{B}\right)\left(\hat{A} \otimes \hat{I}\right) = \hat{A} \otimes \hat{B} \tag{4.250}$$

Theorem 18. *(Kochen–Specker theorem) Quantum information is contextual, namely, it is not possible for all quantum mechanical observables of a quantum system to have predetermined values ahead of time and independent of the apparatus used to measure those observables.*

Proof. The Kochen–Specker theorem holds for quantum systems whose Hilbert space $\mathcal{H}$ is with dimension $n \geq 3$ [295]. Consider a single qutrit, such as a spin-1 particle, and measurement of the squared spin-1 observables (Section 4.13) along 33 different rays in the real 3-dimensional space $\mathbb{R}^3$ [295, 384], as shown in Figure 4.9. If the center of the cube is at $(0, 0, 0)$, the 33 rays are collinear with the following 33 vectors:

$$\vec{r_1} = (1,0,0) \quad \vec{r_2} = (0,1,0) \quad \vec{r_3} = (0,0,1) \quad \vec{r_4} = (0,1,1) \quad \vec{r_5} = (0,-1,1)$$

$$\vec{r_6} = (1,0,1) \quad \vec{r_7} = (-1,0,1) \quad \vec{r_8} = (1,1,0) \quad \vec{r_9} = (-1,1,0)$$

$$\vec{r_{10}} = \left(0, \frac{1}{\sqrt{2}}, 1\right) \quad \vec{r_{11}} = \left(0, 1, -\frac{1}{\sqrt{2}}\right) \quad \vec{r_{12}} = \left(0, -\frac{1}{\sqrt{2}}, 1\right) \quad \vec{r_{13}} = \left(0, 1, \frac{1}{\sqrt{2}}\right)$$

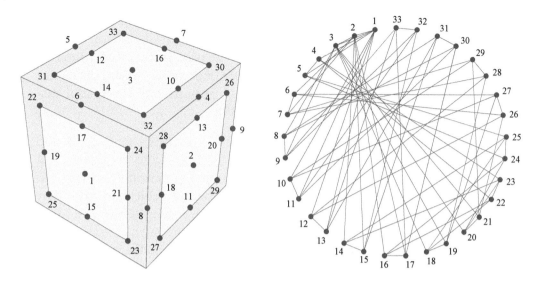

Figure 4.9   The proof of the Kochen–Specker theorem in 3-dimensional Hilbert space requires 33 rays in real 3-dimensional Euclidean space $\mathbb{R}^3$. Each ray passes through a point on the surface of a cube connected with the cube center. The cube has sides of length 2, whereas the inner squares have lengths $\sqrt{2}$. Quantum measurements of the squared spin-1 observables along any three orthogonal (perpendicular) rays always satisfy the 1, 0, 1 rule. The orthogonality relationships between different rays are shown on the circle graph where every two orthogonal rays are connected with a line.

$$\vec{r}_{14} = \left(\frac{1}{\sqrt{2}}, 0, 1\right) \quad \vec{r}_{15} = \left(1, 0, -\frac{1}{\sqrt{2}}\right) \quad \vec{r}_{16} = \left(-\frac{1}{\sqrt{2}}, 0, 1\right) \quad \vec{r}_{17} = \left(1, 0, \frac{1}{\sqrt{2}}\right)$$

$$\vec{r}_{18} = \left(\frac{1}{\sqrt{2}}, 1, 0\right) \quad \vec{r}_{19} = \left(1, -\frac{1}{\sqrt{2}}, 0\right) \quad \vec{r}_{20} = \left(-\frac{1}{\sqrt{2}}, 1, 0\right) \quad \vec{r}_{21} = \left(1, \frac{1}{\sqrt{2}}, 0\right)$$

$$\vec{r}_{22} = \left(1, -\frac{1}{\sqrt{2}}, \frac{1}{\sqrt{2}}\right) \quad \vec{r}_{23} = \left(1, \frac{1}{\sqrt{2}}, -\frac{1}{\sqrt{2}}\right) \quad \vec{r}_{24} = \left(1, \frac{1}{\sqrt{2}}, \frac{1}{\sqrt{2}}\right)$$

$$\vec{r}_{25} = \left(1, -\frac{1}{\sqrt{2}}, -\frac{1}{\sqrt{2}}\right) \quad \vec{r}_{26} = \left(-\frac{1}{\sqrt{2}}, 1, \frac{1}{\sqrt{2}}\right) \quad \vec{r}_{27} = \left(\frac{1}{\sqrt{2}}, 1, -\frac{1}{\sqrt{2}}\right)$$

$$\vec{r}_{28} = \left(\frac{1}{\sqrt{2}}, 1, \frac{1}{\sqrt{2}}\right) \quad \vec{r}_{29} = \left(-\frac{1}{\sqrt{2}}, 1, -\frac{1}{\sqrt{2}}\right) \quad \vec{r}_{30} = \left(-\frac{1}{\sqrt{2}}, \frac{1}{\sqrt{2}}, 1\right)$$

$$\vec{r}_{31} = \left(\frac{1}{\sqrt{2}}, -\frac{1}{\sqrt{2}}, 1\right) \quad \vec{r}_{32} = \left(\frac{1}{\sqrt{2}}, \frac{1}{\sqrt{2}}, 1\right) \quad \vec{r}_{33} = \left(-\frac{1}{\sqrt{2}}, -\frac{1}{\sqrt{2}}, 1\right)$$

There are 16 triads of mutually orthogonal vectors: 1-2-3, 1-4-5, 1-10-11, 1-12-13, 2-6-7, 2-14-15, 2-16-17, 3-8-9, 3-18-19, 3-20-21, 4-22-23, 5-24-25, 6-26-27, 7-28-29, 8-30-31, 9-32-33, and 24 dyads of orthogonal vectors: 10-27, 10-29, 11-30, 11-32, 12-26, 12-28, 13-31, 13-33, 14-23, 14-25, 15-31, 15-32, 16-22, 16-24, 17-30, 17-33, 18-22, 18-25, 19-27, 19-28, 20-23, 20-24, 21-26, 21-29 (Fig. 4.9).

The squared spin-1 observables measured along three mutually perpendicular directions in real 3-dimensional space always obey the experimentally verifiable 1, 0, 1 rule (Section 4.13). To prove that one cannot consistently assign values according to the 1, 0, 1 rule to all 33 rays, let us attempt coloring individual rays in white (W) for 0 and black (B) for 1.

First, by rotation symmetry of the rays 1-2-3, we can always assign 1W, 2B and 3B as the starting point of the analysis. From 1W would also follow 4B and 5B. Next, to satisfy the 1, 0, 1 rule for 4B and 5B, we have to assign two 0s for rays 22, 23, 24, 25, which can be done in one of four possible ways:

22W, 24W: implies that 3B-18B-19W and 3B-20B-21W. From 19W, 21W it follows that 6W-26B-27B and 7W-28B-29B. 6W and 7W are orthogonal and contradict the 1, 0, 1 rule, hence the assignment of 22W, 24W leads to a contradiction.

22W, 25W: implies that 2B-14B-15W and 2B-16B-17W. From 15W, 17W it follows that 8W-30B-31B and 9W-32B-33B. 8W and 9W are orthogonal and contradict the 1, 0, 1 rule, hence the assignment of 22W, 25W leads to a contradiction.

23W, 24W: implies that 2B-14B-15W and 2B-16B-17W. From 15W, 17W follows the 8W, 9W contradiction.

23W, 25W: implies that 3B-18B-19W and 3B-20B-21W. From 19W, 21W follows the 6W, 7W contradiction.

Because there are no other ways to satisfy 4B and 5B, it follows that the 33 rays cannot have predetermined values of 0 or 1 that obey the 1, 0, 1 rule. In other words, when the squared spin-1 observables are measured along certain rays, the spin-1 particle makes up an answer 0 or 1 on the fly, without having had that answer already existing as a physical property. Notably, the proof does not rely on a specific initial quantum state of the qutrit. The contextuality revealed by the Kochen–Specker theorem implies that the values of quantum physical observables are made up answers generated at the time of the measurement. $\qquad\square$

The Kochen–Specker theorem has a simpler proof for $n \geq 8$. Consider a system of three qubits whose quantum state resides in an 8-dimensional Hilbert space formed as a tensor product of the 2-dimensional Hilbert spaces of the three individual qubits. Five sets of mutually commuting observables could be measured, one set at a time, as shown in Figure 4.10, such that each observable in the sets has only two eigenvalues +1 and −1 [340]. Each set of commuting observables contains four observables expressible with the Pauli spin matrices $\hat{\sigma}_x$ and $\hat{\sigma}_y$ (Section 4.12). The products of all four observables produce the result $-\hat{I}$ for a single set, and the result $\hat{I}$ for the remaining four sets, as follows

$$(\hat{\sigma}_x \otimes \hat{\sigma}_x \otimes \hat{\sigma}_x)(\hat{\sigma}_y \otimes \hat{\sigma}_y \otimes \hat{\sigma}_x)(\hat{\sigma}_y \otimes \hat{\sigma}_x \otimes \hat{\sigma}_y)(\hat{\sigma}_x \otimes \hat{\sigma}_y \otimes \hat{\sigma}_y) = -\hat{I} \qquad (4.251)$$

$$(\hat{\sigma}_x \otimes \hat{\sigma}_x \otimes \hat{\sigma}_x)(\hat{\sigma}_x \otimes \hat{I} \otimes \hat{I})(\hat{I} \otimes \hat{\sigma}_x \otimes \hat{I})(\hat{I} \otimes \hat{I} \otimes \hat{\sigma}_x) = \hat{I} \qquad (4.252)$$

$$(\hat{\sigma}_y \otimes \hat{\sigma}_y \otimes \hat{\sigma}_x)(\hat{\sigma}_y \otimes \hat{I} \otimes \hat{I})(\hat{I} \otimes \hat{\sigma}_y \otimes \hat{I})(\hat{I} \otimes \hat{I} \otimes \hat{\sigma}_x) = \hat{I} \qquad (4.253)$$

$$(\hat{\sigma}_y \otimes \hat{\sigma}_x \otimes \hat{\sigma}_y)(\hat{\sigma}_y \otimes \hat{I} \otimes \hat{I})(\hat{I} \otimes \hat{\sigma}_x \otimes \hat{I})(\hat{I} \otimes \hat{I} \otimes \hat{\sigma}_y) = \hat{I} \qquad (4.254)$$

$$(\hat{\sigma}_x \otimes \hat{\sigma}_y \otimes \hat{\sigma}_y)(\hat{\sigma}_x \otimes \hat{I} \otimes \hat{I})(\hat{I} \otimes \hat{\sigma}_y \otimes \hat{I})(\hat{I} \otimes \hat{I} \otimes \hat{\sigma}_y) = \hat{I} \qquad (4.255)$$

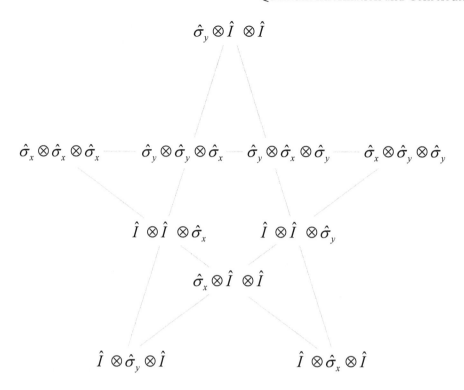

Figure 4.10   The proof of the Kochen–Specker theorem in eight dimensions requires five sets of mutually commuting observables. Each set contains four observables lying along the five legs of a five-pointed star.

In each set, the order of multiplication of the observables does not matter because they commute with each other (Definition 4.14). As a consequence of their commutation, the observables also have a complete set of common eigenvectors (Theorem 11), which allows simultaneous measurements to be performed.

Suppose now that it is possible to assign a predetermined value of +1 or −1 to each observable. If we denote the assigned value of an observable $\hat{\sigma}$ as $v(\hat{\sigma})$ we will obtain the system of equations

$$v\left(\hat{\sigma}_x \otimes \hat{\sigma}_x \otimes \hat{\sigma}_x\right)v\left(\hat{\sigma}_y \otimes \hat{\sigma}_y \otimes \hat{\sigma}_x\right)v\left(\hat{\sigma}_y \otimes \hat{\sigma}_x \otimes \hat{\sigma}_y\right)v\left(\hat{\sigma}_x \otimes \hat{\sigma}_y \otimes \hat{\sigma}_y\right) = -1 \qquad (4.256)$$

$$v\left(\hat{\sigma}_x \otimes \hat{\sigma}_x \otimes \hat{\sigma}_x\right)v\left(\hat{\sigma}_x \otimes \hat{I} \otimes \hat{I}\right)v\left(\hat{I} \otimes \hat{\sigma}_x \otimes \hat{I}\right)v\left(\hat{I} \otimes \hat{I} \otimes \hat{\sigma}_x\right) = 1 \qquad (4.257)$$

$$v\left(\hat{\sigma}_y \otimes \hat{\sigma}_y \otimes \hat{\sigma}_x\right)v\left(\hat{\sigma}_y \otimes \hat{I} \otimes \hat{I}\right)v\left(\hat{I} \otimes \hat{\sigma}_y \otimes \hat{I}\right)v\left(\hat{I} \otimes \hat{I} \otimes \hat{\sigma}_x\right) = 1 \qquad (4.258)$$

$$v\left(\hat{\sigma}_y \otimes \hat{\sigma}_x \otimes \hat{\sigma}_y\right)v\left(\hat{\sigma}_y \otimes \hat{I} \otimes \hat{I}\right)v\left(\hat{I} \otimes \hat{\sigma}_x \otimes \hat{I}\right)v\left(\hat{I} \otimes \hat{I} \otimes \hat{\sigma}_y\right) = 1 \qquad (4.259)$$

$$v\left(\hat{\sigma}_x \otimes \hat{\sigma}_y \otimes \hat{\sigma}_y\right)v\left(\hat{\sigma}_x \otimes \hat{I} \otimes \hat{I}\right)v\left(\hat{I} \otimes \hat{\sigma}_y \otimes \hat{I}\right)v\left(\hat{I} \otimes \hat{I} \otimes \hat{\sigma}_y\right) = 1 \qquad (4.260)$$

Multiplying all five equations gives +1 on the right-hand side because each operator value v appears exactly twice, and $v^2 = (\pm 1)^2 = 1$. On the left-hand side, however, the product is $-1 \times 1 \times 1 \times 1 \times 1 = -1$, due to the negative sign in the first equation. Thus, the system of equations results in a logical contradiction, $1 = -1$.

From the obtained contradiction it follows that the premise is false, hence it is not possible to assign a predetermined value to each observable in all sets of mutually commuting observables for Hilbert space with dimension $n \geq 8$. Again, the proof of the Kochen–Specker theorem does not rely on a specific initial quantum state.

In the case of three qubits, the Kochen–Specker theorem could be converted into a proof of Bell's theorem without inequalities [222, 340]. Consider the case in which an experimenter prepares three qubits in the Greenberger–Horne–Zeilinger state

$$|\Psi\rangle = \frac{1}{\sqrt{2}}(|\uparrow_z\rangle|\uparrow_z\rangle|\uparrow_z\rangle + |\downarrow_z\rangle|\downarrow_z\rangle|\downarrow_z\rangle) \tag{4.261}$$

where $\uparrow$ and $\downarrow$ indicate the eigenstates with $+1$ or -1 eigenvalues of a corresponding spin-$\frac{1}{2}$ observable (Section 4.12). Each qubit is then sent to a distant location so that measurements could be performed without sufficient time for communication between any of the qubits with classical signals whose speed is bound by the speed of light in vacuum c. Explicit matrix multiplication in the $|\uparrow_z\rangle, |\downarrow_z\rangle$ basis confirms that $|\Psi\rangle$ is an eigenstate of $\hat{\sigma}_y \otimes \hat{\sigma}_y \otimes \hat{\sigma}_x$, $\hat{\sigma}_y \otimes \hat{\sigma}_x \otimes \hat{\sigma}_y$ and $\hat{\sigma}_x \otimes \hat{\sigma}_y \otimes \hat{\sigma}_y$ with eigenvalue -1, and an eigenstate of $\hat{\sigma}_x \otimes \hat{\sigma}_x \otimes \hat{\sigma}_x$ with eigenvalue $+1$.

$$
\left(\hat{\sigma}_y \otimes \hat{\sigma}_y \otimes \hat{\sigma}_x\right)
\begin{pmatrix}
0 & 0 & 0 & 0 & 0 & 0 & 0 & -1 \\
0 & 0 & 0 & 0 & 0 & 0 & -1 & 0 \\
0 & 0 & 0 & 0 & 0 & 1 & 0 & 0 \\
0 & 0 & 0 & 0 & 1 & 0 & 0 & 0 \\
0 & 0 & 0 & 1 & 0 & 0 & 0 & 0 \\
0 & 0 & 1 & 0 & 0 & 0 & 0 & 0 \\
0 & -1 & 0 & 0 & 0 & 0 & 0 & 0 \\
-1 & 0 & 0 & 0 & 0 & 0 & 0 & 0
\end{pmatrix}
\begin{pmatrix} \frac{1}{\sqrt{2}} \\ 0 \\ 0 \\ 0 \\ 0 \\ 0 \\ 0 \\ \frac{1}{\sqrt{2}} \end{pmatrix}
= -1
\begin{pmatrix} \frac{1}{\sqrt{2}} \\ 0 \\ 0 \\ 0 \\ 0 \\ 0 \\ 0 \\ \frac{1}{\sqrt{2}} \end{pmatrix}
\tag{4.262}
$$

$$
\left(\hat{\sigma}_y \otimes \hat{\sigma}_x \otimes \hat{\sigma}_y\right)
\begin{pmatrix}
0 & 0 & 0 & 0 & 0 & 0 & 0 & -1 \\
0 & 0 & 0 & 0 & 0 & 0 & 1 & 0 \\
0 & 0 & 0 & 0 & 0 & -1 & 0 & 0 \\
0 & 0 & 0 & 0 & 1 & 0 & 0 & 0 \\
0 & 0 & 0 & 1 & 0 & 0 & 0 & 0 \\
0 & 0 & -1 & 0 & 0 & 0 & 0 & 0 \\
0 & 1 & 0 & 0 & 0 & 0 & 0 & 0 \\
-1 & 0 & 0 & 0 & 0 & 0 & 0 & 0
\end{pmatrix}
\begin{pmatrix} \frac{1}{\sqrt{2}} \\ 0 \\ 0 \\ 0 \\ 0 \\ 0 \\ 0 \\ \frac{1}{\sqrt{2}} \end{pmatrix}
= -1
\begin{pmatrix} \frac{1}{\sqrt{2}} \\ 0 \\ 0 \\ 0 \\ 0 \\ 0 \\ 0 \\ \frac{1}{\sqrt{2}} \end{pmatrix}
\tag{4.263}
$$

$$\left(\hat{\sigma}_x \otimes \hat{\sigma}_y \otimes \hat{\sigma}_y\right) \qquad |\Psi\rangle \qquad = -1 \quad |\Psi\rangle$$

$$\begin{pmatrix} 0 & 0 & 0 & 0 & 0 & 0 & 0 & -1 \\ 0 & 0 & 0 & 0 & 0 & 0 & 1 & 0 \\ 0 & 0 & 0 & 0 & 0 & 1 & 0 & 0 \\ 0 & 0 & 0 & 0 & -1 & 0 & 0 & 0 \\ 0 & 0 & 0 & -1 & 0 & 0 & 0 & 0 \\ 0 & 0 & 1 & 0 & 0 & 0 & 0 & 0 \\ 0 & 1 & 0 & 0 & 0 & 0 & 0 & 0 \\ -1 & 0 & 0 & 0 & 0 & 0 & 0 & 0 \end{pmatrix} \begin{pmatrix} \frac{1}{\sqrt{2}} \\ 0 \\ 0 \\ 0 \\ 0 \\ 0 \\ 0 \\ \frac{1}{\sqrt{2}} \end{pmatrix} = -1 \begin{pmatrix} \frac{1}{\sqrt{2}} \\ 0 \\ 0 \\ 0 \\ 0 \\ 0 \\ 0 \\ \frac{1}{\sqrt{2}} \end{pmatrix} \qquad (4.264)$$

$$\left(\hat{\sigma}_x \otimes \hat{\sigma}_x \otimes \hat{\sigma}_x\right) \qquad |\Psi\rangle \qquad = +1 \quad |\Psi\rangle$$

$$\begin{pmatrix} 0 & 0 & 0 & 0 & 0 & 0 & 0 & 1 \\ 0 & 0 & 0 & 0 & 0 & 0 & 1 & 0 \\ 0 & 0 & 0 & 0 & 0 & 1 & 0 & 0 \\ 0 & 0 & 0 & 0 & 1 & 0 & 0 & 0 \\ 0 & 0 & 0 & 1 & 0 & 0 & 0 & 0 \\ 0 & 0 & 1 & 0 & 0 & 0 & 0 & 0 \\ 0 & 1 & 0 & 0 & 0 & 0 & 0 & 0 \\ 1 & 0 & 0 & 0 & 0 & 0 & 0 & 0 \end{pmatrix} \begin{pmatrix} \frac{1}{\sqrt{2}} \\ 0 \\ 0 \\ 0 \\ 0 \\ 0 \\ 0 \\ \frac{1}{\sqrt{2}} \end{pmatrix} = +1 \begin{pmatrix} \frac{1}{\sqrt{2}} \\ 0 \\ 0 \\ 0 \\ 0 \\ 0 \\ 0 \\ \frac{1}{\sqrt{2}} \end{pmatrix} \qquad (4.265)$$

Thus, the product in Eq. (4.251) is definitely −1. Now, assuming that the three qubit system has predetermined classical correlations that are locally set at the moment of creation of the state $|\Psi\rangle$, one can deduce that it is impossible for the remaining four sets of measurements in Eqs. (4.252), (4.253), (4.254) and (4.255) to give all products equal to +1. However, quantum experiments show that all of the remaining four sets of measurements do have products equal to +1 [66]. Therefore, the quantum correlations have to be nonlocal.

For completeness, we note that quantum mechanics allows simultaneous measurement of the spin product observables $\hat{\sigma}_y \otimes \hat{\sigma}_y \otimes \hat{\sigma}_x$, $\hat{\sigma}_y \otimes \hat{\sigma}_x \otimes \hat{\sigma}_y$, $\hat{\sigma}_x \otimes \hat{\sigma}_y \otimes \hat{\sigma}_y$ and $\hat{\sigma}_x \otimes \hat{\sigma}_x \otimes \hat{\sigma}_x$ that provides definite values for these observables, but importantly does not provide any information for the individual spin values for each of the qubits. On the other hand, it is possible to measure the individual spin values for each of the qubits together with only one of the four spin product observables. This is a manifestation of quantum complementarity according to which it is not possible for noncommuting observables to have definite eigenvalues simultaneously.

The quantum mechanical prediction for the set of measurements containing the observables $\hat{\sigma}_y \otimes \hat{\sigma}_y \otimes \hat{\sigma}_x$, $\hat{\sigma}_y \otimes \hat{I} \otimes \hat{I}$, $\hat{I} \otimes \hat{\sigma}_y \otimes \hat{I}$ and $\hat{I} \otimes \hat{I} \otimes \hat{\sigma}_x$ gives four different experimental outcomes, all of which could occur with probability of $\frac{1}{4}$. We have already shown that the measurement of $\hat{\sigma}_y \otimes \hat{\sigma}_y \otimes \hat{\sigma}_x$ will always give outcome −1 according to Eq. (4.262). What we need is to calculate the measurement outcomes of the individual spin values using change of basis (Section 4.12) as follows

$$|\Psi\rangle = \frac{1}{4}\Big[\big(|\uparrow_y\rangle+|\downarrow_y\rangle\big)\big(|\uparrow_y\rangle+|\downarrow_y\rangle\big)\big(|\uparrow_x\rangle+|\downarrow_x\rangle\big)$$

$$-\big(|\uparrow_y\rangle-|\downarrow_y\rangle\big)\big(|\uparrow_y\rangle-|\downarrow_y\rangle\big)\big(|\uparrow_x\rangle-|\downarrow_x\rangle\big)\Big]$$

$$= \frac{1}{4}\Big[|\uparrow_y\rangle|\uparrow_y\rangle|\uparrow_x\rangle+|\uparrow_y\rangle|\downarrow_y\rangle|\uparrow_x\rangle+|\downarrow_y\rangle|\uparrow_y\rangle|\uparrow_x\rangle+|\downarrow_y\rangle|\downarrow_y\rangle|\uparrow_x\rangle$$

$$+|\uparrow_y\rangle|\uparrow_y\rangle|\downarrow_x\rangle+|\uparrow_y\rangle|\downarrow_y\rangle|\downarrow_x\rangle+|\downarrow_y\rangle|\uparrow_y\rangle|\downarrow_x\rangle+|\downarrow_y\rangle|\downarrow_y\rangle|\downarrow_x\rangle$$

$$-|\uparrow_y\rangle|\uparrow_y\rangle|\uparrow_x\rangle+|\uparrow_y\rangle|\downarrow_y\rangle|\uparrow_x\rangle+|\downarrow_y\rangle|\uparrow_y\rangle|\uparrow_x\rangle-|\downarrow_y\rangle|\downarrow_y\rangle|\uparrow_x\rangle$$

$$+|\uparrow_y\rangle|\uparrow_y\rangle|\downarrow_x\rangle-|\uparrow_y\rangle|\downarrow_y\rangle|\downarrow_x\rangle-|\downarrow_y\rangle|\uparrow_y\rangle|\downarrow_x\rangle+|\downarrow_y\rangle|\downarrow_y\rangle|\downarrow_x\rangle\Big]$$

$$= \frac{1}{2}\Big[|\uparrow_y\rangle|\downarrow_y\rangle|\uparrow_x\rangle+|\downarrow_y\rangle|\uparrow_y\rangle|\uparrow_x\rangle+|\uparrow_y\rangle|\uparrow_y\rangle|\downarrow_x\rangle+|\downarrow_y\rangle|\downarrow_y\rangle|\downarrow_x\rangle\Big]$$

From the calculation it can be seen that the probability amplitudes for individual spin outcomes whose product is +1 annihilate each other through destructive interference. The products of the individual spin values for all of the possible outcomes that remain through constructive interference is −1, which, when multiplied by the −1 outcome of $\hat{\sigma}_y\otimes\hat{\sigma}_y\otimes\hat{\sigma}_x$, gives a composite product of +1 for all four observables. A similar argument holds for the other two sets of commuting observables involving $\hat{\sigma}_y\otimes\hat{\sigma}_x\otimes\hat{\sigma}_y$ or $\hat{\sigma}_x\otimes\hat{\sigma}_y\otimes\hat{\sigma}_y$, since

$$|\Psi\rangle = \frac{1}{2}\Big[|\uparrow_y\rangle|\downarrow_x\rangle|\uparrow_y\rangle+|\downarrow_y\rangle|\uparrow_x\rangle|\uparrow_y\rangle+|\uparrow_y\rangle|\uparrow_x\rangle|\downarrow_y\rangle+|\downarrow_y\rangle|\downarrow_x\rangle|\downarrow_y\rangle\Big]$$

$$= \frac{1}{2}\Big[|\uparrow_x\rangle|\downarrow_y\rangle|\uparrow_y\rangle+|\downarrow_x\rangle|\uparrow_y\rangle|\uparrow_y\rangle+|\uparrow_x\rangle|\uparrow_y\rangle|\downarrow_y\rangle+|\downarrow_x\rangle|\downarrow_y\rangle|\downarrow_y\rangle\Big]$$

For the set $\hat{\sigma}_x\otimes\hat{\sigma}_x\otimes\hat{\sigma}_x$, $\hat{\sigma}_x\otimes\hat{I}\otimes\hat{I}$, $\hat{I}\otimes\hat{\sigma}_x\otimes\hat{I}$ and $\hat{I}\otimes\hat{I}\otimes\hat{\sigma}_x$ we have

$$|\Psi\rangle = \frac{1}{4}\Big[\big(|\uparrow_x\rangle+|\downarrow_x\rangle\big)\big(|\uparrow_x\rangle+|\downarrow_x\rangle\big)\big(|\uparrow_x\rangle+|\downarrow_x\rangle\big)$$

$$+\big(|\uparrow_x\rangle-|\downarrow_x\rangle\big)\big(|\uparrow_x\rangle-|\downarrow_x\rangle\big)\big(|\uparrow_x\rangle-|\downarrow_x\rangle\big)\Big]$$

$$= \frac{1}{4}\Big[|\uparrow_x\rangle|\uparrow_x\rangle|\uparrow_x\rangle+|\uparrow_x\rangle|\downarrow_x\rangle|\uparrow_x\rangle+|\downarrow_x\rangle|\uparrow_x\rangle|\uparrow_x\rangle+|\downarrow_x\rangle|\downarrow_x\rangle|\uparrow_x\rangle$$

$$+|\uparrow_x\rangle|\uparrow_x\rangle|\downarrow_x\rangle+|\uparrow_x\rangle|\downarrow_x\rangle|\downarrow_x\rangle+|\downarrow_x\rangle|\uparrow_x\rangle|\downarrow_x\rangle+|\downarrow_x\rangle|\downarrow_x\rangle|\downarrow_x\rangle$$

$$+|\uparrow_x\rangle|\uparrow_x\rangle|\uparrow_x\rangle-|\uparrow_x\rangle|\downarrow_x\rangle|\uparrow_x\rangle-|\downarrow_x\rangle|\uparrow_x\rangle|\uparrow_x\rangle+|\downarrow_x\rangle|\downarrow_x\rangle|\uparrow_x\rangle$$

$$-|\uparrow_x\rangle|\uparrow_x\rangle|\downarrow_x\rangle+|\uparrow_x\rangle|\downarrow_x\rangle|\downarrow_x\rangle+|\downarrow_x\rangle|\uparrow_x\rangle|\downarrow_x\rangle-|\downarrow_x\rangle|\downarrow_x\rangle|\downarrow_x\rangle\Big]$$

$$= \frac{1}{2}\Big[|\uparrow_x\rangle|\uparrow_x\rangle|\uparrow_x\rangle+|\downarrow_x\rangle|\downarrow_x\rangle|\uparrow_x\rangle+|\uparrow_x\rangle|\downarrow_x\rangle|\downarrow_x\rangle+|\downarrow_x\rangle|\uparrow_x\rangle|\downarrow_x\rangle\Big]$$

The products of the individual spin values for all of the possible outcomes is +1, which, when multiplied by the +1 outcome of $\hat{\sigma}_x\otimes\hat{\sigma}_x\otimes\hat{\sigma}_x$, gives again a composite product of +1 for all four observables. Thus, the quantum mechanical predictions are incompatible with the predictions of any classical model that is noncontextual or local.

Theorem 19. *(No-deleting theorem) Quantum information contained in an unknown quantum state $|\Psi\rangle$ cannot be deleted.*

Proof. The *no-deleting theorem* states that given two copies of an unknown quantum state, it is impossible to delete one of the copies against the other [367, 368, 369]. In the case of two qubits, there is no linear operator $\hat{U}$ such that

$$\hat{U}|0_A\rangle|0_B\rangle = |0_A\rangle|d_B\rangle \tag{4.266}$$

$$\hat{U}|1_A\rangle|1_B\rangle = |1_A\rangle|d_B\rangle \tag{4.267}$$

$$\hat{U}\left(\alpha|0_A\rangle + \beta|1_A\rangle\right)\left(\alpha|0_B\rangle + \beta|1_B\rangle\right) = \left(\alpha|0_A\rangle + \beta|1_A\rangle\right)|d_B\rangle \tag{4.268}$$

where α and β are arbitrary complex coefficients such that the input quantum states are normalized, namely, $\alpha^*\alpha + \beta^*\beta = 1$.

From the linearity of the operator $\hat{U}$, together with Eqs. (4.266) and (4.267), we get

$$\hat{U}\left(\alpha|0_A\rangle + \beta|1_A\rangle\right)\left(\alpha|0_B\rangle + \beta|1_B\rangle\right)$$
$$= \hat{U}\left[\alpha^2|0_A\rangle|0_B\rangle + \beta^2|1_A\rangle|1_B\rangle + \alpha\beta\left(|0_A\rangle|1_B\rangle + |1_A\rangle|0_B\rangle\right)\right]$$
$$= \alpha^2|0_A\rangle|d_B\rangle + \beta^2|1_A\rangle|d_B\rangle + \alpha\beta\hat{U}\left(|0_A\rangle|1_B\rangle + |1_A\rangle|0_B\rangle\right) \tag{4.269}$$

Equations (4.268) and (4.269) are consistent if $\alpha = 0, 1$. But for $\alpha = 0, 1$ the state $\alpha|0\rangle + \beta|1\rangle$ is reduced to one of the basis states $|0\rangle$ or $|1\rangle$, hence it does not represent a general input state for the quantum deleting operation. On the other hand, assuming $\alpha, \beta \neq 0, 1$ and equating terms on the right-hand sides of Eqs. (4.268) and (4.269) gives us

$$\alpha|0_A\rangle|d_B\rangle = \alpha^2|0_A\rangle|d_B\rangle + \alpha\beta u_1|0_A\rangle|d_B\rangle); \qquad u_1 = \frac{1-\alpha}{\beta} \neq 0 \tag{4.270}$$

$$\beta|1_A\rangle|d_B\rangle = \beta^2|1_A\rangle|d_B\rangle + \alpha\beta u_2|1_A\rangle|d_B\rangle); \qquad u_2 = \frac{1-\beta}{\alpha} \neq 0 \tag{4.271}$$

Thus, the action of $\hat{U}$ upon $(|0_A\rangle|1_B\rangle + |1_A\rangle|0_B\rangle)$ should be

$$\hat{U}\left(|0_A\rangle|1_B\rangle + |1_A\rangle|0_B\rangle\right) = \left[\frac{1-\alpha}{\beta}|0_A\rangle + \frac{1-\beta}{\alpha}|1_A\rangle\right]|d_B\rangle \tag{4.272}$$

But such an action of the operator $\hat{U}$ contradicts linearity because it generates different outputs for the sum $(|0_A\rangle|1_B\rangle + |1_A\rangle|0_B\rangle)$ depending on the context of which input states are intended for deletion, as can be directly verified by plugging in $\alpha_1 = \frac{3}{5}, \beta_1 = \frac{4}{5}$ and comparing the result with $\alpha_2 = \frac{4}{5}, \beta_2 = \frac{3}{5}$. Therefore, deletion of unknown quantum states is impossible. Of course, one can achieve swapping of the second qubit with another qubit from the environment. Swapping, however, is not a genuine deleting procedure and represents discarding of the second qubit into the environment without actually deleting it [367, 368, 369].   □

Theorem 20. *(No-teleportation theorem) Quantum information contained in an unknown quantum state* $|\Psi\rangle$ *cannot be converted into classical information.*

Proof. The no-cloning theorem forbids the conversion of an unknown quantum state into a sequence of classical bits, because if the quantum information contained in a qubit were convertible into classical information encoded in a sequence of classical bits, then we would have been able to read those classical bits and clone the original qubit as many times as we like. ☐

The impossibility of unambiguously determining an unknown quantum state by a single measurement, together with the impossibility of converting quantum information into a sequence of classical bits, has been collectively referred to as the *no-teleportation theorem* [366]. The name of the latter theorem is derived from the impossibility of *teleporting* any qubit by merely moving classical bits around. Here, we have formulated two separate Theorems, 15 and 20, because *reading* is not the same as *conversion*, even though these two operations may be related. Also, Theorem 15 is weaker because only a single measurement is allowed, whereas Theorem 20 is stronger as there is no limitation on the number of allowed measurements.

As a caveat, we should point out that the *no-teleportation theorem* should not be confused with *quantum teleportation*, which is a physically achievable operation provided that one has access to both a classical channel for transfer of classical information and a quantum channel for sharing of quantum entangled qubit pairs [42, 57, 495]. In the process of quantum teleportation, the sender called Alice performs a joint measurement on one of a pair of quantum entangled qubits and the unknown qubit $|\psi\rangle$ to be teleported, then Alice reports the outcome of her measurement to the receiver called Bob using the classical channel of information, and finally, Bob performs a unitary operation based on the classical information obtained from Alice so that the second qubit of the quantum entangled pair is transformed into the state $|\psi\rangle$. At the end of the quantum teleportation, there is only one copy of $|\psi\rangle$ left, because the original qubit in state $|\psi\rangle$ has been destroyed by the measurement performed by Alice [358, pp. 26–28].

Theorem 21. *(Holevo's theorem) Given n qubits that carry quantum information in the 2^n quantum amplitudes of the state vector $|\Psi(n)\rangle$, the amount of classical information that can be accessed by an external observer can be only up to n classical bits [260, 261].*

Proof. Only orthogonal quantum states could be unambiguously distinguished through quantum measurement, as we have shown in the proof of Theorem 13. For n qubits, the Hilbert space $\mathcal{H}$ is 2^n dimensional. Therefore, the maximal number of possible orthogonal states in $\mathcal{H}$ and the maximal number of non-degenerate eigenvalue outcomes of any quantum observable $\hat{A}$ on $\mathcal{H}$ is also 2^n. Since each quantum measurement has a single outcome, registering one out of 2^n possible outcomes delivers n bits of classical information. Because the original quantum state cannot be cloned and is irreversibly corrupted by the measurement process, n bits is the maximal amount of classical information that can be accessed by an external observer from a quantum system prepared in a pure quantum state $|\Psi(n)\rangle$. Thinking in terms of the capacity of a quantum channel to transmit classical information, it could be said that encoding of a message consisting of n bits of classical information in one of 2^n orthogonal pure quantum states $|\psi_i\rangle$ would allow unambiguous determination of the state $|\psi_i\rangle$ at the end of the quantum channel and recovery of the n bits of the message with absolute certainty provided that one measures a quantum observable $\hat{A} = \sum_{i=1}^{n} \lambda_i |\psi_i\rangle\langle\psi_i|$ whose eigenvalues are non-degenerate, namely, $\lambda_i \neq \lambda_j$ for $i \neq j$. If the quantum channel, however, transmits a classical message encoded in one of 2^n mixed quantum states described by density matrices $\hat{\rho}_i(n)$, there will be no measurement outcome of any quantum observable $\hat{A}$ on $\mathcal{H}$ that will occur with absolute certainty, hence one cannot unambiguously determine what the original classical message was, or in other words, fewer than n bits are received at the end of the quantum channel.   □

The latter theorem is named after Alexander Holevo, who proved in 1973 a stronger inequality setting an upper bound on the capacity of mixed state quantum channels to transmit classical information [260, 261], from which follows not only that it is not possible to communicate more than n classical bits of information by the transmission of n qubits alone, but also that from the transmission of n qubits in a maximally mixed quantum state cannot be extracted any useful information.

The quantum no-cloning and no-deleting theorems (together with all results implied by these theorems) are intertwined with the fact that quantum states are not observable (Theorem 15). In quantum physics, we can always transform a known quantum state into another desired output quantum state. Therefore, if we had means to determine an *unknown* quantum state, we would have been able to clone or delete any qubit. Indeed, cloning and deleting of bits in classical physics is always a combination of observing the unknown classical state followed by transformation into a desired output state [437]. Thus, the unobservability of the quantum states appears to be the most prominent feature of quantum physics, and in Chapter 6 we will show how this quantum property is pertinent to the main problems of consciousness.

Part III

A quantum information theory of consciousness

CHAPTER 5

Consciousness in classical physics

5.1 Physical boundary of consciousness in classical physics

The laws of classical physics (Chapter 3) do not provide a theoretical description of the physical boundaries of individual conscious minds because consciousness does not enter explicitly in any of the equations or the axioms describing classical mechanics (Section 3.12) or classical electrodynamics (Section 3.14). Classically, there are only four fundamental observable physical quantities (mass, charge, length and time) that determine all future dynamics given any initial conditions for the state of physical objects in the universe. Since classical physics does not identify consciousness with any of these fundamental physical quantities, consciousness does not exist in the classical world unless new physical axioms are introduced to specify what consciousness is and how it relates to the physical world.

Eliminative materialism, also known as *eliminativism*, embraces the absence of consciousness in the world of classical physics as if it were a real fact rather than a defect of the theory, and argues that we are hallucinating having experiences while in fact we have none [122, 123, 124, 126]. Because hallucinations are a form of conscious experiences, such a viewpoint is self-defeating and logically inconsistent. Alternatively, one could try to supplement the axioms of classical physics with new rules that specify how consciousness is generated by or emerges from a collection of classical particles. Every extension of classical physics with new physical laws that introduce consciousness into the physical world will be referred to as a *classical theory of consciousness*. Next, we will discuss the problems faced by the classical theories of consciousness and will identify the origin of these problems.

Example 5.1. *(Mind duplication) Suppose that the conscious mind is generated by a collection of classical physical particles that are all located within a 3-dimensional spatial region with certain volume (Fig. 5.1). We can draw a 2-dimensional boundary enclosing that region and claim that this is the boundary of the conscious mind in space at a certain point in time. From our everyday experience, we know that we are able to move around. Motion implies that the mind boundary changes its position relative to the environment and remains around the collection of particles that generate the mind. Therefore, the mind boundary cannot be fixed in space and time; it should be able to move around together with the collection of particles that generate the conscious experiences. Next, consider a classical process that takes individual particles from within the mind boundary, throws them in the environment, and replaces them with identical particles. In such a case, it seems reasonable to assume that the mind boundary stays the same due to the fact that the new particles replace those that are thrown away. Otherwise, the mind boundary will not be enclosing a single connected 3-dimensional volume, but will be represented by a scrambled disconnected collection of smaller volumes*

143

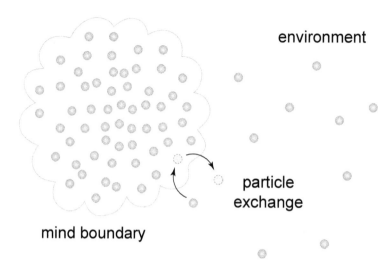

Figure 5.1   The mind boundary problem in classical physics. If consciousness is generated by a certain collection of classical particles, what exactly is the rule that specifies the boundary between the mind and its environment? Particles that build up a mind could be exchanged one by one with particles from the environment and the discarded particles could be used to duplicate exactly the original mind at a different location.

surrounding the old particles that are thrown into the environment. Finally, instead of just discarding the old particles into the environment, consider a classical process that arranges these particles in exactly the same state as they were in the collection that produces the original mind. The output will be two identical collections of particles that differ only by their location in space (Theorem 7). Paradoxically, the original mind will also be duplicated so that it is now located at two different places, each with its own mind boundary. Thus, if minds can be cloned they can no longer have a single closed boundary but should be able to occupy a discontinuous collection of spatial volumes. Because the origin of this problem lies in the ability of classical information to be cloned (Theorem 7), it cannot be avoided in any theory of consciousness that obeys the laws of classical physics.

The possibility of extending the axioms of classical physics with additional rules that specify what the mind is does not imply that such an extension will be satisfactory. As a matter of fact, every attempt to explicitly provide an additional physical rule that defines consciousness with the use of biochemical, electromagnetic or other classical processes runs into insurmountable problems when it comes to predicting where the mind boundary is and where the rest of the world begins. The origin of these problems stems from the fact that there is no difference between the constituent atoms that build the living and the non-living matter. In other words, all classical physical rules inevitably run into the problem that they are unable to prevent the diffusion and penetration of the mind into the surrounding environment.

From the biological sciences, we know that humans are born before their brains and minds are fully developed. Each newborn baby needs a daily consumption of food and drinks that provide energy and building blocks (such as atoms combined in molecules) for the brain. Some of the obtained atoms are then incorporated into the growing brain, and participate in the development and generation of consciousness. Conversely, some of the atoms that were incorporated into the brain and were involved in the generation of consciousness, get excreted as waste products from the body and no longer participate in the generation of consciousness. As an example, consider the water molecule H_2O composed of two hydrogen atoms and one oxygen atom. Over 80% of the brain is composed of water [290]. Water molecules constantly move in and out of neurons, get lost through breathing, perspiration or excretion, and are constantly resupplied by the water we drink. While we rarely contemplate on the need to constantly drink and excrete water in order to stay alive and conscious, this basic physiologic activity poses serious questions: Why do the water molecules not produce conscious experiences in the rivers and the oceans, but only when inside the neurons of the brain? Why do your conscious experiences not diffuse out in the water when you are swimming? Where exactly is the boundary between the part of the brain that directly generates your consciousness and the rest of the brain? In Section 1.1, we have provided experimental and clinicopathological evidence that our minds are generated by the brain cortex but we have not provided any theoretical rule that outlines the mind boundary. Next, we will show that classically there is no satisfactory physical mechanism that sets the boundary of the mind.

Example 5.2. *(Mind expansion into the surrounding environment) In classical physics, the possibility of mind expansion into the surrounding environment is far from unreasonable given the wondrous natural example provided by* weakly electric fish *(Fig. 5.2). Brain neurons input, process and communicate information via electric signals generated by opening or closing of voltage-gated ion channels (Fig. 1.8). The electric signals propagate in the brain* electrolyte *solution composed of water and dissolved salts. Salts provide the essential charged ions, such as Na^+, K^+, Ca^{2+} or Cl^-, whose motion conducts electricity. Since water in rivers and oceans is a natural electrolyte, weakly electric fish have evolved the ability to generate electric fields in the surrounding water with an electric organ and receive electric signals with specialized electroreceptors. The creation and detection of electric fields allows them to perceive not only the surrounding world [505], but to communicate with other weakly electric fish in a social context [449]. Therefore, unlike humans, whose neurons are electrically connected only within the neural network of the individual brain, in weakly electric fish their neurons are connected by electric signals that propagate through the surrounding environment. In such a case, is there a single social mind composed of several bodies of weakly electric fish? The idea that there is a single social mind inhabiting several bodies of weakly electric fish is paradoxical. Yet, if the electric signals in the brains of these fish are not fundamentally different from the electric signals in the surrounding water, there is nothing to set the boundaries of individual conscious minds and prevent their extension into the surrounding water.*

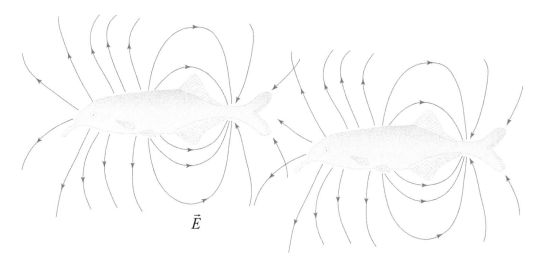

Figure 5.2 The elephantnose fish is a species of African freshwater weakly electric fish that generates an electric field $\vec{E}$ with its electric organ, and then processes the input from its electroreceptors to locate nearby objects in the surrounding water, including other weakly electric fish.

Example 5.3. *(Minds within minds) Individual human beings possess single conscious minds, but is it not possible that the human population of a whole city like New York also has its own New York conscious mind? And what about the possession of a conscious mind by a whole ocean [313] or a whole planet? The possibility of minds existing within other minds is paradoxical because if minds existed within other minds, it would not be clear which mind did what and whose free will were to blame for a given action. Hence, for a self-consistent theory of consciousness that includes free will and causally potent consciousness, it would be necessary to explicitly rule out scenarios such as overlapping of mind boundaries or existence of minds within other minds. Unfortunately, in classical physics every collection of physical particles has its own classical physical state from which follows that general identity between mental states and physical states cannot be valid. Indeed, if every mental state were a physical state and every physical state were a mental state, with the use of the axioms of classical physics one could pick up randomly constituent physical particles from different minds and conclude that they form another mind due to the fact that they have their own classical physical state. Therefore, to rule out the existence of minds within minds, one needs to deny the general identity between mental states and physical states. But if only some physical states were identified with mental states, then an inherently dualistic physical world is produced in which some physical states would be mental and some physical states would not be mental.*

5.2   Binding of consciousness in classical physics

Cognition performed by conscious agents relies on the input, storage, transmission and processing of classical information. Sensory perception, understanding, learning, reasoning and problem solving are branded as *cognitive processes* because they are associated with conscious experiences that can be attributed to a conscious mind. For example, when a student memorizes a poem, we refer to the process as *learning* since the student is capable of experiencing the meaning of the poem. When a computer stores a text file with the poem, however, we refer to the process as *storage of information* since the computer is incapable of consciously experiencing the contents of the stored file. Linguistically, it is awkward to state that the computer "learns" as we fill its hard drive with files, but not all verbs denoting cognitive processes sound awkward when applied to inanimate systems. As an example, making a chess move by either a human chess player or a computer chess program is referred to as *problem solving* regardless of who makes the move. Further terminological confusion could arise when classical information theory is used to define cognitive processes in an operational way that drops the requirement for associated conscious experiences. It is then possible to attribute "distributed cognition" to social systems such as the people in an emergency department or to a network of computers connected through the internet [215, 216], but such a metaphorical usage of the term "cognition" would require subsequent division into two varieties: "cognition associated with conscious experiences" performed by human agents and "cognition not associated with conscious experiences" performed by artificial machines or societies of biological organisms. To avoid terminological clumsiness, here we will use the term *cognition* only in the narrow sense that refers to processes that are associated with conscious experiences and are attributable to a conscious mind [58]. Consequently, cognitive processes such as sensory perception, learning and problem solving could be easily understood as corresponding to input, storage and processing of classical information by the brains of conscious agents. What is really difficult to explain in classical terms is what generates the conscious experiences in the first place and how these conscious experiences are bound together into a single mind.

We communicate daily with the members of our family or with our friends. From that personal experience, we introspectively know that two communicating brains do not produce a single united mind, otherwise we would have had direct access to the conscious experiences of others. Thus, the binding problem on one side is to explain what binds our conscious experiences into a single united whole, whereas on the other side is to explain why the experiences that constitute other conscious minds do not get bound with the experiences that constitute our own conscious mind (Section 1.2).

One possible solution to the binding problem would be to find a physical process that operates within each brain, but breaks down between two different brains. An alternative solution would be to find physical rules that both define the boundaries of individual minds and forbid the occurrence of minds within minds, and then require that all conscious experiences produced within the boundary of a single mind necessarily have to be bound together. The prospects of finding a classical answer, however, are dim because if consciousness were a form of classical information, then the instantaneous binding of conscious experiences would violate the physical limit for transmission of classical information that is the speed of light in vacuum c (Section 3.18).

At the microscopic level, the human brain contains 86 ± 8 billion neurons interconnected into a complex neural network [21]. As discussed in Section 1.7, each of these neurons is enclosed by an electrically excitable plasma membrane that is able to respond to various stimuli with the generation of electric spikes (Fig. 1.8). Within the network, each neuron collects and processes electric inputs from its neighbors mainly through its dendrites and cell body, and then sends electric outputs through its axon that targets the dendrites and cell bodies of other neurons. Because the neural network as a whole is able to perform complex computational tasks that cannot be accomplished by any of the individual neurons alone, several attempts to explain the binding of conscious experiences have been previously made using the theory of neuronal networks [336]. Five popular proposals are binding by convergence, binding by assembly, binding by synchrony, binding through integrated information or binding through electromagnetic fields.

5.2.1 Neural convergence

Convergence of neuronal inputs as a putative binding mechanism could be explained as follows. Suppose that you are looking at a green ball. Certain neurons in your primary visual cortex will react to the shape, whereas other neurons will react to the color of the visual object. If the firing of a neuron N_1 that detects the shape of a "ball" and a neuron N_2 that detects the color "green" converge onto a neuron N_3 that fires as a result of the convergent input, it could be speculated that the firing of neuron N_3 corresponds to the concept of a "green ball" (Fig. 5.3).

Unfortunately, if followed to the extreme, the idea of convergence will predict that at any instant of time our conscious mind should be produced by the firing of a single neuron that binds all sensory inputs that make up one's mental picture. This is paradoxical because we would not be able to explain how we could see things that we have never seen before in our lives. To see a new thing, we would need to have a virgin neuron that has never fired before. Moreover, because each new day we experience things that we have never experienced before, our brains need to have an immense pool of virgin neurons that have never fired before. In the brain, however, there are no such virgin neurons. Instead, experiments have shown that electric firing is essential both for the maturation of newborn neurons and for keeping these neurons in a healthy state [212, 385].

Another objection to the convergence idea comes from the possibility of having multiple inputs to the dendrites of a single neuron that could summate to produce an electric spike even if only a couple of the inputs were delivered. Suppose that we want to bind three inputs corresponding to "sour" + "green" + "apple," but let any two of the inputs be sufficient to pass the threshold of −55 mV for firing an electric spike. In such a case, the neuron would also fire under two inputs

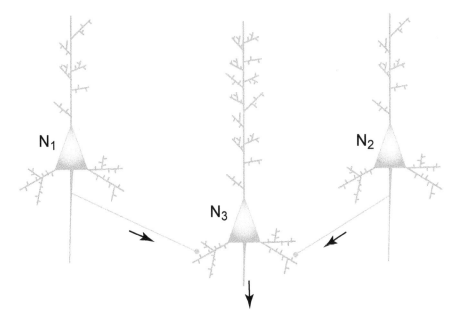

Figure 5.3   Illustration of the binding by convergence hypothesis. If neurons N_1 and N_2 fire electric spikes that converge onto neuron N_3, thereby causing neuron N_3 to fire as well, it could be speculated that neuron N_3 has bound together the information provided by both neurons N_1 and N_2.

such as "sour" + "apple" or "green" + "apple." Since the neuronal spike will be exactly the same in cases such as "sour" + "green" + "apple" or "sour" + "apple" or "green" + "apple" there would be nothing to distinguish which of the possible meanings were bound together by the firing neuron. In reality, our consciousness binds not just three but thousands of inputs. Therefore, the scenario with summation of multiple synaptic inputs that substantially exceed the spike threshold of -55 mV should occur frequently for those neurons that fire in the brain cortex.

5.2.2   Neural assembly

Binding of conscious experiences by assembly seems to avoid the necessity of a pool of virgin neurons that have never fired before. Instead, it hypothesizes that the information is encoded in the firing of an assembly of neurons. This is analogous to the storage of information in a written text: no single letter contains all of the information, but the letters in the text collectively convey that information. Unfortunately, this does not solve the binding of conscious experiences, because it is hard to find a physical law that specifies what a "neuronal assembly" means in the context of binding of consciousness. For example, one could select randomly neurons from different individual brains and ask what conscious experience that particular ensemble of neurons gives rise to. Somehow, the assembly of neurons needs to be considered only within a single brain, and the ensemble of neurons should not be picked up randomly.

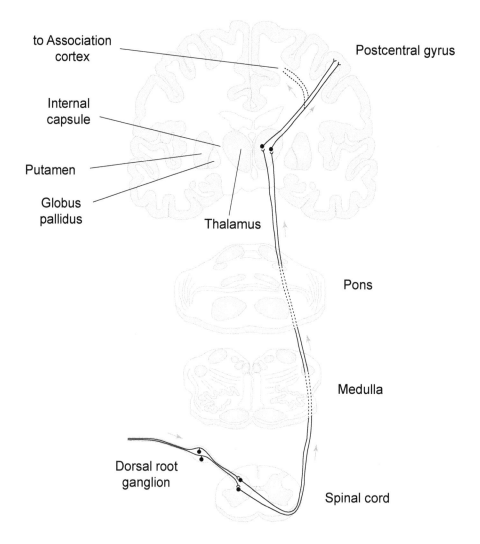

to Association cortex

Postcentral gyrus

Internal capsule

Putamen

Globus pallidus

Thalamus

Pons

Medulla

Dorsal root ganglion

Spinal cord

Figure 5.4   A sensory neural pathway to the brain cortex. The lateral spinothalamic tract transmits pain electric signals from the periphery toward the brain cortex where the pain is consciously experienced. The cell bodies of the first order neurons of the tract reside in the dorsal root ganglia. The first order neurons project peripheral processes to the tissues in the form of free nerve endings that are sensitive to molecules indicative of cell damage. The central processes of these neurons enter the spinal cord in an area at the back of the posterior horn where they synapse onto second order neurons. The axons of second order neurons decussate across the anterior white commissure and ascend in the contralateral lateral spinothalamic tract through the medulla oblongata, pons and midbrain, until synapsing in the ventral posterolateral nucleus of the thalamus. The third order neurons in the thalamus will then project through the internal capsule to the main somatosensory cortex (postcentral gyrus) and other associative cortical regions. The spinal cord, medulla and pons are represented with their transverse sections, whereas the thalamus and cortex are shown in frontal slice.

Indeed, we have already provided as an example the split-brain human subjects that host two minds, rather than one, as a result of surgical severing of their corpus callosum (Section 1.2). Because the corpus callosum contains axons of cortical pyramidal neurons that project from one brain hemisphere to the opposite hemisphere, it follows that the binding of consciousness should be accomplished through the axons that make synapses onto dendrites of target neurons. The existence of intact synaptic connections between neurons, however, is not sufficient for binding of conscious experiences. There is a need for a nonlocal physical mechanism that shares instantaneously a given experience with all of the neurons participating in the generation of the conscious mind.

Example 5.4. *(Anatomical connection is not sufficient for binding of consciousness) Incoming sensory information from our bodies or the surrounding world is transmitted through neural pathways that contain anatomically connected sequences of neurons. Sensory neurons from the spinal cord extend their axons and make synaptic connections with neurons inside the thalamus. Then, thalamic neurons project further to the brain cortex where they make synaptic connection with cortical neurons. In Section 1.1, we have discussed the work by William H. Dobelle showing that the electric impulses are experienced consciously only after they arrive in the brain cortex, not before that. Spinal anesthesia also nicely illustrates the fact that the sensory signal is not experienced while it is on its way toward the cortex. The application of local anesthetics at a certain level of the spinal cord causes analgesia by blocking the propagation of pain electric impulses toward the cortex (Fig. 5.4), through inhibition of voltage-gated sodium channels (Fig. 1.8) located in the plasma membranes of neurons that comprise the lat-*eral spinothalamic tract. *The human subject undergoing spinal anesthesia is conscious but does not feel the pain from the surgery. If the anatomical connectedness were sufficient for the pain signal to be consciously experienced, the pain electric signal should have been experienced immediately as it enters the peripheral nervous system, and there would have been no need for the electric impulses to propagate from the peripheral nervous system toward the brain cortex. Thus, being anatomically connected is not enough for binding of conscious experiences.*

Example 5.5. *(Electric firing is not sufficient for binding of consciousness) Functional connectedness through electric firing is also inadequate for explaining the binding of conscious experiences. At clinical concentrations, inhaled volatile anaesthetics such as halothane, isoflurane and sevoflurane, cause loss of general sensation and reversibly erase consciousness during certain medical and surgical procedures. With the experiences gone during general anesthesia our consciousness ceases to exist, even though the neurons in the brain cortex are still able to fire under sensory stimulation, as shown by in vivo electrophysiological recordings in both rodents [272, 308] and primates [307]. If electric firing could occur without generating conscious experiences, then there is no reason to expect that electric firing itself is able to bind conscious experiences together.*

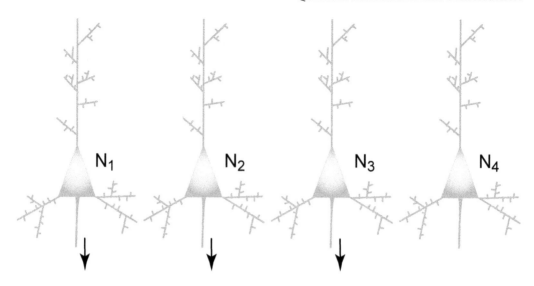

Figure 5.5   Illustration of the binding by synchrony hypothesis. If neurons N_1, N_2, and N_3 fire electric spikes synchronously whereas neuron N_4 does not fire a spike (or fires with a delay), it could be speculated that only the information processed by neurons N_1, N_2, and N_3 was bound together.

5.2.3   Neural synchrony

Binding of conscious experiences by synchrony [432] could be viewed as a refinement of the "assembly" proposal (Fig. 5.5). Claims have been made that groups of neurons that fire synchronously are producing and binding the conscious experiences at that particular moment of time [506]. Unfortunately, there are multiple issues with the latter proposal as well.

First, it is technically challenging to answer how exactly the synchrony of spikes is defined. For example, is there a time window within which the electric spikes are considered bound together, or the synchrony must be sharp and occurring at a precise time point? Also, since neurons differ in size and shape, it must be answered at what place in the neuron the spike is measured. Is it in the soma, or further down the axon?

Second, the firing of pyramidal neurons in the visual cortex during general anesthesia seems to invalidate the hypothesis that electric firing, synchronous or not, is directly generating consciousness or binding conscious experiences together [272, 307, 308].

Third, for every possible definition of what synchrony is, one may ask why one's experiences do not get bound with the experiences of somebody else standing nearby whose neurons happen to fire synchronously at the same time.

Fourth, "synchrony" or "simultaneity" of events is not an absolute concept according to Einstein's relativity theory (Section 3.18). It is difficult to see how a relative concept can be adequate for explaining an absolute phenomenon such as the binding of one's conscious experiences.

5.2.4   Integrated information

Binding of conscious experiences through integrated information is an elaborate proposal that takes into consideration both the structure and the actual functional state of the neural network [362, 482, 483, 484, 485, 486, 487, 488]. Although calculating integrated information in the real brain is computationally hard and there is a zoo of competing integrated information measures [479], here we will focus only on certain conceptual issues in the theory developed by Giulio Tononi and colleagues [486, 487].

Mathematically, the neural network could be represented by a *graph*, where each neuron corresponds to a vertex and synaptic connections between neurons are depicted by *directed edges* (Fig. 5.6). The functional state of a network composed by n neurons is then expressed by a string of n bits, where 0s indicate neurons that do not fire, and 1s indicate neurons that fire. In general, the weights of the synaptic connections w_i between neurons could vary. Inhibitory synapses could have negative weights, whereas excitatory synapses could have positive weights. In the following exposition, we will set the weights of all synapses equal to $w = \frac{1}{2}$, which means that all synapses are excitatory and each neuron needs to receive at least two simultaneous synaptic inputs in order to fire. In order to be able to calculate the amount of integrated information generated by a neural network, we would need several mathematical preliminaries.

Definition 5.1. *(Potential repertoire) The* potential repertoire *of a neural network composed of n neurons, each of which could be either firing or not, is represented by a uniform probability distribution of 2^n possible network states. The uniform probability distribution expresses complete uncertainty about the previous state of the neural network.*

Definition 5.2. *(Actual repertoire) The* actual repertoire *of a neural network composed of n neurons, each of which has a certain synaptic connectivity and integrates synaptic inputs by a specified mechanism, is represented by a probability distribution of only those network states that could have led to the current state of the neural network in a single step of time that executes all functional connections once. Thus, given the graph of a neural network and its current state, it may be possible to logically attribute zero probability to some states from the potential repertoire and exclude them from the actual repertoire.*

Definition 5.3. *(Information gain) The* Kullback–Leibler divergence *measures in bits the information gain when one revises one's beliefs from the prior discrete probability distribution P_1 to the posterior discrete probability distribution P_2*

$$D_{KL}(P_2\|P_1) = \sum_i P_2(i)\log_2 \frac{P_2(i)}{P_1(i)} \tag{5.1}$$

The *effective information* generated by a neural network could be calculated as the information gain when one revises one's beliefs from the discrete probability distribution P_1 representing the potential repertoire to the discrete probability

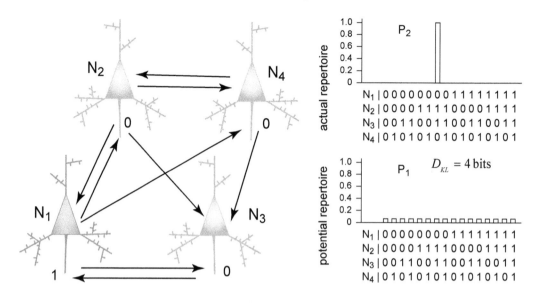

Figure 5.6   The effective information generated by a neural network composed of four neurons N_1, N_2, N_3 and N_4 is the Kullback–Leibler divergence between the actual and the potential repertoires of the network. The current state of the network is 1000. The actual repertoire is composed of a single state 0110, which is the only state that could have led to the current state in a single step of time.

distribution P_2 representing the actual repertoire of the network. As an example, consider the neural network shown in Figure 5.6. From the graph of the network, it could be deduced that only the state 0110 could have led to the current state 1000 in a single step of time. The information gain is $D_{KL} = 4$ bits. This effective information, however, is not yet a measure of integration. In order to determine whether there is an integration, one needs to analyze what happens with the amount of effective information generated when the network is partitioned into disconnected parts.

Let us suppose that the neural network is partitioned into two disconnected parts, as shown in Figure 5.7. In such a case, the external inputs to each disconnected part have to be treated as extrinsic noise, meaning that every possible combination of external inputs could occur with the same probability. For the first partition composed of neurons N_1 and N_2, there are two external synaptic inputs that could deliver with equal probability one of four synaptic firing patterns: 00, $0\frac{1}{2}$, $\frac{1}{2}0$, $\frac{1}{2}\frac{1}{2}$. Because neuron N_1 is currently firing, the previous states of the network should have been such that neuron N_2 is firing, namely, 01 or 11. Because there are two synaptic firing patterns $\frac{1}{2}0$ and $\frac{1}{2}\frac{1}{2}$ that transform the state 01 into the current state 10 and only one synaptic firing pattern $\frac{1}{2}0$ that transforms the state 11 into 10, the first partition of the neural network could have been in the state 01 with probability $\frac{2}{3}$ and in the state 11 with probability $\frac{1}{3}$. The analysis is similar for the partition composed of neurons N_3 and N_4, but it needs to take into account 16 synaptic firing patterns: 0000, $000\frac{1}{2}$, $00\frac{1}{2}0$, $00\frac{1}{2}\frac{1}{2}$, $0\frac{1}{2}00$, $0\frac{1}{2}0\frac{1}{2}$, $0\frac{1}{2}\frac{1}{2}0$, $0\frac{1}{2}\frac{1}{2}\frac{1}{2}$, $\frac{1}{2}000$,

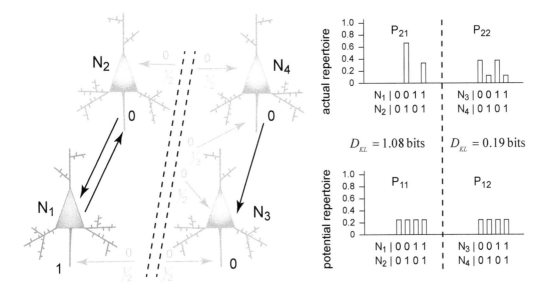

Figure 5.7   Effective information generated by two disconnected partitions of the neural network containing either neurons N_1 and N_2, or neurons N_3 and N_4. The first partition generates $D_{KL} = 1.08$ bits of effective information, whereas the second partition generates $D_{KL} = 0.19$ bits.

$\frac{1}{2}00\frac{1}{2}$, $\frac{1}{2}0\frac{1}{2}0$, $\frac{1}{2}0\frac{1}{2}\frac{1}{2}$, $\frac{1}{2}\frac{1}{2}00$, $\frac{1}{2}\frac{1}{2}0\frac{1}{2}$, $\frac{1}{2}\frac{1}{2}\frac{1}{2}0$, $\frac{1}{2}\frac{1}{2}\frac{1}{2}\frac{1}{2}$. The neuron states 00 and 10 are transformed into the current state 00 by nine synaptic firing patterns: 0000, $000\frac{1}{2}$, $00\frac{1}{2}0$, $0\frac{1}{2}00$, $0\frac{1}{2}0\frac{1}{2}$, $0\frac{1}{2}\frac{1}{2}0$, $\frac{1}{2}000$, $\frac{1}{2}00\frac{1}{2}$, $\frac{1}{2}0\frac{1}{2}0$, whereas the neuron states 01 and 11 are transformed into 00 by just three synaptic firing patterns: 0000, $000\frac{1}{2}$, $00\frac{1}{2}0$ due to the $\frac{1}{2}$ synaptic input contributed from the firing neuron N_4 to neuron N_3. Hence, the second partition of the neural network could have been in the states 00 or 10 with probability $\frac{3}{8}$ and in the states 01 or 11 with probability $\frac{1}{8}$. Considered separately, the two partitions generate $D_{KL} = 1.08$ and $D_{KL} = 0.19$ bits of effective information that do not sum up to the $D_{KL} = 4$ bits of information generated by the whole unpartitioned neural network.

In order to calculate how much effective information is generated by the whole network above and beyond the two network partitions, one should calculate the information gain D_{KL} between the actual repertoire P_2 of the whole network and the actual repertoire $P_{21} \otimes P_{22}$ of the partitioned network (Fig. 5.8). Still, this is not yet a measure of integrated information. One needs to further consider all possible partitions of the neural network and find the minimal amount of effective information that is generated by the whole network but cannot be accounted for by the network parts. The decomposition of a network into minimal parts, which leaves the least amount of effective information unaccounted for, is referred to as the *minimum information partition* of the network. After laborious analysis of all possible partitions, it could be shown that the bipartition shown in Figure 5.8 is actually the minimum information partition of the neural network.

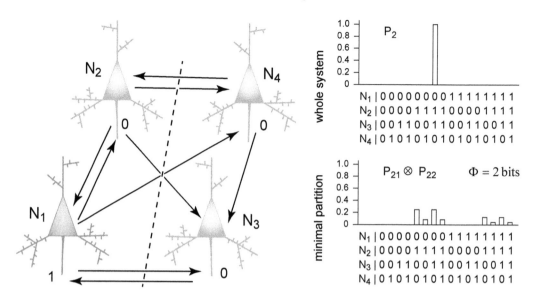

Figure 5.8 The minimum information partition of the neural network separates neurons N_1 and N_2 from N_3 and N_4. The Kullback–Leibler divergence between the actual repertoire of the whole network and the actual repertoire of the minimum information partition of the network measures in bits the amount of integrated information Φ.

Definition 5.4. *(Integrated information) The* integrated information Φ *is the information gain between the actual repertoire of a network and the actual repertoire of the network minimum information partition. Thus, Φ is the least amount of effective information generated by the network, but unaccounted for by the network parts [486].*

Because the integrated information Φ depends only on the number and complexity of functional connections between different vertices in the graph that represents the physical system, many different physical systems ranging from electrically wired transistors to pump-equipped pipe-connected water tanks could generate the same amount of integrated information Φ provided that the functional organization is replicated. While taking the road to panpsychism is not problematic in itself, there are several conceptual issues that require special attention.

First, classical communication between people does not lead to binding of conscious experiences. However, if a large number of brains, each possessing a separate mind, are engaged in replicating the functional graph of a working neural network (Fig. 5.9), the generation of integrated information Φ would predict the emergence of a super-mind or Über-mind that necessarily has to shut down the conscious experiences of the individual minds in order to prevent the paradoxical occurrence of minds within minds [487]. Even though the integrated information theory of consciousness has been supplemented with an exclusion principle, according to which only complexes that specify local maxima of integrated information Φ^{max} correspond to conscious minds, it is possible through deliberate design to construct a network of brains that generates a higher amount of Φ compared with the amount of Φ generated by each brain individually. For example, corti-

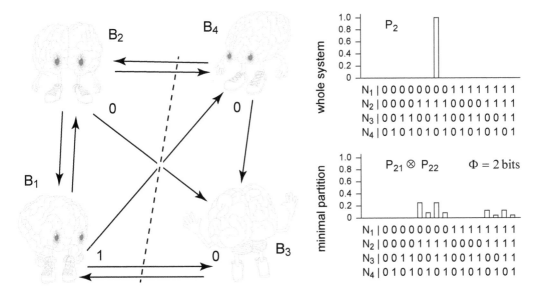

Figure 5.9   Classical communication between brains, each of which has its own mind, also generates integrated information Φ. If binding of conscious experiences were due to integrated information, a network of communicating brains would have been able to generate a super-mind or Über-mind in which all individual brains suddenly lose their own minds as Φ reaches a certain value.

cal neurons have on average about 7000 synapses [363], which means that the graph representing the neural network of a human brain would also have about 7000 incoming edges per vertex. To maximize the amount of Φ generated by a network of brains, however, one does not need to stick to the graph of a real brain but could hugely increase the number of incoming and outgoing edges so that say one million brains are allowed to send and receive messages from any other brain in the network. Tononi and Koch admit that at the point where the Φ^{max} of the interacting brains would exceed Φ^{max} of the individual brains, their individual conscious mind would disappear and be replaced by a new Über-mind that subsumes the individual brains [487]. Here, we point out that experimental test with one million volunteers potentially exchanging e-mails through the internet is not unfeasible, and we predict that if such a test is ever performed it would show no emergence of a new Über-mind, thereby disproving the binding of conscious experiences through integrated information.

Second, the anatomical organization of the brain into two cerebral hemispheres communicating with each other through the corpus callosum, greatly constrains the amount of integrated information Φ^{max} attainable by both hemispheres together. Tononi and Koch admit that if the 200 million callosal fibers through which the two cerebral hemispheres communicate with each other were severed progressively, there would be a moment at which conscious experience would go from being a single one to suddenly splitting into two separate experiencing minds, as we know to be the case with split-brain patients. This would be the point at which Φ^{max} for the whole brain would fall below the value of Φ^{max} for the left and for the right hemisphere taken by themselves [487]. Noteworthy, the latter prediction appears to have been inadvertently tested and falsified in human subjects that have undergone surgical control of epilepsy. Extensive commissurotomy that severs the anterior commissure and a major portion of the corpus callosum but spares its posterior, rounded end called the *splenium*, does not produce split-brain syndrome [220]. Thus, the cerebral hemispheres seem to bind conscious experience if but a small fraction of the corpus callosum remains intact, contrary to what one would expect if binding of conscious experiences were due to integrated information.

5.2.5 EEG waves

Physical fields provide an alternative approach to the binding problem of consciousness. Instead of starting from the functional behavior of discrete localized physical entities such as neurons and then trying to find a binding mechanism, the field approach starts with delocalized physical entities such as classical fields and then tries to find a mechanism that localizes the fields in the brain. An example is the creation of electric fields from localized physical charges (Fig. 3.15). Each individual charge q_i generates an electric field $\vec{E}_i$ that extends in space according to Coulomb's law, given by Eq. (3.84). At any point x, the total electric field is the vector sum of all individual electric fields

$$\vec{E}_{\text{total}}(x) = \sum_i \vec{E}_i(x) \tag{5.2}$$

Because individual electric fields are delocalized and naturally sum with each other, it may appear that fields are better suited to solving the binding problem of consciousness. The field extension into the environment, however, leads to the converse problem of localizing consciousness within the brain. In particular, the total electric field generated by the electrophysiological activity of neurons in the brain propagates through nearby tissue, including the bones of the skull, and can be recorded by electrodes attached to the scalp in the form of an *electroencephalogram* (Fig. 5.10).

Electroencephalography (EEG) is routinely used for monitoring of the brain function. In physiological conditions, EEG studies have revealed five categories of normal EEG wave patterns that can be classified by their frequency range. Gamma rhythms (30–80 Hz) exhibit the highest correlation with conscious processes and are thought to modulate perception, cognition and motor function. EEG recordings have further shown that the power of gamma rhythms is decreased during general anesthesia and increased during awakening from anesthesia [298]. Beta rhythms (14–29.9 Hz) are associated with active thinking, arousal, alertness and concentration. Alpha rhythms (8–13.9 Hz) appear when the eyes are closed during relaxation, light trance, pre-sleep or pre-waking. Theta rhythms (4–7.9 Hz) occur during meditation, drowsiness or dreaming sleep, whereas delta rhythms (0.1–3.9 Hz) are found during dreamless sleep.

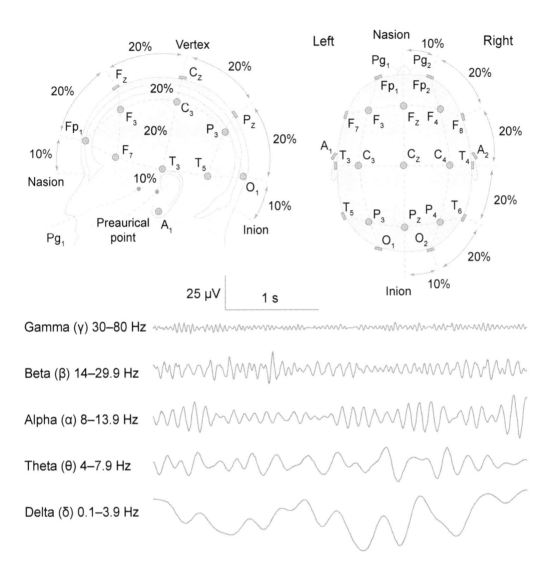

Figure 5.10   Electroencephalography (EEG) records the electrical activity of the brain using electrodes attached to the scalp. The international 10-20 electrode placement system assigns alphabetical and numerical abbreviations to identify the location of each electrode: A, auricular; C, central; F, frontal; Fp, frontal pole; O, occipital; P, parietal; Pg, pharyngeal; T, temporal. The nasopharyngeal electrodes Pg_1 and Pg_2 are used only in special cases. The five categories of normal EEG wave patterns are shown together with their characteristic frequency range. Gamma rhythms modulate perception and consciousness, and can be used to monitor depth of anesthesia. Beta rhythms are associated with concentration, arousal, alertness and cognition. Alpha rhythms appear during relaxation, light trance, pre-sleep or pre-waking. Theta rhythms occur in dreaming sleep and delta rhythms in dreamless sleep.

Despite of the correlations observed between the power of certain EEG rhythms and different types of mental states, identification of consciousness with classical electromagnetic fields in the brain leads to *paradoxes*. First, the electromagnetic fields extend in the universe at large, which exacerbates the physical boundary problem and makes it hard to explain why do we experience ourselves as agents localized in space (Section 5.1). Second, the heart is also an electrically active organ, as exemplified by *electrocardiography* (ECG) recordings, but does not support consciousness (Section 1.1). Third, EEG recordings during general anesthesia do not show a lack of electric activity of the brain, but rather a change in the power of different EEG rhythms. Thus, it appears that it is not the electric field itself, but rather the electric field dynamics that is relevant for consciousness. Further elaboration on this argument leads down the slippery road of classical functionalism with epiphenomenal consciousness, which will be thoroughly discussed next in Section 5.3.1.

5.3   Causal potency of consciousness in classical physics

In Sections 5.1 and 5.2, we have discussed the apparent problem that conscious experiences do not appear anywhere in the fundamental equations or axioms of classical physics. As a result, one needs to provide additional rules that define which collections of physical particles comprise individual minds and explain how those particles generate conscious experiences. Theoretical construction of such additional rules can be done in one of two ways: postulating that consciousness is a product of a part of the brain (functional approaches), or postulating that consciousness is identical with a part of the brain (reductive approaches). Here, we will show that functional and reductive approaches have opposing implications for the causal potency of conscious experiences and the ability of human consciousness to evolve.

5.3.1   Classical functionalism implies epiphenomenalism

Functionalism or *functional approaches to consciousness* recognize that *unconscious brains* do exist and define the conscious mind as a function or a functional product of the physical brain. Indeed, awake human brains generate conscious experiences, whereas anesthetized brains or dead brains do not generate any conscious experiences. Since we know that at least some brain states produce conscious experiences [431] and consciousness can be turned on or off by a variety of factors (including low blood glucose levels, changes in brain pH, general anesthetics or brain concussion), it might seem natural to assume that consciousness is somehow a product of the brain functioning [147]. Modern medical practice also seems to support the functional view: If a person is unconscious in clinical death, the resuscitation efforts are predominantly focused on ensuring steady oxygen supply to the brain for preventing neuronal death with the hope that consciousness may be regained as long as the brain neurons are kept alive.

Example 5.6. *(Crushed brain experiment) Even though the brain is the seat of consciousness and is responsible for the generation of our feelings and sensations, it does not have pain receptors. As a result, during certain types of neurosurgical operations, the skull is open and the surgeon operates while the patient is conscious. Keeping the patient awake and conscious is particularly important when the surgeon needs to remove a tumor or an epileptogenic piece of brain cortical tissue. By first stimulating the cortical tissue, the patient can report his experiences, and therefore help the surgeon to assess whether the tissue is healthy or not. The purpose of the surgery is to remove only the tumor or the epileptogenic cortical tissue, without anything else, otherwise the patient can lose some of his cognitive abilities. At this point, one can imagine a gruesome thought experiment in which the brain is crushed with a hammer or scrambled by a mixer, leaving the patient permanently unconscious. Because the composition of the intact conscious brain and the crushed unconscious brain is exactly the same in terms of chemical atoms, classically one could conclude that consciousness must be somehow a product of the functional organization of the brain rather than its chemical content.*

Depending on what kind of brain function is supposed to generate conscious experiences, the functional approaches to consciousness can be divided into different schools.

Behaviorism postulates that all statements about mental states and conscious processes should be equivalent in meaning to statements about behavioral dispositions. For example, anger should not be equated with a certain posture of facial expression, loud voice, or the responses of smooth muscles and glands such as sweating or salivating. Instead, anger should be characterized by a set of certain dispositions for future behavior such as an increased probability of striking, insulting, or otherwise inflicting injury and a lowered probability of aiding, favoring, comforting or making love [441]. Among the most infamous claims of behaviorism is the statement that there is no difference between two states of mind (including experiences, desires or beliefs) unless there is a demonstrable difference in the behavior associated with each of those states.

Computationalism postulates that the conscious mind is a computer program or a software, and the brain is a hardware that runs the mind software. A central claim of the theory is that the mind is an information processing system and that thinking is a form of computing. Importantly, computationalism does not simply state that the human mind is able to compute, or that the human mind resembles a computing system, rather it postulates that the computation performed by the brain is what generates the conscious experiences. Among the most infamous claims of computationalism is the statement that the brain substrate itself is irrelevant for the generation of consciousness; all that matters is the computation that is performed. An immediate drawback of computationalism is the loss of distinction between the mind as a phenomenon and the computer simulation of that phenomenon. In other words, if the mind is a running software, the simulation of the mind would also be a mind since it is a running software too.

Example 5.7. *(Piecemeal replacement of conscious brain neurons with silicon chips)*
David Chalmers constructed a thought experiment in which the neurons in a human
brain are gradually replaced by silicon chips and argued that any functional isomorph
of a conscious system should have qualitatively identical experiences [82]. As a first
step in the experiment, only a single neuron is replaced with a silicon chip that per-
forms precisely the same local function as the neuron. In particular, given certain in-
puts, the silicon chip calculates the appropriate electrical and chemical outputs that
are then delivered to the neighboring neurons. Since the chip is designed to have the
right input/output function, the replacement will make no difference to the functional
organization of the system as a whole. As a second step, another neighboring neuron is
replaced with a silicon chip. As before, the functional organization will be kept intact,
but it will be possible to merge the two silicon chips into a single one that has boosted
processing power. Further proceeding with the outlined piecemeal replacement of larger
and larger groups of neighboring neurons with silicon chips will ultimately convert the
whole brain into a silicon chip. Each system in the sequence will be functionally isomor-
phic to a human brain, but while the system at one end of the spectrum has a conscious
mind, the system at the other end will be essentially a copy of a silicon robot. If func-
tionalism is correct, it follows that this silicon robot, including all intermediate hybrid
systems, should also possess the same conscious mind as the human brain. Still, there
is no good reason to make us believe that functionalism is correct. Neurological and
neurosurgical data show that removing pieces of brain cortex do not cause unconscious-
ness, but gradually deteriorate the cognitive abilities of the remaining conscious mind.
Therefore, if the silicon chip replacement is performed with an embodied brain, the con-
sciousness produced by the remaining piece of brain will not fade away, but rather will
be incapacitated in its control over the body and will probably experience a feeling that
part of the body has been hijacked by an external agent (that will be the silicon chip).
Furthermore, even if the silicon chip were also generating conscious experiences, these
experiences need not be bound together with the conscious experiences generated by the
brain.

Computationalism implies that computer programs that are run in silicon
chips should be able to generate consciousness experiences. Moreover, it promises
that in the near future human consciousness could be detached from its brain
substrate and uploaded into silicon memory chips analogously to how computer
programs can be stored and executed in personal computers. A notable propo-
nent of such a viewpoint is the theoretical physicist Stephen Hawking, who was
interviewed at the Cambridge Film Festival in 2013, after the premiere of a new
biographic movie about his life. Asked about whether a person's consciousness can
live on after the person dies, seemingly using the words *brain* and *mind* with their
meanings interchanged, Hawking replied

> I think the brain is like a programme in the mind, which is like a computer,
> so it's theoretically possible to copy the brain onto a computer and so pro-
> vide a form of life after death. [44, 96, 241]

Representationalism postulates that representation of a certain kind suffices for the generation of conscious experiences, where the kind of representation needs to be specified as a physical function without any recourse to fundamental mental properties. For example, light rays reflected from external objects enter the eyes and deliver visual information for the shape and color of those objects. The visual images detected by the retina are encoded in the form of electric spikes that are subsequently delivered to the brain cortex. Thus, the objects that we consciously see are represented by electric excitations of neurons in the brain. The central claim of the theory is not simply that our visual conscious experiences represent the objects whose light images enter the eyes, but rather that the act of representing of those light images through physical processes in the brain (such as propagation of electric impulses in the network of neurons) is itself generating the visual conscious experiences.

Emergentism postulates that conscious experiences are emergent properties of the brain that are not identical with, reducible to, or deducible from the other physical properties of the brain [61, p. 59]. Among the most infamous claims of emergentism is that cumulative quantitative changes lead to qualitative leaps [160, p. 27]. Thus, the human consciousness is a product of the material brain [314, p. 55] that emerges when the brain structure and functional organization reaches a certain, sufficiently high, level of complexity [314, p. 75]. In essence, emergentism puts the emphasis not on *what* function the brain performs, but on *how complex* that function is. Despite the apparent resort to complexity, emergentism is dangerously close to pseudoscience because consciousness is able to miraculously pop into existence as if conjured by a magic trick.

Example 5.8. *(Atomic magic wand for switching emergent consciousness on or off) Because the chemical atoms are essentially identical for both the sentient brain and the insentient matter, we can start a recipe for building up a sentient brain from insentient matter. According to emergentism, at a certain threshold level of complexity our brain product will start to feel. Therefore, while following the recipe for building up a sentient brain we are in principle able to pause at a stage at which adding or removing a single atom will give or take the mind of the system. Thus, emergentism implies that a single atom can act as a magic wand for switching consciousness on or off.*

Having briefly outlined several functional approaches to consciousness, we will now prove that all functional approaches consistent with the deterministic physical laws of classical physics reduce conscious experiences to the level of an *epiphenomenon* that is a physical phenomenon produced by and accompanying the brain processes, but itself having no causal influence upon those brain processes [269]. When confronted with the question of why do we have conscious experiences, evolutionary biologists would typically reply that *consciousness* is a useful thing to have and it provides conscious organisms with a competitive edge in the struggle for survival [259]. The evolutionary answer, however, would contradict directly the epiphenomenal nature of consciousness, since epiphenomena do not have any causal influences in the physical world and cannot be selected for by natural processes [207, 278, 279].

Theorem 22. *All functional approaches to consciousness (including emergentism) that are consistent with deterministic physical laws reduce conscious experiences to the level of a causally ineffective byproduct (epiphenomenon). Because epiphenomenal consciousness is utterly useless, it cannot evolve and its presence in humans and other animals is evolutionary unexplainable.*

Proof. Classically, the brain (or any other physical system that can generate conscious experiences) is composed of physical particles (such as atoms, ions and molecules) whose physical properties, including mass, charge, position and velocity, determine with certainty what will happen at any future moment of time. Therefore, knowing precisely where each physical particle is, with all of its physical properties, allows exact calculation of which neuron in the brain will fire and which neuron will not, and provides an exact prediction of how the brain will react to an incoming information. In other words, the determinism that is characteristic to the laws of classical physics implies that if we knew in detail the initial physical state of the brain, the behavioral responses would also have been exactly known. Thus, the behavioral responses are predetermined with absolute certainty, even though we may actually not know what they are. Consider now the premise that the brain somehow generates or produces conscious experiences. Since these conscious experiences do not enter in any of the physical equations that predict the behavioral responses, it follows that the conscious experiences are just an epiphenomenon that does not and cannot have any causal impact on the dynamics of the brain and the behavior. Without a causal impact upon behavior, however, conscious experiences cannot have any survival value and cannot be selected for by natural selection.   □

Epiphenomenalism clashes seriously with the theory of evolution, because our conscious experiences seem to be adequately matched to the neural responses and the corresponding behavior. For example, the neural responses to detrimental factors are always associated with unpleasant feelings and avoiding behavior. If conscious experiences are causally effective, the adequate matching between the unpleasant experiences, the avoiding behavior and the negative influence of detrimental factors upon the organism becomes evolutionary explicable, since the animals that would have enjoyed detrimental factors would not have avoided dangers, thereby dying out in the competition with other organisms that do not enjoy detrimental factors [278]. If conscious experiences were causally ineffective, however, one would expect that in nature there are some organisms that avoid detrimental factors due to the organization of their neural processes but consciously experience pleasant feelings when being injured. Even though we do not have direct access to the conscious experiences of other organisms, through our own introspection we could convince ourselves that we never enjoy detrimental and never dislike beneficial factors. Hence, epiphenomenalism is false.

5.3.2   Classical reductionism implies trivial immortality

Reductionism, also known as *reductive materialism, type physicalism, type identity theory, mind–brain identity theory* or *identity theory of mind*, asserts that the mental events can be grouped into types that can be identified with types of physical events in the brain. In classical physics, the addition of novel causally effective ingredients is forbidden by the *determinism* of the physical laws. In a deterministic theory, everything that is causally effective must be taken as a fundamental physical quantity in the dynamical equations that govern the behavior of the physical system because only those physical quantities that enter in the physical equations can make a difference for the future time dynamics of the system. In order to bypass the restrictions imposed by determinism, the reductive approaches to consciousness identify the mind with some physical quantity that generates real physical forces in the physical world. For consistency, however, reductionism should predict that the physical forces attributed to the mind are non-zero for conscious brain states and zero for unconscious brain states.

Postulating that the mind *is* the brain (or a subsystem of the brain such as the brain cortex) makes the evolution of the mind equivalent to the evolution of the brain, thereby avoiding epiphenomenalism and eliminating all paradoxes related to the causal potency of consciousness. The general mind–brain identity thesis, however, is demonstrably false because *anesthetized brains* or *dead brains* do not give rise to conscious experiences (Section 1.6). If the mind and the brain were identical, then it would have been impossible to separate them. Hence, having *unconscious brain states* such as anesthetized brains or dead brains is an empirical fact that contradicts the general mind–brain identity thesis. Starting from the premise that the brain *is* the mind, one could also deduce that all of the fundamental classical physical quantities (mass, charge, length and time) are properties of the mind. But if all physical quantities are mind properties, it would follow that the physical world is composed of conscious minds. Thus, one ends up with either *panpsychism*, asserting that the physical reality is fundamentally mental and there is no such thing as insentient matter, or *idealism*, asserting that insentient physical properties such as mass, charge, length and time are nonentities or just an illusion [47, p. 58].

Refining the general mind–brain identity thesis could be done by postulating that the mind is identical with either a single physical property of the brain (such as the brain charge, brain electromagnetic field, etc.) or a combination of several such properties. This type of reasoning leads to different *reductive approaches to consciousness*, among which is the extreme case of the general mind–brain identity theory stating that all of the physical properties of the brain are also properties of the mind. Identifying the mind not with the whole brain (or a whole brain subsystem), but with a single physical property of the brain could be speculated to provide a mechanism for turning the mind off by setting the value of the relevant physical quantity to *zero*. In Section 5.1, we have discussed weakly electric fish and the possibility of an *electromagnetic theory of consciousness* that identifies the mind with a non-zero electromagnetic field generated by the brain. Such a theory, however, refutably predicts that consciousness can never be turned off in the presence of external electromagnetic fields such as Earth's magnetic field or radio emissions (due to cell phones, wireless networks, television, etc.), since the electromagnetic field inside the brain is not zero in those cases. Similar empirical refutation can be provided for every theory that identifies the mind with any other non-zero physical quantity. Since the values of all classical physical quantities can always be made non-zero with the use of available physical devices, it follows that consciousness in a dead brain could be resurrected by adding extra electromagnetic fields, electric currents, or any other physical quantity identified with the mind. Remarkably, in 1803 Giovanni Aldini (1762–1834) attempted resurrecting a hanged criminal called George Forster using electric currents. According to eyewitnesses, as a result of the electric stimulation indeed Forster's eye opened, his right hand was raised and clenched, and his legs moved [373, pp. 380–381]. Now we know that all these motions were not a result of resurrecting Forster's mind, but were caused by direct electric action upon his muscles. Modern medicine has further accumulated data showing that addition of any classical physical quantity is not sufficient to resurrect consciousness in a dead brain. Thus, both the temporal loss of consciousness during general anesthesia and the mortality of the conscious mind accompanying the death of the brain, seriously constrain the theoretical modeling of the mind–brain relationship and provide serious grounds for rejecting the mind–brain identity thesis.

Theorem 23. *All reductive approaches to consciousness that reduce consciousness to a fundamental classical physical quantity imply trivial immortality that is achievable by the current technology.*

Proof. In order to explain how general anesthetics turn off conscious experiences during anesthesia, the theory needs to provide a range of values for the fundamental physical quantity within which consciousness is generated and outside of which consciousness fails to be generated. But then consciousness can always be revived by physical means that bring the relevant physical quantity within the prescribed theoretical range, which makes consciousness immortal given a trivial support from a suitably selected classical machine that is currently available.   □

Complicated type identity approaches in which there are both conscious and unconscious brain states, based not on a defined physical rule but on an improvised list of statements, are just a form of disguised *emergentism* and do not compress any information, contrary to what scientific theories should do. As an example, consider the list:

> 1. *Awake human brain states are conscious.*
> 2. *Crushed human brain states are unconscious.*
> 3. *Human brain states under general anesthesia are unconscious.*
> 4. *Physical states of rocks are unconscious.*
> 5. *Physical states of computers are unconscious.*
> 6. *Healthy brain states of dolphins are conscious.*
> 7. ....

Such a list does not provide any scientific explanation of what consciousness is. Moreover, a simple comparison between a conscious healthy brain and the same brain brought to the unconscious state through crushing it with a hammer could also easily bring back the problems of functionalism (Section 5.3.1). In particular, if the physical composition of the intact brain and the crushed brain is the same, then consciousness must have been related to the functional organization of the brain since classically there is little else left to explain the loss of consciousness after the crushing (Example 5.6).

5.4   Free will in classical physics

The laws of classical physics are *deterministic* in nature. This means that given the detailed description of the brain at any given time, the physical laws determine with absolute certainty the brain state at any other future moment of time. Deterministic laws imply, however, that *free will* is impossible, because in order to be able to choose between alternative future courses of action one should be first provided with at least two different options to choose from. Thus, without the ability to choose, there can only be *will*, but not *freedom*.

5.4.1   Debunking compatibilism

Historically, the compatibility between free will and determinism has been the subject of a prolonged controversy. *Compatibilism* is the position that free will and determinism are logically compatible, whereas *incompatibilism* is the alternative position that free will and determinism are logically incompatible. Further division among incompatibilists gives rise to *hard determinists* who believe in determinism but not in free will, and *libertarians* who believe in free will but not in determinism.

Definition 5.5. *(Hard determinism)* Hard determinism *is the position that because we are living in a universe that is governed by deterministic physical laws, we are agents without free will.*

Definition 5.6. *(Libertarianism)* Libertarianism *is the position that because we are agents with free will, we cannot be living in a universe that is governed by deterministic physical laws.*

Thomas Hobbes (1588–1679) argued that we have the *will* to do things, but since the will is always caused by something else, it cannot be free. Consequently, he regarded *free will* as a logical contradiction [256, p. 275]. Hobbes is credited with being the modern inventor of compatibilism, but his claims that determinism and voluntary actions are compatible should be understood only in the narrow context where the term *voluntary action* stands for desired or willed action [256, p. 274], but not an action that could have been done otherwise [256, p. 275]. Modern hard determinists make the latter distinction explicit in the statement that if determinism is true, it follows that we cannot have free will and necessarily our subjective experience of being capable to choose otherwise has to be an *illusion* [121, 125, 239, 514]. Such a conditional statement, however, does not prove that we lack free will. A libertarian can as easily claim the converse, namely, that because we have free will, it follows that the universe cannot be deterministic. If ultimately one has to *choose* between determinism and free will, the rational choice would be to believe the evidence from our conscious experience in favor of free will. Indeed, because our only access to the surrounding world is through our senses, it would be irrational to think that we are capable of producing a correct physical theory of the universe solely through guesswork if we had fundamentally flawed and untrustworthy senses.

To avoid possible narrowing of the meaning of the term *voluntary action* down to *willed action*, instead of *freely willed action*, compatibilists have made attempts to prove that an agent endowed with free will is able to choose freely even though he could not have done otherwise. Below, we will discuss two of the most famous arguments in favor of compatibilism, and we will explain where exactly each of these arguments fails.

Example 5.9. *(Locke's man in a locked room argument) John Locke (1632–1704) attempted to show that our belief in free will could be based on an illusion sustained by our ignorance. Suppose that a man wakes up in a room that, unknown to him, is locked from the outside. He then chooses to stay in the room, believing that he has freely chosen to do so. Because the man really had no available the option of not staying in the room, he could not have done otherwise. Nonetheless, the ignorance of the man of the real situation he is in sustains the illusion of freedom [320, p. 153].*

Critical analysis of Locke's argument reveals that whereas it is true that ignorance is able to sustain illusory beliefs for a short period of time, it is not true that genuine free will is compatible with determinism. The fault of the argument resides in the fact that it is based on a conditional premise, namely, that the man in the room happens to choose to stay in the room. Since the man could also have chosen not to stay in the room, there is a possibility and non-zero probability that the man tries to open the locked door and then realizes that he has been locked in. Indeed, running the experiment multiple times or letting the man stay in the room sufficiently long so that he finally attempts to exit the room will inevitably result in the realization by the man that the room is locked.

In reality, we can directly test whether compatibilism is true. For example, we can try moving one of our limbs thousands of times in various directions. If we had free will but we were living in a deterministic universe, it would have been the case that for a substantial percentage of our trials we would have been willing to move our limb in a direction that is forbidden by the actual physical state of the brain and the physical laws. In those particular trials, we would have discovered by experience that we want to move our limb in one direction but it either does not move or goes in another direction. Thus, compatibilism is testable through introspection and experimental data unambiguously falsifies it: when we perform voluntary actions, our limbs always move in the direction that we have chosen.

Example 5.10. *(Frankfurt's argument for compatibilism) Harry Frankfurt attempted a refinement of Locke's argument. Suppose that an external agent possesses two advanced physical devices capable of the following: The first device performs precise measurements upon the neurons in your brain and from the experimental data deduces your current mental state together with your future decisions. The second device rearranges the firing of your neurons in such a way that you will do a certain action* A *only in case you have chosen not to do the action* A. *Then, the external agent could use these devices to observe your mental states, monitor your conscious choices, and put you into a situation in which if you have chosen to do* A *you would have done it on your own will, despite the fact that you could not have done otherwise. Hence, it appears that one could act by his own free will, even if one is unable to choose otherwise [193, 333].*

Frankfurt's argument appears to be a refined, but faulty, attempt at establishing the credibility of compatibilism in the framework of a classical theory of mind. The first problem is that it is inappropriate to apply deterministic physical laws or classical information theorems (such as Theorem 6) to mental states and free will, otherwise the argument will be circular [207]. In particular, such reasoning can only reiterate that if determinism can be applied to conscious states and free will, then the free will has to be consistent with determinism. But since from Definition 1.3 of free will follows that the free will is not consistent with determinism, we can conclude that determinism cannot and should not be applied to conscious states and free will. The quantum information theorems (Section 4.20) also imply that one cannot indiscriminately assume that the mental states are classical bits with free will, and then derive intended conclusions about the mental states or free will from the postulated classical bit properties, as these already presuppose the validity of all principles behind classical physics, including determinism. Thus, caution needs to be exerted before taking for granted that physical information can be read, copied or deleted. The second problem in Frankfurt's argument is that it also relies on a conditional premise. Consequently, given enough time, sooner or later you will notice that someone or something is hindering your free will. Exactly as in Locke's locked room scenario, you may choose not to perform the action A, only to find to your own surprise that there really is no such an option as you inevitably end up doing the action A.

5.4.2   Determinism implies moral nonresponsibility

The view that the physical laws need to be deterministic is widely held by physicists. Determinism is attractive because given powerful computers one could use the equations of the physical laws to calculate and predict with absolute certainty the future behavior of the surrounding world. Exact predictions are needed for flying spaceships to other planets, for building skyscrapers, for constructing bridges, for manufacturing machines, and for curing human diseases. When applied to ourselves as conscious minds, however, determinism deprives us of our free will and allows external agents to take control over us. Thus, determinism makes it hard to answer who is controlling who, thereby forcing us into a state of moral nonresponsibility for our actions.

Clarence Darrow (1857–1938) was a famous trial lawyer who successfully defended murderers from receiving the death penalty. When Darrow defended persons accused of murder, he argued that his clients were not morally responsible for their actions because of the deterministic physical laws that govern the world. A standard argument used by Darrow goes as follows:

> Every one knows that the heavenly bodies move in certain paths in relation to each other with seeming consistency and regularity which we call law. If instead of the telescope we use the microscope, we find another world so small that the human eye cannot otherwise see it, but fully as wonderful as the one revealed by the telescope. No one attributes free will or motive to the material world. Is the conduct of man or the other animals any more subject to whim or choice than the action of the planets? [105]

Determinism implies, as Darrow correctly noticed, that the criminals are not morally responsible for what they do because their actions are the result of what other people have done to them. But these other people, too, cannot be morally responsible for what they have done because their actions, in turn, are the product of what had earlier been done to them. The argument works like a line of falling dominoes. Thus, determinism infamously leads to the *domino theory of moral non-responsibility*, where no one is responsible for anything.

Despite its devastating implications for the existence of free will, determinism has been cherished by many prominent scientists. In 1931, Albert Einstein wrote

> If the moon, in the act of completing its eternal way around the earth, were gifted with self-consciousness, it would feel thoroughly convinced that it was traveling its way of its own accord on the strength of a resolution taken once and for all. So would a Being, endowed with higher insight and more perfect intelligence, watching man and his doings, smile about man's illusion that he was acting according to his own free will. [263, p. 172]

Einstein was not bothered by the lack of free will, because this apparently helped him to accept calmly the misdeeds of others. Humans had already caused the insane atrocities of World War I and Germany was on its way to electing Nazis to rule the country, when in 1932, Einstein recorded his credo stating

> I do not believe in free will. Schopenhauer's words: 'Man can do what he wants, but he cannot will what he wills,' accompany me in all situations throughout my life and reconcile me with the actions of others, even if they are rather painful to me. This awareness of the lack of free will keeps me from taking myself and my fellow men too seriously as acting and deciding individuals, and from losing my temper. [156]

The British philosopher Bertrand Russell (1872–1970) thought that nobody believed in free will in practice, and argued that the free will doctrine was born out of religious necessity to attribute sin to others:

When a man acts in ways that annoy us we wish to think him wicked, and we refuse to face the fact that his annoying behaviour is a result of antecedent causes which, if you follow them long enough, will take you beyond the moment of his birth, and therefore to events for which he cannot be held responsible by any stretch of imagination. No man treats a motor-car as foolishly as he treats another human being. When the car will not go, he does not attribute its annoying behaviour to sin; he does not say: 'You are a wicked motor-car, and I shall not give you any more petrol until you go.' [418, p. 34]

Different scholars may hold incompatible views on certain scientific problems. Scientific criticism and the clash of ideas are thought to help the growth of science due to the fact that true ideas survive whereas false ideas die out. If we lacked free will, however, science would have been a meaningless enterprise and scientific argumentation would have been futile. In a deterministic world, whether a scientist will write or read a scientific paper, will accept or reject a scientific argument, or will understand or misconstrue a scientific theory, would have been preordained. Thus, entering into a scientific discussion would have been useless if we were not able to choose whether and how to revisit our beliefs whenever they are found to be deficient.

5.4.3 Instability and chaos cannot rescue free will

Instability of the solutions for the corresponding physical states leads to an enhanced sensitivity of the system dynamics to infinitesimal perturbations. An example of an *unstable state* is the state of a spherical ball centered on top of a perfectly symmetric hill of a potential field (Fig. 5.11). Without any external perturbation, the ball will stay on top of the hill, keeping its delicate balance. If there is even an infinitesimal sideways push, however, the ball will be set in motion and it will roll down the hill.

Dynamical trajectories of physical systems can also be unstable. For example, consider the dynamics of the physical state $y(t)$ of a system governed by the differential equation

$$\frac{d}{dt}y(t) = \cos[\pi y(t)t] \tag{5.3}$$

To solve this equation one needs a boundary condition such as the initial value of the state $y(0)$ at time $t = 0$ [41, p. 198]. At sufficiently large time t, the solutions of the equation asymptotically approach a discrete set of trajectories given by

$$y \sim \left(n + \frac{1}{2}\right)\frac{1}{t} + \frac{(-1)^n}{\pi}\left(n + \frac{1}{2}\right)\frac{1}{t^3}, \qquad \text{(as } t \to \infty) \tag{5.4}$$

where $n = 0, \pm1, \pm2, \pm3, \ldots$ The dynamic trajectories with *even* n are stable, whereas those with *odd* n are unstable [40]. The unstable solution corresponding to $n = 1$ is plotted in Figure 5.11. Its value at $t = 0$ is given by $y(0) = 1.602572\ldots$ An infinites-

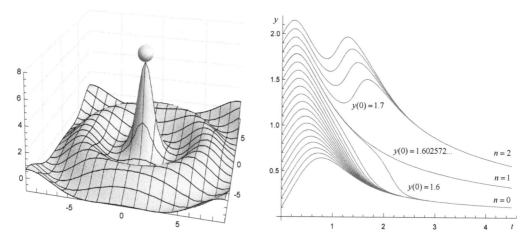

Figure 5.11   Instability in deterministic physical systems. On the left is shown an unstable physical state of a spherical ball that is centered on top of a perfectly symmetric hill of a potential field. In such a case, even an infinitesimal sideways push is able to set the ball rolling down the potential hill. On the right is shown a numerical example of an unstable solution with $n = 1$ of the nonlinear differential equation $\frac{d}{dt}y(t) = \cos[\pi y(t)t]$ corresponding to the boundary condition $y(0) = 1.602572\ldots$ An infinitesimal decrement or increment of the $y(0)$ value will cause the trajectory to be attracted toward the stable solutions with $n = 0$ or $n = 2$, respectively, for $t > 4$.

imal decrease of the $y(0)$ value will cause the unstable trajectory to be attracted toward the stable solution with $n = 0$, whereas an infinitesimal increase will cause attraction toward the stable solution with $n = 2$.

In mathematics, *chaos* is the field of study of dynamical systems that are governed by nonlinear deterministic equations whose solutions are very sensitive to the initial conditions. This very high sensitivity to the initial conditions of the system is referred to as the *butterfly effect*: the flap of the wings of a butterfly in South America can cause a hurricane hitting the coast of North America several weeks later [324]. Because *chaotic systems* are very sensitive to the initial conditions, even very small differences in the values of those conditions could lead to widely diverging predictions for the state of the physical system at a later time. As a result, the chaotic dynamics would appear to be unpredictable even for future moments of time that are not too distant from the present.

Examples of chaotic systems include global weather, fluid turbulence, the double pendulum, the stock market, etc. The presence of instabilities with sensitive dependence on the initial conditions alone, however, is not what makes these systems chaotic. For chaos, the time evolution of the system through phase space should appear to be quite random. In particular, the system should have *dense periodic orbits* in phase space, and the dynamics should be *topologically mixing*, that is, any given region of the phase space should eventually overlap with any other given region. Deterministic continuous systems need at least a phase space of dimension 3 in order to support chaos. In 1963, Lorenz proposed a parametric sys-

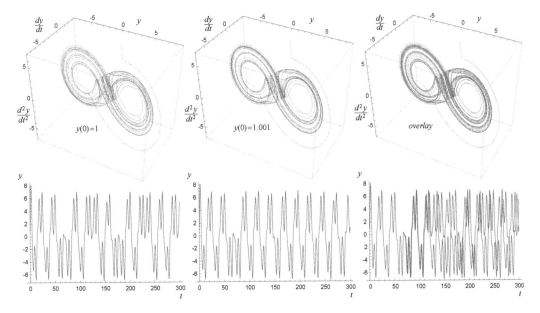

Figure 5.12   Chaotic behavior of a jerky Lorenz-like system governed by Eq. (5.5) with fixed initial conditions $\frac{d}{dt}y(0) = 2$ and $\frac{d^2}{dt^2}y(0) = 4$, but slightly different $y(0)$ that is $y(0) = 1$ (blue line) or $y(0) = 1.001$ (red line). The behavior of the system for these two cases appears to be similar for time $t < 120$, but widely diverges for $t > 120$.

tem of equations describing the dynamics of a physical system in a 3-dimensional phase space that exhibited chaos for a certain range of the parameters [323]. Here, we will provide an example of a *jerky Lorenz-like system* described by a single third order differential equation [448]

$$\frac{d^3}{dt^3}y(t) = -\frac{d^2}{dt^2}y(t) - \frac{d}{dt}y(t) + 6\arctan[y(t)] - 2y(t) \tag{5.5}$$

where t is time, $y(t)$ is the position, $\frac{d}{dt}y(t)$ is the velocity, $\frac{d^2}{dt^2}y(t)$ is the acceleration, and $\frac{d^3}{dt^3}y(t)$ is the jerk of the system. The initial values of $y(0)$, $\frac{d}{dt}y(0)$ and $\frac{d^2}{dt^2}y(0)$ at time $t = 0$ completely determine the dynamics of the system inside the 3-dimensional phase space. The extreme sensitivity of the chaotic system to a small variation of a single initial parameter is illustrated in Figure 5.12 where we have plotted the dynamical trajectories for time $t \in [0, 300]$ in two cases with $y(0) = 1$ or $y(0) = 1.001$, for fixed $\frac{d}{dt}y(0) = 2$ and $\frac{d^2}{dt^2}y(0) = 4$. It can be seen that the behavior of the system for the two cases widely diverges for $t > 120$.

In practice, usually the short term prediction of a system with chaotic dynamics is satisfactory, whereas the long term prediction utterly fails (Fig. 5.12). One possible cause of the long term unpredictability is the *measurement error of the initial state* of the system due to technical limitations of the measurement instruments used. Another possible cause of the long term unpredictability is the *accumulation of very small rounding errors in numerical calculations* used to predict the future state of the chaotic system. Because both sources of long term unpredictability are related to our limited ability to handle irrational numbers, unpredictability is not an inherent feature of the deterministic chaotic system. Thus, our inability to predict the long term behavior of chaotic systems is irrelevant for the existence of free will. Determinism forbids free will regardless of how good our ability to work with irrational numbers is.

Example 5.11. *(Physical laws are always enforced with perfect precision and never fail) All fundamental physical laws in scientific theories are postulated to hold for all possible physical situations at all times. This implies that all real objects in the universe behave exactly as the mathematical equations of physical laws predict, even if the actual physical values for various physical quantities are given by irrational numbers that can only be expressed as infinite aperiodic decimal expansions. In other words, the physical universe appears to have this astonishing mathematical ability to calculate with irrational numbers without ever making rounding errors and without approximating the solutions predicted by the physical laws.*

Some physicists, including Stephen Hawking, believe that determinism is a part of the grand design of the universe, and accept that free will is an illusion. But then they redefine and corrupt the term *free will* to denote nothing but *our inability to use the deterministic physical laws to predict human behavior.*

> It is hard to imagine how free will can operate if our behavior is determined by physical law, so it seems that we are no more than biological machines and that free will is just an illusion. While conceding that human behavior is indeed determined by the laws of nature, it also seems reasonable to conclude that the outcome is determined in such a complicated way and with so many variables as to make it impossible in practice to predict. [...] In the case of people, since we cannot solve the equations that determine our behavior, we use the effective theory that people have free will. [242]

Hawking admits that by his redefinition of free will, we should attribute free will to all complex systems whose future dynamics is beyond our capacity to predict. Thus, the weather, the volcanoes or the stock market would have free will as long as we lack sufficiently powerful measuring devices and computers to help us predict those phenomena. Entertainingly, once we have the technology to do successful predictions, the weather, the stock market or the volcanoes would lose their free will.

> We would therefore have to say that any complex being has free will–not as a fundamental feature, but as an effective theory, an admission of our inability to do the calculations that would enable us to predict its actions. [242]

It should be emphasized that redefining words does not and cannot solve theoretical problems [395, p. 28]. Free will is our ability to make genuine choices among two or more available alternative courses of action (Definition 1.3) and the free will problem is to explain whether and how physical laws grant us such an ability. If someone wants to change the meaning of the phrase *free will*, we can bring back the original free will problem by simply formulating it in terms of our ability to make genuine choices and dispensing with any reference to *freedom* or *will*. Thus, the deficiency of Hawking's redefinition can be pinpointed: Either we are able to make genuine choices or we are not, and our inability to solve differential equations and handle irrational numbers has nothing to do with our ability to make such choices.

5.4.4   Belief in free will and human conduct

Determinism implies that free will is impossible. Yet, we feel that we are able to make choices and it is up to us to decide what to do in our lives. Thus, determinists cannot simply claim that there is no free will; rather they have to face the challenge of explaining how we can be so terribly deluded in thinking that we possess something that does not exist. Children can be fooled into believing that there are fairies in the forest, but once they are told that fairies do not exist, the false belief can be eradicated. In contrast, our belief in free will is much stronger than the belief in fairies and is reinforced continuously by our daily exercise of free will. We can immediately test whether we have free will by closing and opening our eyes or moving one of our limbs up and down. Such a test will readily convince us that we are freely controlling our actions. Thus, in order to explain the origin of such an apparently strong and persistent human delusion in the existence of free will, one would need extraordinary evolutionary evidence.

Because entities without free will are also subject to Darwinian evolution, the evolution of human consciousness itself does not necessarily imply that humans have to be endowed with free will. Advocates of the viewpoint that free will is an illusion [514], however, cannot easily argue that the *belief in free will* has been evolutionarily advantageous and selected for by natural selection. Accumulated experimental data suggests that the belief in free will enhances counterfactual thinking [6], improves error detection [414], promotes gratitude [330], reduces lying and cheating [504], decreases ethnic/racial prejudice [534] and contributes to prosocial, morally virtuous behavior [32]. The interpretation of these data is consistent with the common sense view that the belief in free will affects our behavior exactly because we do possess free will. In contrast, the determinism of classical physics that is used to justify the lack of free will implies that if conscious experiences and beliefs are considered products of brain function, then they are necessarily causally ineffective epiphenomena (Section 5.3.1). Therefore, it is self-contradictory and irrational to hold simultaneously the views that consciousness is a product of brain function, that free will is an illusion due to the deterministic laws of classical physics, and that the belief or disbelief in free will can have any behavioral effects.

Theorem 24. *If consciousness is a product of brain functioning and physical laws are deterministic, then the conscious belief in free will cannot have any behavioral effects. Any association between the belief in free will and behavioral effects can only be coincidental, not causal.*

Proof. We have already proved that if conscious experiences, desires and beliefs are products of brain functioning and determinism holds, then consciousness is an epiphenomenon (Theorem 22). Therefore, the *brain processes* that generate the belief in free will may affect behavior, whereas the *belief in free will* as a form of phenomenal conscious experience cannot affect behavior in any way whatsoever. As a result, any association between the conscious belief in free will and certain behavioral effects would be a mere coincidence, not a causal relation. □

In psychology, one cannot directly measure conscious experiences, desires or beliefs, but it is possible to obtain some form of quantification using standardized questionnaires and verbal reports from the individuals under study. The assumption is that the individuals are trustworthy while grading and honest when reporting their experiences. Combining psychological questionnaires with other physical measurements makes it possible for one to obtain correlations between reported mental states and physical events. The existence of correlations between mental and physical events, however, does not imply any causal relationship. In order to interpret some experimental data as supportive of a certain causal link, one always needs an underlying theory whose postulates are already accepted as being true. In particular, evidence that the belief in free will can have measurable effects upon human conduct [6, 317, 330, 414, 504, 534] can only be produced by a theory in which conscious experiences are able to exert causal effects upon the brain or the surrounding world. The causal efficacy of our belief in free will is not sufficient to disprove determinism as a fundamental physical principle, but necessarily implies that classical functional approaches to consciousness are empirically inadequate due to their epiphenomenal character. Consequently, reductive (type identity) approaches are left as the only alternative option in classical physics for constructing theories of consciousness in which our mental lives are able to change the course of human history. Unfortunately, reductionism brings us back to the problem of *unconscious brains* (Section 5.3.2).

5.5   Inner privacy of consciousness in classical physics

Classically, the disturbing effect of any measurement could be made arbitrarily small. Hence, it is possible for an observation not to change the state of the physical system that is observed. Because different classical observables commute with each other, they can be measured either simultaneously or one by one in any order and still give the same measurement results. Thus, at least in principle, one can measure all physical observables of a classical system and deduce the complete state of the system. The knowledge of the current state of the system could be then used to calculate the state of the system at any future moment of time.

Observability of the state of every classical physical system poses a serious challenge to any classical theory of consciousness because the *conscious experiences* are *unobservable*. We are all aware that conscious experiences are only accessible from a first-person perspective and that we have no access to somebody else's experiences. The first-person accessibility of consciousness implies that we can never be sure whether any other animal species is conscious, or even whether another human being is conscious. Classical physics, however, leads to a paradox, namely, if everything that describes the state of a physical system is observable (Theorem 6), how is it then possible that the conscious experiences are not observable?

Communicability of classical information poses another challenge to classical theories of consciousness. If everything about our conscious experiences were completely communicable, we would have been able to explain to others what it is like to experience the redness of a red rose or the smell of vanilla. Here, it is important to note that we do not question our ability to answer yes-or-no questions of the type "Are you consciously seeing a red rose?" or "Are you experiencing the smell of vanilla?" What is questioned is our ability to communicate the phenomenal nature of qualia, namely, what it is like to have certain experiences from the first-person perspective.

Example 5.12. *(Mary's room argument) Suppose that Mary, a brilliant neuroscientist, has been raised and lived all her life in a black and white room where she was surrounded only by black and white objects. She has been well educated by reading all important books in neurophysiology and knows everything about vision. For example, she knows how the incoming light of different wavelengths activates different red, green or blue cones in the retina, how this color information is converted into electric signals and delivered to the visual brain cortex where color is consciously experienced. Suppose that after having all the neurophysiological knowledge, Mary is allowed to see a red rose. By experiencing the redness of the red rose, did Mary learn something new? Intuitively it seems that by experiencing the first red object in her life, Mary acquires some kind of new knowledge that she did not have before. But how can this be if in the physical world there is only one kind of information, namely, classical information that can be perfectly well communicated and learned from books? Mary's room thought experiment was proposed in 1982 by Frank Jackson [274], and is now widely referred to as the* knowledge argument, *since it highlights the fact that there seem to be two kinds of knowledge, objective knowledge that can be communicated to others and subjective knowledge that cannot [275, 328].*

Classical approaches to consciousness have tried to address the Mary's room argument by arguing that *being* in a certain mind/brain state is not the same as *knowing* what the mind/brain state is. Mary has to *be* in a brain state of experiencing the redness of the rose in order to experience her first red object. Before Mary experiences the redness of the red rose she knew in what brain state she had to be in, but she never was in that brain state while living in the black and white room. Such a statement is valid, but it completely misses the point of the knowledge argument, which is about communicability of knowledge of what it is like to experience something. After Mary experiences her first red object, she should form a memory trace in her brain that would inform her of what it is like to see red colored objects. If that memory of the red color were classical information, it would have been convertible into a string of bits, 0s and 1s, and could have been written in the books that Mary studied. Therefore, the main claim of the knowledge argument is that memories of phenomenal experiences could not be converted into communicable classical information. Otherwise, by simply sending strings of bits, 0s and 1s, we would have been capable of making people born blind know what seeing is and people born deaf know what hearing is.

Accepting that classical physics is the correct description of the physical world implies that the information about all existing things is communicable. Consequently, one is forced to conclude that subjective, first-person conscious experiences are impossible. Indeed, the philosopher Daniel Dennett has argued that we are deluded about having conscious experiences and that we as conscious minds do not exist [122]. Entertaining as it may be, it would be irrational to believe the arguments produced by a non-existing mind or a mind deluded about its own existence. Furthermore, our memories of having had conscious experiences should not exist too because we do not exist, but then how are we able to remember things? Since classical physics leads to paradoxes, it is rational to discard classical physics and study other empirically corroborated physical theories that support the kinds of information irreducible to classical information. Indeed, we will show in Section 6.6 that quantum physics and quantum information theory are able to explain the origin of the inner privacy of our conscious lives.

5.6   Mind–brain relationship in classical physics

We exist as *conscious minds* that are composed of *conscious experiences*. Introspectively, we do not perceive ourselves as being built of atoms, molecules or neurons. When we decide to close our eyes or to move one of our limbs, we do not have any idea which neuron in our brain is firing to deliver the appropriate electric signal to our muscles. Yet, if we undergo open skull neurosurgery, what can be observed from a third-person point of view is our *brain* that is a pinkish-gray, walnut-shaped, jelly-like substance. If a piece of the brain is observed under a microscope it could be further established that the brain is composed of *neurons* (Fig. 1.8). At the microscopic scale, the size of various neuronal compartments is in the order of a micrometer (10^{-6} m). Further zooming in reveals that neurons are composed of *biomolecules* (such as proteins, lipids or nucleic acids) whose size is of the order of a nanometer (10^{-9} m). Biomolecules are composed of *atoms* of chemical elements (Fig. 3.2) that are built from elementary physical particles such as the *electron* whose classical size is of the order of a femtometer (10^{-15} m). Reconciling the subjective, first-person picture of our conscious experiences with the third-person objective picture of our brain requires an explanation of how exactly the mind and the brain relate to each other.

5.6.1 Idealism

Idealism postulates that only minds exist, whereas the material objects, such as brains, are just illusions. Introspection shows unambiguously that we as conscious minds are composed from experiences or ideas. The third-person view of our material brain, however, appears to be devoid of any mental properties. Since the third-person view of our material brain is different from the first-person view of our conscious mind, it follows that the mind or at least some aspects of the mind are not observable by external observers. Idealists go further than that, however, and consider third-person observations as misleading, unreliable, and altogether false descriptions of what our minds are. Thus, idealism is a theory of mind, but it is not a physical theory because it bans all knowledge about the physical world that is concisely compressed in classical mechanics (Section 3.12), electrodynamics (Section 3.14) or Einstein's relativity (Section 3.18). After rejecting the physical laws, typically *idealism* is complemented with a divine creator who ensures the interaction between the existing minds and produces the apparent consistency of physical laws in our experiences. The divine creator can also violate the physical laws in the form of miracles and can subject us to misfortunes in order to test our moral strength.

5.6.2 Eliminativism

Eliminativism, or *eliminative materialism*, postulates that the brain is built from classical physical particles that obey deterministic physical laws, whereas conscious minds do not exist and are just illusions [122, 126]. Thus, eliminativism is a physical theory, but it is not a theory of mind since it bans the existence of conscious experiences. Denying that we exist or denying that we have conscious experiences is a self-defeating viewpoint. In Chapter 3, we have seen that conscious experiences do not enter any of the basic axioms of classical physics. Thus, uncritical application of classical physics to consciousness could indeed lead to the paradoxical conclusion that we are hallucinating having experiences while we have none. However, arriving at a paradox or contradiction should logically lead to the rejection of the false premise, which in this particular case is the validity of the laws of classical physics and not the existence of consciousness. Thus, if we are rational, we should keep consciousness and discard classical physics. Nonetheless, some researchers who are not familiar with how the proof by contradiction works, choose to accept the resulting contradiction as if it reflected a true fact about our consciousness, as exemplified by the works of Daniel Dennett [122, 126]. Remarkably, rather than being embarrassed by writing nonsense, Dennett attempts to shift the burden of proof to those who may criticize him:

> if you believe you have conscious experiences that you don't in fact have—then it is your beliefs that we need to explain, not the nonexistent experiences! [126, p. 45]

5.6.3 Functionalism

Functionalism postulates that the brain is a computing hardware, whereas the conscious mind is a functional product or software run by the brain. If the mind is a functional product of the brain, however, it follows that the brain is dispensable and conscious experiences should also be supported by personal computers or other silicon chip devices. One of the most important problems of functionalism is that functions can be constructed using the dynamics of entities that already possess minds. Thus, running a *function* that generates a mind, using physical entities that already possess minds, will lead to highly paradoxical occurrence of *minds within minds.*

Example 5.13. *(Ned Block's China brain argument) Suppose that the whole nation of China is asked to simulate the information processing of a brain under incoming pain stimuli. Let each Chinese person act as a single neuron and communicate with other people (that also simulate neurons) using mobile phones connected through satellites. Assuming that the China brain perfectly simulates the information processing of pain stimuli in a real brain [52, p. 279], it is not easy to answer who is experiencing the pain. There is no global China mind who feels the pain, therefore functionalism is false.*

Dennett has insisted that the China brain will produce real pain experiences attributable to an emergent China mind, even though none of the Chinese individuals participating in the simulation experiences pain. For Dennett, the problem occurs because we are unable to imagine how the emergent mind occurs [122, p. 438]. We would like to stress, however, that imagining that a mind could emerge is less of a problem compared to the possibility of accepting the existence of minds within minds. For example, it is strange to think that the Chinese people taking part in the simulation of the China brain are blissfully unaware that they are causing pain sensation in an emergent China mind. And it would be even stranger to think that while you are drinking your coffee you might be inadvertently causing excruciating pain suffered by an emergent cosmic mind.

Example 5.14. *(Searle's Chinese room argument) Suppose that a person who does not understand the Chinese language is locked in a room that contains a library of books that provide a complete set of rules for matching questions written in Chinese to answers also written in Chinese. If the person in the Chinese room is given questions written outside of the room, he will be able to reply to these questions based on the rules found in the books. Apparently, however, even though the person may be replying correctly to the questions, he will not understand Chinese and will not know what his replies mean, by simply following the given set of rules [430]. Executing a certain set of rules does not lead to understanding, therefore functionalism is false.*

Dennett again claims that the problem occurs because we are unable to imagine how the whole Chinese room, including the person and the books, could have its own emergent mind [122, p. 438]. The essence of the argument, however, is that the existence of minds within minds is a ridiculous prediction contradicting directly our experiences. If minds within minds were possible, then it would not have been clear whose free will is acting in the world and the concept of moral responsibility would have been inconsistent. Also, if a global emergent mind were operating and we were part of it, that global mind would have had a direct control over us despite that we feel in control of our own actions. But if we had to dismiss our own experiences about ourselves as untrustworthy, then the whole prospect of doing natural science would have been hopeless.

5.6.4   Reductionism

Reductionism postulates that the conscious mind is identical to (a part of) the brain. Thus, the relationship between the mind and the brain turns out to be the *identity* relation. Unfortunately, the mind–brain identity approach (or any variant of it) faces the apparently insurmountable problem of explaining how *unconscious brains* such as *dead brains* or *anesthetized brains* are possible (Section 5.3.2).

5.6.5   Dualism

Dualism postulates that both the brain and the mind exist but they are made of different substances. Thus, the brain is made of matter, whereas the mind is composed of conscious experiences. The main problem of dualism is to explain whether and how the mind and the brain are able to interact with each other.

Interactionism is the common sense view asserting that material events can cause conscious experiences and those experiences can subsequently cause material events. For example, a stone falling on someone's foot (material event) can cause a pain sensation (mental event) that can further cause yelling and screaming (material event) that can be heard by other people (mental event), and so on. Thus, *interactionism* is the natural form of dualism that we would expect based on our past experiences (Fig. 5.13). In classical physics, however, any interaction between substances has to be modeled by physical forces that can be calculated using mathematical equations. Unfortunately, it is difficult to consistently expand the classical world with additional mental physical quantities. For example, the mental states are unobservable, the phenomenal nature of conscious experiences is not communicable, and the conscious minds possess free will, whereas the brain states are observable, communicable, and lacking free will as they evolve deterministically. In essence, a physical world built up only from classical information (Section 3.19) cannot support *consciousness* and any form of interactionism.

Psycho-physical parallelism is a pseudoscientific attempt to rescue dualism by asserting that the brain states and the mental states do not need to interact at all, since they are set in pre-established harmony by a divine creator (Fig. 5.13).

Epiphenomenalism is another unsuccessful form of dualism asserting that the brain states produce mental states, but the mental states are unable to causally affect the brain states or the material world (Fig. 5.13). In Section 5.3.1, we have shown that functional approaches to consciousness based on deterministic physical laws imply epiphenomenalism, and we have argued that epiphenomenalism is incompatible with evolutionary theories of the human mind.

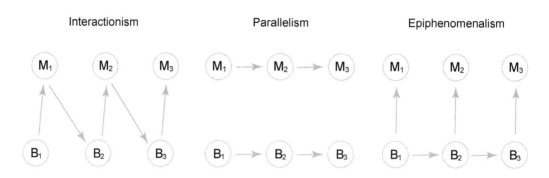

Figure 5.13   Varieties of interactions between the mind and the brain in different types of dualism. Interactionism asserts that the brain states produce mental states, which may in turn select the future brain states. Psycho-physical parallelism asserts that the brain states and the mental states do not interact with each other but are set in pre-established harmony by a divine creator. Epiphenomenalism asserts that the brain states generate the mental states, which do not affect causally the brain states. B_1, B_2, B_3 represent brain states; M_1, M_2, M_3 represent mental states; time flows horizontally; arrows indicate the direction of the causal influence.

5.6.6   Panpsychism

Panpsychism postulates that all physical particles possess primordial mental features or *psyche*. Thus, the physical world is composed of conscious minds. In contrast to idealism, however, the brain and the classical physical laws are not considered illusions imposed upon the mind by a divine creator. Rather, the physical laws govern the physical interaction between existing minds and allow natural evolution of complex minds from simpler minds. William James (1842–1910) was among the first scientists to argue that if new qualities such as *sentience* cannot emerge magically out of quantity or any form of motion, then the theory of evolution demands the existence of sentient "mind dust" in order to assemble sentient minds [279, p. 146]. Thus, if the natural evolution is to work smoothly, consciousness in some shape must have been present at the very origin of things [279, p. 149]. Panpsychism makes it aesthetically pleasing to look at a natural scenery with mountains, trees, wild animals in the open spaces, fish in the rivers, birds in the sky and realize that all that reality is built up from the same mental ingredients, just differently organized. Since the mental ingredients are already there, they do not emerge out of the organization; rather the content of the assembled mental states depends on the organization. Unfortunately, the nature of classical information (Section 3.19) makes it impossible to incorporate additional mental features in a world governed by the deterministic laws of classical physics. In particular, the axioms of classical mechanics (Section 3.12) would be violated by the existence of conscious experiences that are not observable or by the ability of conscious minds to use their free will in order to make choices among two or more alternative future courses of action.

5.7   The hard problem of consciousness in classical physics

David Chalmers argued that the *hard problem of consciousness* is to provide an explanation for the fact that we are sentient beings who have experiences rather than being just *mindless machines* who react the way we do but having no experiences at all [83, 84]. To answer the question of why we have conscious experiences, one cannot simply resort to classical physics because the classical physical laws have nothing to say on whether a given material body has any conscious experiences or not. Consequently, classical physics cannot tell us where the boundary of one mind is and where another mind starts (Section 5.1), what binds conscious experiences within a single mind (Section 5.2), and why we are different from mindless machines. The construction of a classical theory of consciousness through addition of new axioms that define what consciousness is and how it relates to the physical world (Section 5.6) is also problematic due to the severe restrictions imposed by the theorems that pertain to classical information (Section 3.19).

Evolutionary biologists assert that we have conscious experiences, because consciousness is a useful thing to have and organisms endowed with consciousness have an edge in the struggle for survival. The utility of something, however, does not imply that we need to necessarily have it. For example, it might be very useful to be able to fly but we cannot fly. Evolution does not and cannot aim for certain end products because it operates by random mutations and natural selection of the fittest organisms. Yet, consciousness could not have been an accidental product of the brain that has been subsequently selected for by natural selection, because the deterministic laws of classical physics together with the assumption that consciousness is a functional product of the brain imply epiphenomenal conscious experiences that are not causally effective and not subject to natural selection (Theorem 22). Therefore, if consciousness is to evolve, it should have already been present in some simple primordial form in some (or maybe all) physical particles at the moment of their creation due to the fundamental physical laws of the universe and this primordial consciousness should have been causally effective. Attributing mental features to all physical particles leads to *panpsychism*.

The formulation of the hard problem of consciousness is comprehensible only within the framework of classical functional approaches to consciousness, and it seems that the mindless machine argument is tailored to disprove functionalism. The two important ingredients that cause the hard problem are the determinism of classical physics (Section 3.2) and the inner privacy of consciousness (Sections 1.5 and 5.5). All classical physical quantities of the brain and the surrounding world are observable. The initial values of these physical observables completely determine the brain dynamics for all times through deterministic equations that concisely express the physical laws. Because consciousness is subjective, private and unobservable, it has to be added by some new physical law as an additional product of the brain. In Theorem 22, we have proved that such an additional product of the brain will be epiphenomenal. If consciousness is an epiphenomenon, however, it will be physically useless and the hard problem to explain why we possess consciousness at all will be exacerbated. A classical world that has identical physical laws but no law that endows us with inner consciousness would be a mindless machine world that behaves exactly as our world. In particular, our mindless machine twins would evolve exactly as successfully in their world as we do in our world. Thus, the hard problem can be understood to originate from the lack of distinction between a deterministic world with inner consciousness and a deterministic mindless machine world without any consciousness. Indeed, if consciousness is epiphenomenal you can add it to or remove it from a physical world without any harm. The lack of harm, however, implies that there can be no good explanation of why we possess consciousness at all.

Classical reductionism avoids the hard problem by postulating that either the brain and the mind are identical or that certain physical observables of the brain are identical to the mind. If the mind is identical to a physical quantity, then it will be causally effective in the physical world and it would not be possible for one to construct a mindless machine world by leaving out the physical laws related to consciousness. For example, consider an electromagnetic theory of consciousness in which the mind is identical with a non-zero electromagnetic field. Duplication of all classical physical laws without Maxwell's laws of electrodynamics (Section 3.14) could produce a world without consciousness, but this world will be quite different from our world because it will lack electromagnetic phenomena. Thus, the causal efficacy of the mind (due to the identification of the mind with physical phenomena) avoids the hard problem by precluding the construction a mindless machine world whose dynamics mirrors the dynamics of our world. The price that has to be paid by classical reductionism, however, is also too high. Identifying the unobservable, subjective conscious experiences with any number of objective physical observables makes the theory of consciousness both logically inconsistent and empirically inadequate.

The fundamental principles of classical physics underlying the properties of classical information (Section 3.19) are the culprit to blame for all paradoxes related to consciousness. Fortunately, in the beginning of the 20th century it became clear that classical physics is fundamentally inadequate to explain the appearance of the surrounding physical world, including simple phenomena such as the working of the light bulb. Next, we will show how quantum physics provides room for construction of a paradox-free theory of consciousness.

CHAPTER 6

Consciousness in quantum physics

6.1 Axioms of quantum information theory of consciousness

Applying classical or quantum information theorems to consciousness cannot be done without having explicit rules stating *what* consciousness is and *how* consciousness enters into the physical description of the world. The theoretical construction of new axioms usually employs the Lakatos method of *proofs and refutations* that goes from the problems at hand toward crafting axioms that resolve those problems without any unwanted byproducts [305]. Once the axioms of the theory are crafted and tested for hidden logical inconsistencies or generation of unwanted byproducts, however, the theory is presented and its explanatory power is assessed in the opposite way that goes from the axioms toward proving theorems that describe *how* the physical world is and explain *why* the physical world behaves the way it does.

By the admittedly high standards of successful laws of physics, at present we do not have any remotely satisfactory lawlike description of consciousness [292]. The following three axioms attempt to remedy this situation by explicitly constructing a physical theory of consciousness that allows one to apply quantum information theorems to consciousness, thereby addressing all of the main problems presented in Chapter 1.

Axiom 6.1.1. *To each individual conscious mind corresponds a single non-factorizable (quantum entangled) state vector $|\Psi\rangle$ that resides in a subspace of the Hilbert space of the universe $\mathcal{H}_U$, and to each non-factorizable state vector corresponds a single mind.*

Axiom 6.1.2. *Every factorizable state vector $|\Psi\rangle = |\psi_1\rangle \otimes |\psi_2\rangle \otimes \ldots \otimes |\psi_k\rangle$ that resides in a subspace of the Hilbert space of the universe $\mathcal{H}_U$ represents a collection of k minds, where the individual (non-factorizable) minds are given by $|\psi_1\rangle, |\psi_2\rangle, \ldots, |\psi_k\rangle$.*

Axiom 6.1.3. *For composite non-factorizable (quantum entangled) quantum systems with a state vector $|\Psi'\rangle$ there is an energy threshold $\mathcal{E}$ at which objective reduction and disentanglement of the individual subsystems could occur in the form $|\Psi'\rangle \to |\psi_1'\rangle \otimes |\psi_2'\rangle \otimes \ldots \otimes |\psi_n'\rangle$, with probability for the actualized outcome given by the Born rule $|\langle\Psi'\|\psi_1'\rangle \otimes |\psi_2'\rangle \otimes \ldots \otimes |\psi_n'\rangle|^2$. The energy threshold $\mathcal{E}$ is a free parameter to be determined empirically, but its maximal possible value is bounded by $\mathcal{E} \leq \frac{\hbar}{t_P}$, where t_P is the Planck time.*

The axiomatization introduced here contains only three new axioms, yet it is sufficient to provide a complete theory of consciousness because consciousness is not introduced as a new mental field but identified with an already well-defined quantum physical entity that is the state vector $|\psi\rangle$. Exactly because the quantum information theory of consciousness is a form of an identity theory, all axioms

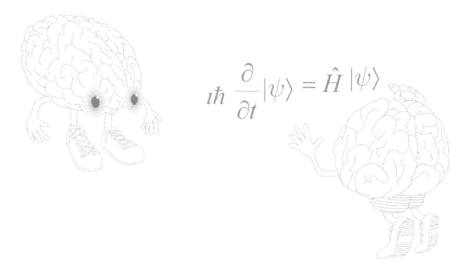

Figure 6.1   Identifying conscious states with non-factorizable quantum state vectors $|\psi\rangle$ that obey the Schrödinger equation $i\hbar\frac{\partial}{\partial t}|\psi\rangle = \hat{H}|\psi\rangle$ implies that all theorems that hold for quantum information should hold for consciousness too. Thus, the quantum information theory of consciousness is a form of an identity theory, but one that is irreducible to the classical mind–brain identity theory.

that hold for quantum systems (Section 4.15) and all theorems that hold for quantum information (Section 4.20) hold for consciousness too (Fig. 6.1). Consequently, quantum information theory can be used to address classical paradoxes related to consciousness [207]. For example, one can easily show that the piecemeal replacement of brain neurons with silicon chips (as described in Example 5.7) cannot keep the conscious experiences intact. From Axiom 6.1.1 it follows that in order to keep the consciousness unchanged, one needs to keep the quantum wave function ψ of the brain unchanged. Since different chemical atoms have distinctive quantum wave functions, however, replacing real neurons containing carbon based organic molecules with functionally equivalent inorganic silicon chips will substitute carbon C wave functions for silicon Si wave functions. Hence, conscious experiences cannot remain the same after the chip replacement is carried out.

6.2   Physical boundary of consciousness in quantum physics

Consciousness exists in the universe. Because physics is supposed to study all existing things, it follows that consciousness has to be identified with a well-defined mathematical entity inside the physical theory [207]. Quantum physics provides a good candidate for such a mathematical entity, namely, the quantum state vector $|\psi\rangle$ in the Hilbert space $\mathcal{H}$. Indeed, the quantum information contained by the state vector $|\psi\rangle$ is not observable (Theorem 15) and cannot be converted into classical bits of information (Theorem 20). Furthermore, the fabric of the physical state vector $|\psi\rangle$ is made of complex-valued quantum probability amplitudes

that determine the probabilities for possible future courses of action whose actu-
alization allows for choices to be made. Thus, the state vector $|\psi\rangle$ seems to possess
many of the properties that a mental state should have. To precisely identify what
a conscious mind is, however, Axiom 6.1.1 imposes several important restrictions
specifically designed to avoid the existence of minds within minds.

There are two types of state vectors: entangled state vectors that cannot be expressed as a tensor product and non-entangled state vectors that can be expressed
as a tensor product. In 1935, Albert Einstein, Boris Podolsky (1896–1966) and
Nathan Rosen (1909–1995) showed that not every physical system has its own
state vector [158]. For example, a singlet state composed of two entangled spin-$\frac{1}{2}$
particles can be expressed by a state vector

$$|\psi_{AB}\rangle = \frac{1}{\sqrt{2}}(|\uparrow_A\rangle|\downarrow_B\rangle - |\downarrow_A\rangle|\uparrow_B\rangle) \tag{6.1}$$

but neither of the component particles A or B has its own state vector. Only if the
composite state of particles A and B is a tensor product state

$$|\psi_{AB}\rangle = |\psi_A\rangle \otimes |\psi_B\rangle \tag{6.2}$$

will the component particles A and B have their own state vectors. In the tensor
product state $|\psi_{AB}\rangle = |\psi_A\rangle \otimes |\psi_B\rangle$, however, it is not only that the particles A and B
have state vectors $|\psi_A\rangle$ in $\mathcal{H}_A$ and $|\psi_B\rangle$ in $\mathcal{H}_B$, but the composite system also has a
state vector $|\psi_{AB}\rangle$ in $\mathcal{H}_{AB} = \mathcal{H}_A \otimes \mathcal{H}_B$.

If the state vector $|\psi\rangle$ is to be identified with the mind of the system, and we
want to avoid the *minds within minds* problem (Examples 5.13 and 5.14), we need
to specify that only non-factorizable (entangled) state vectors correspond to individual minds. To show that factorizable state vectors should not have minds,
let us suppose the opposite and affirm that all state vectors $|\psi\rangle$ correspond to
conscious minds. Then, any state vector $|\psi_{AB}\rangle$ that is expressible as a tensor
product $|\psi_{AB}\rangle = |\psi_A\rangle \otimes |\psi_B\rangle$ would also have a mind. However, since any collec-
tion of n non-interacting quantum systems, each of which has an independent
state vector $|\psi_i\rangle$, forms a composite system that has a tensor product state vector
$|\Psi_{1,2,...,n}\rangle = |\psi_1\rangle \otimes |\psi_2\rangle \otimes ... \otimes |\psi_n\rangle$, it would be the case that every *collection of minds*
forms another composite mind. Since this is exactly the minds within minds sce-
nario that we wanted to avoid, we have to reject the assumption that factorizable
states correspond to single minds. Instead, if a state vector is factorizable and can
be expressed as a tensor product, it has to correspond to a collection of minds.
Thus, by decomposing the quantum state vector $|\Psi\rangle$ of the universe into a tensor
product

$$|\Psi\rangle = |\psi_1\rangle \otimes |\psi_2\rangle \otimes ... \otimes |\psi_k\rangle \tag{6.3}$$

we can have a theoretical rule that explicitly specifies the boundaries of in-
dividual minds $|\psi_i\rangle$ ($i = 1, 2, ..., k$) inside the Hilbert space of the universe
$\mathcal{H}_U = \mathcal{H}_1 \otimes \mathcal{H}_2 \otimes ... \otimes \mathcal{H}_k$. Noteworthy, there will be no global cosmic mind $|\Psi\rangle$ if
the quantum state of the universe can always be written as a tensor product state.

Axiom 6.1.1 solves the physical boundary problem of consciousness in the Hilbert space $\mathcal{H}$ (Section 4.6), rather than the classical space and time of classical mechanics (Section 3.12) or the unified spacetime of Einstein's theory of relativity (Section 3.18). This is not surprising, because in Section 5.1 we have already argued that the physical boundary problem of consciousness cannot be solved within classical physics. Indeed, the classical spacetime is built up from observable physical quantities, whereas the Hilbert space $\mathcal{H}$ is built up from quantum probability amplitudes that are similar in character to the subjective, unobservable conscious experiences. Thus, the proposed quantum theory of consciousness interprets the Hilbert space of the universe $\mathcal{H}_U$ as a mental space that at any time supports a large collection of individual minds $|\psi_1\rangle \otimes |\psi_2\rangle \otimes ... \otimes |\psi_k\rangle$. Even though, such a theory may look like an instantiation of Berkeley's idealism, it actually leads to interactionism as we will show in Section 6.7.

6.3   Binding of consciousness in quantum physics

Quantum entangled states such as the singlet state given by Eq. (6.1) exhibit nonlocal features that can be experimentally tested using the Bell inequality given in Eq. (4.230). Classical physical theories are local (Theorem 9) and predict that the Bell inequality can never be violated, whereas quantum theory is nonlocal (Theorem 17) and predicts that the Bell inequality can be violated by measurements performed upon entangled states such as the singlet state. Indeed, let the entangled particles in the singlet state be created in a laboratory and then separated a huge distance away. If after the separation, the spins of both particles A and B are measured along any arbitrary direction z, quantum mechanics correctly predicts that the outcomes of the spin measurements will necessarily be anticorrelated so that if the spin of particle A is up, the spin of particle B will be down, and vice versa. Remarkably, these spin outcomes could not have been predetermined at the instant of particle creation in the laboratory (Theorem 18) because if one decides to measure the second spin along another direction, such as x or y, the spins will not be anti-correlated anymore. Instead, the spins will be precisely correlated by another mathematical function given by the Born rule that depends on both measurement choices regardless of how far apart the measurements are performed [39]. The nonlocal correlations due to quantum entanglement were called "spooky action at a distance" by Einstein, who never believed that experimental tests will agree with the quantum mechanical predictions. Rather he thought that quantum mechanics is an incomplete theory, which will be eventually completed by the addition of hidden "elements of reality" whose existence will restore the locality and determinism characteristic to classical physics. Yet, contrary to Einstein's expectations, all experimental tests performed so far have shown that the Bell inequality is indeed violated and that quantum entangled states exhibit nonlocal character [18, 19, 20, 196]. In other words, even if there are hidden "elements of reality" of which we are currently unaware, these will not bring back the lost classical world.

Quantum entangled states are non-factorizable. Conversely, tensor product quantum states are not entangled. As a result, Axiom 6.1.1 not only sets the boundaries of individual conscious minds, but together with Bell's theorem also implies that *quantum entanglement* provides the nonlocal mechanism for binding of conscious experiences into a single whole. Still, several important caveats need to be made.

First, quantum entanglement is not used here to explain paranormal phenomena such as telekinesis, clairvoyance or telepathy [206, 208], because there is no scientifically reproducible evidence that such paranormal phenomena do exist [3]. Furthermore, many of the stories describing telepathic experiences are vague, hazy, ambiguous, and appear to be the result of guesswork. In contrast, the physiological binding of consciousness is concrete, sharp and unambiguous. In reality, the amount of detail in our conscious perceptions is astounding, so that we are quite sure and very clear about what we hear, what we see and what we feel. Thus, if one uses quantum entanglement in order to explain the clear, rich in detail, binding of conscious experiences, it naturally follows that the same mechanism cannot and should not be used for explaining hazy, vague and irreproducible paranormal phenomena.

Second, quantum entanglement is not used to bind unconscious or subconscious processes. If we apply Axiom 6.1.1 to the medical and experimental evidence pointing to the brain cortex as the seat of consciousness (Section 1.1), we can further deduce that the quantum entanglement of physical particles within the brain cortex provides the physical mechanism that binds the conscious experiences into a single mental picture. But if the proposed quantum theory of consciousness is correct, it follows that the brain cortex could not be quantum entangled with any of the sensory organs, otherwise we would have been able to experience the visual or auditory stimuli immediately as they enter our eyes or ears. Here, we note that alternative proposals for a quantum theory of consciousness such as the Hameroff–Penrose Orch OR model [231, 232, 233] consider quantum entanglement only in relation to subconsciousness (see also Section 6.7.3). For example, the quantum superposition of tubulins inside brain microtubules has been claimed to produce a subconscious quantum state that has to undergo a process of orchestrated objective reduction (Orch OR) in order to generate a flash of conscious experience. Thus, the consciousness in the Hameroff–Penrose Orch OR model occurs at a frequency of 40 Hz as a series of brief discrete conscious events that are separated by longer periods of subconscious quantum activity [233]. In contrast, our Axiom 6.1.1 implies that the quantum activity of the brain cortex represents the continuous time evolution of a conscious mental state.

Third, quantum entanglement is a ubiquitous process resulting from the interaction between quantum physical systems. If for a composite quantum system with time independent Hamiltonian $\hat{H}$, at time $t = 0$ one starts with a collection of k quantum component subsystems in a tensor product state

$$|\psi(0)\rangle = |\psi_1(0)\rangle \otimes |\psi_2(0)\rangle \otimes \ldots \otimes |\psi_k(0)\rangle \tag{6.4}$$

this state will not last for long, and at time t it will unitarily evolve into a quantum entangled (non-factorizable) state $|\psi(t)\rangle$ given by

$$\hat{U}(t)|\psi(0)\rangle = e^{-\frac{i}{\hbar}\hat{H}t}|\psi(0)\rangle = |\psi(t)\rangle \tag{6.5}$$

Thus, the quantum entanglement between all k physical systems will ultimately generate a single conscious mind. The process will not stop there, however, if the collection of k physical systems is not isolated from its environment. Rather, the entanglement with the environment will create a larger and larger entangled physical system, until the whole universe becomes entangled into a single cosmic mind. Axiom 6.1.3 avoids such unbounded growth of the entangled physical system, by introducing a physical process that for sufficiently large quantum systems leads to *objective reduction* of the state vector $|\psi'\rangle$ into a tensor product collection of disentangled localized quantum states

$$|\psi'\rangle \rightarrow |\psi_{1'}\rangle \otimes |\psi_{2'}\rangle \otimes \dots \otimes |\psi_{k'}\rangle \tag{6.6}$$

The reason for requiring objective reduction in the quantum information theory of consciousness is quite different from the motivation to eliminate observable macroscopic superpositions, such as dead or alive Schrödinger's cats (Section 6.5.4). Namely, if no objective reduction occurs, each quantum system will interact and get entangled with its immediate environment. This enlarged quantum system composed of the system and its nearby environment will further interact and get entangled with another external shell of the physically present environment. Thus, quantum entanglement will diffuse out of individual quantum systems into bigger and bigger entangled bubbles of environment (Fig. 6.2), and at the end the whole universe will be in a single entangled state. If the state vector of the universe $|\Psi\rangle$ is not expressible as a tensor product state, however, according to Axiom 6.1.1 it would describe just a single cosmic mind. Because we utilized quantum entanglement to explain the binding of consciousness, but at the same time we know from experience that we are not anything like a cosmic mind [454], it should be concluded that there is a physical mechanism that leads to objective state vector reduction similarly to the proposals by Lajos Diósi [135, 136, 137, 138] and Roger Penrose [376, 377, 382]. The important feature of such an objective state vector reduction is that it produces tensor product states of disentangled quantum physical subsystems. Thus, the existence of objective disentanglement ensures the empirical adequacy of quantum information theory of consciousness and avoids the generation of super-minds or Über-minds, which are classically unavoidable (Section 5.2.4).

Both consciousness and quantum information do exist, but are unobservable. Quantum entanglement, quantum superpositions and quantum interference are utilized by quantum computers to achieve tasks that could not be achieved by classical computers [247]. Exactly because one needs to have a physically existing quantum computer in order to perform classically impossible tasks, it is incorrect to claim that the quantum information is not real. To exist means to be real. Therefore, one cannot object that we are identifying consciousness, which is real, with quantum information that is not real. Quantum information is real and its existence is what makes quantum computers outperform classical computers.

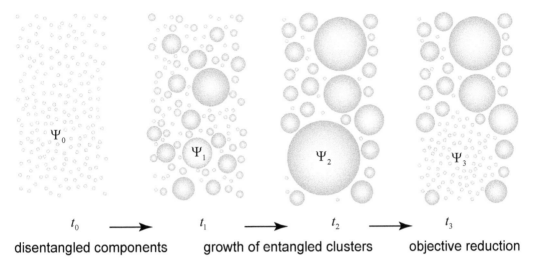

$$t_0 \longrightarrow t_1 \longrightarrow t_2 \longrightarrow t_3$$

disentangled components   growth of entangled clusters   objective reduction

Figure 6.2   Growth of quantum entangled clusters and objective reduction. At time t_0, the quantum state $|\Psi_0\rangle$ is a tensor product of disentangled component quantum subsystems represented as bubbles. As time goes on, quantum entanglement produces composite non-factorizable quantum states represented by larger bubbles. At time t_2, the composite entangled quantum state $|\Psi_2\rangle$ reaches a certain energy threshold $\mathcal{E}$ and undergoes objective reduction that at time t_3 leaves a disentangled collection of quantum components. Different composite quantum entangled systems reach the needed energy threshold at different times, hence undergo objective reductions asynchronously.

Here, it would be instructive to highlight the important differences between the proposed quantum information theory of consciousness and classical reductionism (Section 5.3.2) or classical panpsychism (Section 5.6.6). Both classical approaches face an insurmountable problem when trying to explain the existence of *unconscious brains* or *anesthetized brains*. In particular, once the brain is identified with the conscious mind, it becomes impossible to explain how the consciousness can be turned off. This problem is pronounced in classical panpsychism in which mental properties are always present in all physical particles. Conversely, once a physical system is postulated to be lacking conscious experiences, it is equally hard to explain how the consciousness can be turned on. The quantum information theory of consciousness addresses the problem of unconscious brains using the distinction between a *single mind* and a *collection of minds*.

Definition 6.1. *(Single mind)* Being a single mind *can be understood introspectively as the unified experience of yourself that you have.* Single minds are conscious.

Definition 6.2. *(Collection of minds)* Being within a collection of minds *can be understood through reflection of what it is like to be a participant in a conversation with another human, namely, when talking with a friend you may only guess what it is like to be in your friend's mind, but definitely you do not have direct access to your friend's conscious experiences.* Collections of minds are non-conscious because they do not possess a single mind.

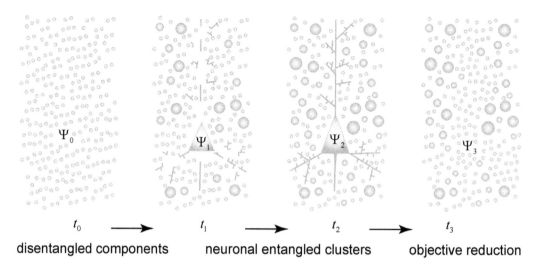

$$t_0 \longrightarrow t_1 \longrightarrow t_2 \longrightarrow t_3$$

disentangled components   neuronal entangled clusters   objective reduction

Figure 6.3   Organized quantum entanglement of neurons inside the brain cortex followed by objective reduction. At time t_0, the quantum state $|\Psi_0\rangle$ is a tensor product of disentangled component quantum subsystems represented as bubbles. As time goes on, neuronal electric activity organizes quantum entanglement, producing composite non-factorizable quantum states inside neurons. At time t_2, the composite neuronal entangled quantum state $|\Psi_2\rangle$ reaches a certain energy threshold $\mathcal{E}$ and undergoes objective reduction that at time t_3 leaves a disentangled collection of quantum components. For simplicity a single neuron is shown; however, the human consciousness would involve millions of quantum entangled cortical neurons.

The quantum information theory of consciousness works as follows: Axiom 6.1.1 endorses a form of panpsychism according to which every non-interacting elementary physical particle will have some simple, primordial conscious experience. Quantum interactions between physical particles will lead to entanglement and growth of quantum entangled clusters (Fig. 6.3). Quantum entanglement will bind together the conscious experiences of the component particles. Each quantum entangled cluster will have a more complicated state vector living in a larger subspace of the Hilbert space of the universe. Since each entangled state vector resides in a Hilbert subspace with a larger number of dimensions, it will correspond to a more sophisticated conscious experience that is within a larger mind. At a certain energy threshold $\mathcal{E}$, the conscious mind has grown so large and sophisticated in its conscious experience that it undergoes objective reduction (Axiom 6.1.3) and disintegrates into a collection of simpler minds (Axiom 6.1.2). The mind cycles of sophistication through growing entanglement followed by mind disintegration through objective reduction and disentanglement will occur at an extremely high frequency greater than 100 GHz [207]. The frequency of mind cycles will be inversely related to the quantum decoherence time resulting from the mechanism causing the objective reductions.

For inanimate objects such as rocks, the quantum information theory of consciousness will predict that the growth of entangled clusters and subsequent objective reductions will occur in a stochastic, disorganized and asynchronous fashion. Thus, the rock will be unconscious because it is a *collection of minds* that stochastically pop in and out of existence. The physical properties of the rock will be the statistical average of a zillion stochastic quantum processes. The same picture will also hold for unconscious brains or dead brains. In contrast, the hallmark feature of the human consciousness occurring in the living brain will be that it is *a single mind* sustained by repeating cycles of binding and disbinding of conscious experiences through entanglement and disentanglement, which are organized in such a way that the conscious "I" is revived and experienced again and again. Since each new conscious "I" is always slightly different from the previous one, we are able to learn and evolve in time, yet due to the memory traces left from the previous "I" the subjective continuity of the self is ensured. Structurally, the neuronal shape, and functionally, the electric firing of the neurons inside the brain cortical network, will orchestrate the growth of quantum entanglement to extend along the plasma membranes of the neuronal projections (dendrites and axons) in an organized string-like fashion (Fig. 6.3) that differs from the stochastic bubble-like mechanism (Fig. 6.2). In essence, the human conscious mind will be a product of organized quantum entanglement within the cortical network of neurons that store one's own memories, whereas the unconscious brain will be a product of disorganized stochastic quantum entanglements that occur in a bubble-like fashion and may involve parts of neurons together with the surrounding tissue including glial, endothelial and blood cells.

The nonlocality of quantum entanglement makes it an ideal physical mechanism for binding of conscious experiences together. Quantum entanglement occurs naturally if the Hamiltonian that enters into the Schrödinger equation contains a few interaction terms, which means that to entangle two quantum systems one does not need a huge informational channel for communication. This explains well why preserving the splenium, a small portion of the corpus callosum, is sufficient to bind conscious experiences between the two cerebral hemispheres, as discussed in Section 5.2.4. Curiously, for a long period of time in the 20th century, quantum entanglement was considered to be not an asset, but a shortcoming of the quantum theory. When Einstein realized that quantum entangled particles do not have their own individual state vectors but exhibit nonlocal correlations that are derived from the state vector of the composite system [158], he argued that quantum mechanics is *incomplete* and needs to be replaced by a better theory that will restore classical realism. Thus, rather than seeing the existence of quantum entanglement as a useful resource, Einstein considered it a mathematical flaw that has to be repaired in the complete physical theory of the universe. To discourage dreaming for the lost classical reality, in Chapter 5 we have purposefully exposed the classical origin of seven of the most important problems of consciousness and explained why the principles of classical physics undermine the construction of a physical theory of consciousness that is consistent with our introspective viewpoint of what we are.

The quantum panpsychism introduced by Axiom 6.1.1 does not endorse super-stitious beliefs such as animism, shamanism or spirit worship. Clearly, physical reality has the potential for generating conscious minds because we are made of physical particles and we need a constant supply of food and drink in order to stay alive and conscious [477, 478]. The question, therefore, is whether the conscious experiences emerge from a completely unconscious matter, or whether conscious experiences are already present in some primordial form in all matter. Classi-cal hard materialists, like Vladimir Lenin (1870–1924) [314], appear to embrace emergentism because they want a sharp demarcation between their worldview and superstition. The attempted demarcation fails, however, because emergentists at-tribute to the physical world the miraculous ability to pop into existence conscious minds where the physical laws predict none. Furthermore, classical emergentism is incompatible with the theory of natural evolution because the emergent con-sciousness is epiphenomenal and cannot be selected for by natural processes as it lacks survival value (Section 5.3.1). In contrast, quantum panpsychism does not require miracles and allows human consciousness to evolve smoothly from "mind dust" (quantum fields and particles) created at the very origin of the universe [279, p. 149].

6.4   Causal potency of consciousness in quantum physics

The causal potency problem originates from determinism: In classical physical theories there are fundamental constituents entering in the mathematical descrip-tion of the physical laws that govern deterministically the dynamics of physical systems. Since only these fundamental constituents are causally potent to affect the deterministic behavior of physical systems, consciousness has to appear as a fundamental constituent of the physical theory as well. If conscious experiences do not enter in the mathematical description of the physical laws, but are pro-duced by the deterministic dynamics of other fundamental physical constituents, then these conscious experiences are an epiphenomenon that cannot have any causal impact on the behavior of the physical systems. Thus, in order to solve the causal potency problem of consciousness one has to identify consciousness with some of the fundamental constituents of the physical theory, or else admit that consciousness is an epiphenomenon. Classically, the fundamental physical quan-tities are the charge q, mass m, length l, and time t. In general, the identification of mind states with brain states does not work because brain states under general anesthesia or dead brains do not produce conscious experiences. If consciousness cannot be identified with the charge, mass, length or time, then what remains is to be a functional product generated by those fundamental physical quantities. Any product that does not enter in the deterministic physical laws, however, is causally ineffective. Similarly, postulation of two kinds of brain states, conscious and un-conscious, would lead to epiphenomenal consciousness, unless the physical laws governing conscious and unconscious brain states are different.

In the quantum information theory of consciousness, one keeps the standard quantum physical laws that are valid for both conscious and non-conscious systems (Section 4.15). The causal potency problem is then elegantly avoided if the quantum wave function of non-factorizable quantum systems is identified with the conscious mental content experienced by the quantum system itself (Axiom 6.1.1). Because the quantum wave function is built up from quantum probability amplitudes ψ that causally determine the probabilities $|\psi|^2$ for various future events to occur, the identification of consciousness with the quantum wave function of the system creates a physical theory in which conscious experiences are intimately linked and causally connected with the possible future choices that can be made. Thus, the causal potency problem is addressed together with the free will problem (to be discussed in detail in Section 6.5).

6.4.1   On the nature of quantum states

Quantum physical states differ qualitatively from classical physical states. Rather than being built up from fundamental physical quantities that have predetermined values, quantum states are built up from complex-valued *quantum probability amplitudes* ψ, whose squared modulus $|\psi|^2$ determines the *quantum probability* for a given fundamental physical quantity to have a particular value. In Chapter 4, we have shown that the quantum physical states behave like *vectors* $|\psi\rangle$ and satisfy the axioms of a Hilbert space $\mathcal{H}$ (Section 4.6) as a consequence of the fact that each quantum state is a solution of the linear Schrödinger equation (Section 4.16). By requirement, all quantum states in $\mathcal{H}$ are normalized so that they have a unit length. The normalized vector $|\psi\rangle$ of any vector $|\tilde{\psi}\rangle$ is

$$|\psi\rangle = \frac{1}{|\tilde{\psi}|}|\tilde{\psi}\rangle = \frac{1}{\sqrt{\langle\tilde{\psi}|\tilde{\psi}\rangle}}|\tilde{\psi}\rangle \tag{6.7}$$

from which it can be directly confirmed that $|\psi\rangle$ has a unit length

$$\langle\psi|\psi\rangle = \frac{1}{|\tilde{\psi}|^2}\langle\tilde{\psi}|\tilde{\psi}\rangle = \frac{|\tilde{\psi}|^2}{|\tilde{\psi}|^2} = 1 \tag{6.8}$$

Definition 6.3. (*U(1) gauge symmetry*) *Multiplying the state vector $|\psi\rangle$ of any quantum system by a pure phase factor $e^{i\theta} \in \mathbb{C}$ does not change the quantum mechanical predictions for any physical observable $\hat{A}$, since the expectation values stay the same $\langle\psi|\hat{A}|\psi\rangle = \langle\psi|e^{-ix}\hat{A}\,e^{ix}|\psi\rangle$. Because the set of all pure phase factors forms a $U(1)$ group, the latter property is referred to as the $U(1)$ gauge symmetry of quantum mechanics.*

The vector nature of quantum states implies that every quantum state $|\psi\rangle$ can be written as a quantum superposition of other quantum states in an infinite number of different ways. Consider, for example, the decomposition of $|\psi\rangle$ into two other mutually orthogonal vectors $|\psi_1\rangle$ and $|\psi_2\rangle$ (Fig. 6.4). The vector $|\psi\rangle$ is expressible as a quantum superposition for an infinite number of values of the parameter α

$$|\psi\rangle = \cos\alpha|\psi_1\rangle + \sin\alpha|\psi_2\rangle \tag{6.9}$$

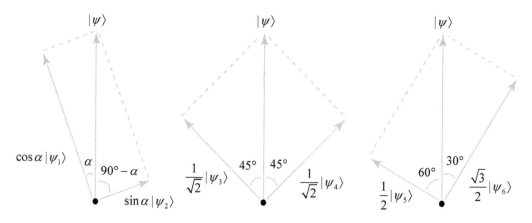

Figure 6.4   Every quantum state $|\psi\rangle$ is a vector in Hilbert space $\mathcal{H}$. Three different decompositions of the vector $|\psi\rangle$ into two other mutually orthogonal vectors illustrate the fact that any quantum state can be represented as a quantum superposition of other states in an infinite number of different ways.

Appreciating the vector nature of quantum physical states is important for the proper understanding of the process of objective reduction given by Axiom 6.1.3. A popular misconception is that during the collapse all quantum superpositions disappear. What actually occurs is a discontinuous quantum jump $|\psi_k\rangle \to |\psi_{k'}\rangle$ from a quantum state $|\psi_k\rangle$ to another quantum state $|\psi_{k'}\rangle$. Both $|\psi_k\rangle$ and $|\psi_{k'}\rangle$ can always be written as quantum superpositions by changing the representation basis. Only in the special case, where the states are written in the k' basis, it will appear that $|\psi_{k'}\rangle$ is not superposed (as it is one of the basis vectors), whereas $|\psi_k\rangle$ may look like a quantum superposition (if it is not one of the basis vectors).

Thus, every quantum state $|\psi\rangle$ could always be thought of as a quantum superposition of other states. Only in very special cases, where $|\psi\rangle$ is among the basis vectors of the representation basis could it be said that the state is not a superposition of the basis vectors. Expressing the state vector $|\psi\rangle$ in any representation basis $\{|k_1\rangle, |k_2\rangle, \ldots, |k_n\rangle\}$ is done by the orthogonal decomposition of the unit operator $\hat{I}$ in that basis

$$|\psi\rangle = \hat{I}|\psi\rangle = \sum_{i=1}^{n} |k_i\rangle\langle k_i\|\psi\rangle = \sum_{i=1}^{n} \langle k_i|\psi\rangle|k_i\rangle = \sum_{i=1}^{n} a_i|k_i\rangle \qquad (6.10)$$

where the coefficients in the superposition are given by $a_i = \langle k_i|\psi\rangle$. Since each complex coefficient $a_i \in \mathbb{C}$ is a quantum probability amplitude, it is directly linked to quantum *indeterminism* through the Born rule (Axiom 4.15.3), namely, $|a_i|^2 = a_i^* a_i$ is the probability of finding the state $|\psi\rangle$ being collapsed into the state $|k_i\rangle$ upon measurement performed in the basis $\{|k_1\rangle, |k_2\rangle, \ldots, |k_n\rangle\}$.

The time evolution of quantum physical systems according to the Schrödinger equation (Axiom 4.15.5) leads not only to superpositions of *states*, but also to superpositions of *events* or *histories* [172, 173, 176, 177, 179]. The quantum interfer-

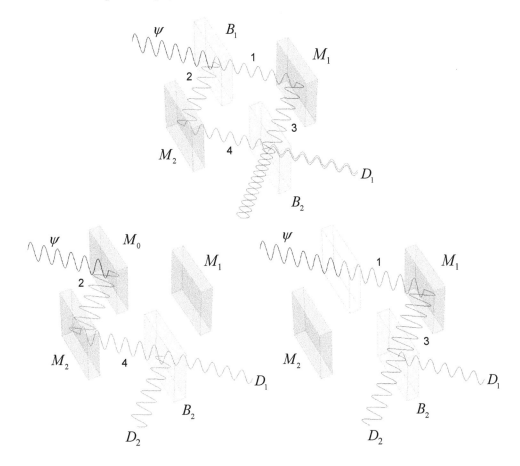

Figure 6.5   Quantum superposition of histories in the two arms of a Mach–Zehnder interferometer leads to constructive interference at detector D_1 and destructive interference at detector D_2. Thus, when both arms of the interferometer are traveled, the photon always goes to D_1 and never to D_2. However, if the photon takes only a single path inside the interferometer, namely, path 1 or path 2, it can go to either detector D_1 or D_2 with an equal probability of $\frac{1}{2}$. B, beam splitter; D, detector; M, mirror; ψ, photon.

ence of single photons inside a Mach–Zehnder interferometer (Fig. 6.5) nicely illustrates how different superpositions of quantum histories could lead to different probability distributions for certain measurement outcomes [132, 133, 134, 421].

Inside the Mach–Zehnder interferometer, only three types of photon events could occur: transmission $\hat{T}$, reflection $\hat{R}$, or beam splitting $\hat{B}$. If $|\psi\rangle$ is the incoming beam, $|\psi_+\rangle$ is the transmitted beam, and $|\psi_-\rangle$ is the reflected beam, the action of the three time evolution operators could be written as

$$\hat{T}|\psi\rangle \;=\; |\psi_+\rangle \tag{6.11}$$

$$\hat{R}|\psi\rangle \;=\; \imath|\psi_-\rangle \tag{6.12}$$

$$\hat{B}|\psi\rangle \;=\; \frac{1}{\sqrt{2}}\left(\hat{T}+\hat{R}\right)|\psi\rangle = \frac{1}{\sqrt{2}}\left(|\psi_+\rangle + \imath|\psi_-\rangle\right) \tag{6.13}$$

After the first beam splitter B_1, the quantum state $|\psi\rangle$ evolves into a superposition

$$\hat{B}|\psi\rangle = \frac{1}{\sqrt{2}}(|1\rangle + i|2\rangle) \tag{6.14}$$

where $|1\rangle$ and $|2\rangle$ are the states indicating the photon presence in the corresponding interferometer arms 1 and 2. Then, after reflection from mirrors M_1 and M_2, the state becomes

$$\hat{M}\frac{1}{\sqrt{2}}(|1\rangle + i|2\rangle) = \frac{1}{\sqrt{2}}(i|3\rangle - |4\rangle) \tag{6.15}$$

Subsequent action of the second beam splitter B_2 leads to constructive interference at detector D_1 and destructive interference at detector D_2

$$\begin{aligned}
\hat{B}\frac{1}{\sqrt{2}}(i|3\rangle - |4\rangle) &= \frac{1}{2}(-|D_1\rangle + i|D_2\rangle) + \frac{1}{2}(-|D_1\rangle - i|D_2\rangle) \\
&= \frac{1}{2}[(-|D_1\rangle - |D_1\rangle) + (i|D_2\rangle - i|D_2\rangle)] \\
&= -|D_1\rangle \tag{6.16}
\end{aligned}$$

The probabilities for detecting the photon by each detector are the expectation values of the corresponding detector projection operators $|D_1\rangle\langle D_1|$ and $|D_2\rangle\langle D_2|$

$$\begin{aligned}
(-1)^2\langle D_1\|D_1\rangle\langle D_1\|D_1\rangle &= 1 \tag{6.17} \\
(-1)^2\langle D_1\|D_2\rangle\langle D_2\|D_1\rangle &= 0 \tag{6.18}
\end{aligned}$$

Thus, the photon always goes to detector D_1 and never to D_2. Since single photons are sent one by one, the quantum interference shows that each photon is able to take at once both available paths through arms 1 and 2 of the Mach–Zehnder interferometer.

Remarkably, any modification of the experimental setup that forces the photon to take either path 1 or path 2 destroys the quantum interference and allows the photon to be detected by either detector D_1 or D_2 with an equal probability of $\frac{1}{2}$ (Fig. 6.5). For example, if the first beam splitter B_1 is removed, the photon will take only path 1

$$\hat{T}|\psi\rangle = |1\rangle \tag{6.19}$$

Then, it will be reflected by mirror M_1

$$\hat{M}|1\rangle = i|3\rangle \tag{6.20}$$

Finally, the action of the second beam splitter B_2 will result in a superposition of the photon being detected by either detector D_1 or D_2

$$\hat{B}i|3\rangle = \frac{1}{\sqrt{2}}(-|D_1\rangle + i|D_2\rangle) \tag{6.21}$$

The probabilities for detecting the photon by each detector are

$$\frac{1}{\sqrt{2}}(-\langle D_1| - \imath\langle D_2|)|D_1\rangle\langle D_1|\frac{1}{\sqrt{2}}(-|D_1\rangle + \imath|D_2\rangle) = \frac{1}{2} \tag{6.22}$$

$$\frac{1}{\sqrt{2}}(-\langle D_1| - \imath\langle D_2|)|D_2\rangle\langle D_2|\frac{1}{\sqrt{2}}(-|D_1\rangle + \imath|D_2\rangle) = \frac{1}{2} \tag{6.23}$$

Similarly, if the first beam splitter B_1 is replaced by a mirror M_0, the photon will take only path 2, will be reflected by mirror M_2, and after the second beam splitter B_2 will also end in a superposition of being detected by either detector D_1 or D_2

$$\hat{M}|\psi\rangle = \imath|2\rangle \tag{6.24}$$

$$\hat{M}\imath|2\rangle = -|4\rangle \tag{6.25}$$

$$-\hat{B}|4\rangle = \frac{1}{\sqrt{2}}(-|D_1\rangle - \imath|D_2\rangle) \tag{6.26}$$

Since the final state in Eq. (6.26) is the complex conjugate of the final state in Eq. (6.21), the probabilities for detecting the photon by each detector are also $\frac{1}{2}$.

The three different scenarios for the Mach–Zehnder setup show that the time evolution of quantum states is extremely sensitive to the boundary conditions of the experiment. The two cases without B_1, in which M_0 is either present or absent, also show that different setups leading to different quantum states can have identical probability distributions for the possible measurement outcomes.

6.4.2   Quantum indeterminism avoids epiphenomenalism

The time evolution of the quantum probability amplitudes according to the Schrödinger equation (Axiom 4.15.5) is deterministic. However, the indeterminism in quantum theory arises from the process of objective state vector reduction (Axiom 6.1.3) that provides the probabilities for observing any of the possible measurement outcomes according to the Born rule (Axiom 4.15.3). Thus, the determinism of the Schrödinger equation allows beforehand calculation of what the probabilities for possible events are, whereas the inherent indeterminism of quantum events allows unpredictable actualization of only one of the possible events. In other words, quantum indeterminism does not arise merely from the mathematical structure of the Schrödinger equation, but from the fact that the physical object that evolves according to the Schrödinger equation is made of *complex-valued quantum probability amplitudes* that are subject to actualization.

Quantum indeterminism is pertinent to the analysis of the crushed brain experiment (Example 5.6). Direct comparison of two brain states with the same chemical composition, but such that one of the states is conscious while the other state is unconscious, suggests that somehow the functional organization of the brain state is relevant for the generation of consciousness. Classically, functionalism leads to epiphenomenalism because the generated conscious experiences are unable to change the already predetermined future dynamics of the brain (Section 5.3.1). In quantum physics, however, the quantum indeterminism in the dynamics of quantum systems provides a multitude of possible outcomes from the measurement process. Since the indeterminism is inherent in quantum systems, there is no compelling reason to conclude that the conscious experiences generated by a quantum system were not causally effective in determining the actualization of a single outcome out of the multitude of available outcomes. Thus, quantum indeterminism could avoid epiphenomenalism in theories claiming that the conscious experiences are a product of the brain. Still, one needs to be extremely careful not to retreat back to classical information theory (Section 3.19). In Section 6.7, we will show how in quantum theory the mind and the brain are able to interact with each other, and will explain in what sense the mind can be viewed as a product of the brain.

6.5 Free will in quantum physics

Free will is our capacity to make choices among at least two possible future alternatives (Definition 1.3). In classical physics, the identification of conscious states with physical states undergoing deterministic dynamics precludes the availability of future alternatives for the mind to choose from and bans genuine free will from existence. In quantum physics, however, the state vector $|\psi\rangle$ and the quantum wave function ψ are composed of *quantum probability amplitudes* that determine the probabilities for possible future outcomes according to the Born rule (Axiom 4.15.3). Therefore, the identification of the state vector $|\psi\rangle$ of non-factorizable quantum physical systems with conscious (mental) states (Axiom 6.1.1) provides an excellent opportunity for solving the free will problem. Since quantum physics is inherently indeterministic, it is able to accommodate free will in the physical process of actualization of one out of many possible future choices.

The quantum probability amplitudes $\psi(x,t)$ determine the probability for possible future outcomes according to the Born rule

$$\text{Prob}(x,t) = \psi^*(x,t)\psi(x,t) = |\psi(x,t)|^2 \tag{6.27}$$

The probability $\text{Prob}(x,t)$ for observing the quantum particle at position x at time t is given by the squared modulus of the probability amplitude $|\psi(x,t)|^2$. Thus, if the position of the quantum particle is measured, the quantum wave function $\psi(x,t)$ uniquely determines the probabilities for finding the particle at space point x at time t. In other words, there are multiple future alternatives for the position of the particle and each such alternative can be actualized with probability $\text{Prob}(x,t)$.

Noteworthy, the quantum state vector $|\psi\rangle$ does not always need to be represented in the position basis as $\psi(x,t)$. If one is interested in the probability for measuring the quantum system being in a state with a certain value for its momentum $p = \hbar k$, it would be necessary to use the momentum representation $\psi(p,t)$ of the wave function (Section 4.11). The Born rule is then applied analogously: The probability $\text{Prob}(p,t)$ for observing the quantum particle having a momentum p at time t is given by $|\psi(p,t)|^2$.

Historically, the fathers of quantum mechanics did not emphasize the Born rule as an axiom, but considered it to be derivable from Heisenberg's uncertainty principle [245, 246, 291]. In the 1920s and 1930s, the uncertainty principle was understood as an empirical regularity (employing no or only a bare minimum of theoretical terms) whose purpose was to help build up a fully fleshed physical theory that is able to explain the empirical data [252]. At present, we already have a powerful Hilbert space axiomatization of quantum theory (Section 4.15), so we derive different Heisenberg uncertainty relations as theorems and no longer need them as a starting point when solving quantum problems [387].

Albert Einstein realized that the unpredictable quantum behavior of an electron appears to be a manifestation of free will; however, he thought that the free will should be banned from physics. In a letter to Max Born (1882–1970) written on 29 April 1924, Einstein expressed his distaste for quantum indeterminism:

> I find the idea quite intolerable that an electron exposed to radiation should choose *of its own free will*, not only its moment to jump off, but also its direction. In that case, I would rather be a cobbler, or even an employee in a gaming-house, than a physicist. [56, p. 82]

Erwin Schrödinger (1887–1961), who discovered the *Schrödinger equation* and shared the 1933 Nobel Prize in Physics for his work on quantum mechanics, doubted that quantum indeterminism is relevant for explaining human free will but for other reasons [424]. Referring to the Born rule for calculating quantum probabilities as Heisenberg's uncertainty principle, he argued that

> the remarkable feeling of responsibility, entails the idea of *choice* between different possibilities for which a clue is sought in the modern views of physics. If that were right, it would mean *either* one of two things. First, that the laws of Nature are after all at "my" mercy. For if my smoking or not smoking a cigarette before breakfast (a very wicked thing!) were a matter of Heisenberg's uncertainty principle, the latter would stipulate between the two events a definite statistics, say 30:70; which I could invalidate by firmness. *Or*, secondly, if that is denied, why on earth do I feel responsible for what I do, since the frequency of my sinning is determined by Heisenberg's principle? [424]

Schrödinger's argument is ungrounded for several reasons. First, our free will does not imply physical *lawlessness*. Because we are unable to perform miracles at will, the physical laws have to be able to set the available alternatives from which we are then allowed to choose. Second, free will does not require violations of the Born rule. Quantum statistics is manifested only when a quantum experiment is repeated multiple times. For each choice, however, the time flow is irreversible and the choice cannot be revisited once it is done. In a series of repeated choosings the overall probability distribution may be given by the Born rule but it is up to us to decide what the exact history of outcomes will be. For example, suppose that you can choose smoking a cigarette with 20/80 statistics for no/yes, and you repeat the experiment five times. Two possible histories are 10111 or 11110, but their consequences might be different. Your free will can still be responsible for all the differences that result from choosing one definite history versus all other physically possible but non-actualized alternative histories. Thus, the existence of probability weights on the possible choices does not invalidate our free will. Third, the Born statistics is not imperative if you repeat the choices only a finite number of times. Suppose that you have to choose in a quantum experiment that will be performed only once in your lifetime with predicted 20/80 no/yes ratio. From the quantum nature of your choices, the outcomes can have only one of the two values, 0 for no or 1 for yes. Thus, the predicted 20/80 ratio does not imply that you cannot in principle choose the less likely outcome of no. If that outcome indeed is chosen, all that can be said is that a rare, but not impossible, event had occurred. Since the Born rule is exact only in the limit when the experiment is repeated an infinite number of times, using your free will for choosing a finite history that does not agree with the calculated probability distribution by the Born rule does not and cannot invalidate the Born rule. Fourth, if the mind choices were able to violate the Born rule, it would have been the case that all available choices are equally likely and we would not have been subject to any biases when making our conscious decisions. Yet, we know that we are biased, and even though we are free to choose, it is more likely for us to choose something that will be pleasant but unhealthy, rather than something that we know will be unpleasant but healthy. Thus, the integration of the Born rule in quantum information theory of consciousness provides an inherent mechanism for biasing our decisions and sheds light on the origin of our personal preferences for choosing some courses of action more frequently than others. Noteworthy, the bias in the probabilities provided by the Born rule does not need to be consciously experienced as an emotional affect, urge, desire or temptation. Indeed, we use daily our free will for all kinds of unimportant things that we do not even remember of doing at the end of the day, and most of the time we perform those activities without having any emotional urge for doing one action instead of another.

6.5.1 Actualization of possibilities and choice making

The availability of future alternatives to choose from is necessary, but not sufficient, to solve the free will problem. What is also needed is a physical process that describes the choice making manifested through the actualization of one of the available possibilities (Axiom 6.1.3). Such a process is only available in collapse models of quantum mechanics and is referred to as the *collapse of the wave function, reduction of the wave packet* or *objective reduction*. Objective reductions are inherently indeterministic and noncomputable, meaning that there exists neither a physical law, nor a conceivable algorithm that could be used to predict with certainty the outcome from the objective reduction [378, 380]. Thus, the process of objective reduction is a physical manifestation of the inherent free will enacted by the quantum physical systems. The ability to make free choices further goes hand in hand with the problems of morality. Since our choices are irreversible once they are made, we are obliged to be responsible for our actions.

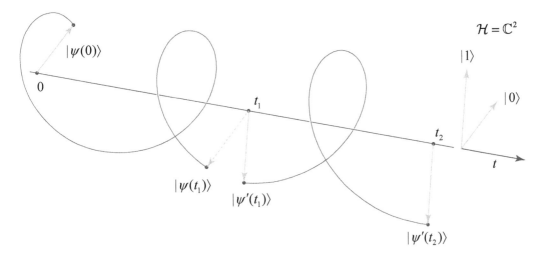

Figure 6.6   Discontinuous quantum jump in the time evolution of a two-level quantum system with energy $E = \hbar\omega$. The time evolution by the Schrödinger equation corresponds to a deterministic continuous rotation with angular frequency ω of the state vector $|\psi(t)\rangle$ in the Hilbert space $\mathcal{H} = \mathbb{C}^2$. The objective reduction of the state vector $|\psi\rangle$ in a certain basis at time t_1 leads to a discontinuous jump $|\psi(t_1)\rangle \to |\psi'(t_1)\rangle$. Due to this quantum jump, the quantum state at time t_2 is given by $|\psi'(t_2)\rangle$ instead of $|\psi(t_2)\rangle$. The discontinuous and unpredictable nature of the quantum jumps gives rise to quantum indeterminism.

Example 6.1. (*Objective reduction is a basis-dependent concept*) *The outcome of objective reduction is a disentangled collection of quantum physical systems that can be described by a tensor product state [372]. Due to the vector nature of quantum states it is always possible to express a quantum state in the form of a quantum superposition by choosing a basis in which the quantum state is not among the basis vectors (Section 6.4.1); the reduction of the wave packet occurs only when the composite system is viewed in a certain basis such that the resulting disentangled state is among the basis vectors. In that basis, the process of objective reduction brings the quantum coherence measures (Section 6.6.4) down to zero. If viewed in another basis, however, the objective reduction may even lead to maximal quantum coherence.*

Example 6.2. (*Quantum jump is a basis-independent concept*) *From the abstract basis-independent view of the time evolution of the quantum state $|\psi\rangle$ inside the Hilbert space $\mathcal{H}$, objective reduction appears to be a discontinuous quantum jump in the dynamics of the state vector $|\psi\rangle$ (Fig. 6.6). Because the outcomes from the discontinuous quantum jumps are unpredictable, the quantum jumps are responsible for the inherent indeterminism manifested by the quantum physical systems. The only way for a possible outcome to be absolutely certain (to occur with 100% probability) is to measure the quantum state in a basis in which $|\psi\rangle$ is one of the basis vectors. Such a measurement, however, will return again $|\psi\rangle$ as the outcome and there will be no discontinuous quantum jump. In other words, objective reduction in a given basis does not occur for states that are already reduced in that basis.*

Situations in which a single quantum measurement outcome occurs with absolute certainty are not impossible. In all such cases the quantum state $|\psi\rangle$ is measured with the use of the projection operator $|\psi\rangle\langle\psi|$, which results with certainty in the state $|\psi\rangle$ as the measurement outcome. One laboratory example is the Mach–Zehnder interferometer in which a single photon passes coherently through both interferometer arms and always hits only one of the detectors (Fig. 6.5). The outcome is predetermined due to the occurrence of constructive or destructive quantum interference in different space regions. Since there are no multiple available choices, the quantum system cannot manifest free will in its behavior. This is not surprising because classical deterministic systems do not exhibit free will, hence no free will should be expected by quantum systems in the domain of physical phenomena where the quantum behavior mimics the classical deterministic behavior. Indeed, when we are subjected to deterministic motion we do not subjectively feel capable of using our free will in order to stop that motion. Thus, the most rational explanation for our feeling of being capable or incapable of exerting our free will in certain situations is that we are actually capable or incapable of exerting our free will in those situations. Next, we will discuss two experiments whose outcomes contradict the classical view that our introspective feeling of possessing free will is an illusion.

Example 6.3. *(Falling from a bridge) Free will cannot be exercised if the outcome of an event is predetermined. Suppose that by an accident you fall from a bridge. For the brief moment of time while you are in a free fall, you may be fully conscious of the situation that you are in but unable to use your free will in order to freeze the motion in midair and avoid hitting the ground. In such a tragic scenario, you will almost certainly not experience the illusion that you are using your own free will in order to keep the fall going. On the contrary, it is likely that you will feel the despair of knowing that you are unable to use your free will in order to change the inevitable outcome.*

Example 6.4. *(Flying in an airplane) Suppose that you are sitting in your seat in a flying airplane. The motion of the airplane is transmitted through the seat, through your body, and then through your skull that exerts certain force upon your brain, setting it in motion. Regardless of how strongly you want it, you are unable to prevent your brain from moving as this will require the airplane to stop its motion in midair. Thus, you will not be able to exert your free will if the outcome (such as the brain motion) is predetermined. Furthermore, if you have ever boarded an airplane, you would already know that during the flight you never experience the illusion of causing your brain to move or causing the airplane to fly merely by the power of your own free will.*

6.5.2   Free will versus superdeterminism

In the proofs of Bell's theorem formalizing quantum nonlocality and the Kochen–Specker theorem formalizing quantum contextuality, we have assumed that the human experimenters had the free will to choose how to set up their measuring apparatuses. Thus, it may appear that the impossibility of local and noncontextual predetermined classical values for noncommuting quantum observables is ultimately dependent on the existence of human free will. Denying quantum nonlocality and contextuality, however, could not be simply based on the negation of the free will assumption. For example, it is not inconceivable that somewhere in the universe 33 identical independent spin-1 particles are measured simultaneously along the 33 rays (Fig. 4.9) for which there cannot be predetermined values for the squared spin-1 components according to the $1, 0, 1$ rule. Or even better, suppose that we set up such a simultaneous measurement of 33 independent spin-1 particles in the laboratory. Then, the free will of the human experimenter becomes irrelevant for proving the quantum contextuality and nonlocality. All that is needed is the ability to prepare identical independent quantum systems in large numbers.

In order to keep the possibility of predetermined classical values, one should postulate a conspirative *superdeterminism*, according to which every time an experimenter wants to prepare 33 identical independent spin-1 particles, nature produces 33 dependent (correlated) spin-1 particles. In other words, superdeterminism postulates the existence of a universe in which all possible events are predetermined at the very origin of the universe in a conspirative way [316]: the processes are fully deterministic but they occur in such a manipulated way that could make us falsely believe that we live in a quantum indeterministic universe. Thus, superdeterminism not only denies free will, it also enforces a universal grand theater whose only purpose is to mislead us. In essence, superdeterminism asks us not to trust the evidence to the contrary, but to blindly trust superdeterminism for the sake of it. As with any other conspiracy theory, superdeterminism is pseudoscientific because it is an overly complicated theory designed merely to immunize itself from experimental falsification. In addition to being able to escape any rational criticism, superdeterminism neither generates new unexpected predictions of the physical theory, nor compresses any information or useful knowledge.

Human free will is an important prerequisite, which if dropped would make the whole scientific enterprise meaningless. Indeed, if scientists did not have the free will to choose what measurements to perform, it would have been impossible to obtain any knowledge from performing experiments. A deterministic universe should already have a plan for everyone's scientific belief regardless of what sham experimentation is accompanying those beliefs. In particular, bad scientists who happen to interpret their data in the wrong way would be just unlucky to be born that way, whereas good scientists who happen to interpret their data in the correct way would be just lucky to be born that way.

6.5.3 Where does free will come from?

John H. Conway and Simon B. Kochen [99, 100] attempted to use human free will as a premise from which to derive the conclusion that the quantum particles outside of one's own brain also possess a certain amount of free will. In particular, they considered two human experimenters with free will who perform measurements of the squared spin-1 components (Section 4.13) of entangled spin-1 particles in the state

$$|\Psi\rangle = |\uparrow_z\rangle|\downarrow_z\rangle + |0_z\rangle|0_z\rangle + |\downarrow_z\rangle|\uparrow_z\rangle \tag{6.28}$$

that are separated far away in space so that classical communication cannot occur during the experiment. Then, Conway and Kochen followed the proof of the Kochen–Specker theorem and argued that the $1, 0, 1$ rule for squared spin-1 components ensures that the spin-1 particles could not have had a predetermined set of values for all 33 quantum observables shown in Figure 4.9. This means that the spin-1 particles make up their answers on the fly, and appear to exhibit some form of free will.

Conway noted that the free will of quantum particles could explain where human free will comes from, namely, from the fact that the human brain is composed of quantum particles [98]. He also explicitly acknowledged that the quantum unitary time evolution according to the Schrödinger equation cannot give rise to free will, and what is needed for free will is an objective physical process that leads to wave function collapse and quantum indeterminism [98].

Remarkable as it may be, the reasoning by Conway and Kochen is deficient in two important respects, both of which are taken care of in our axiomatization (Section 6.1).

First, the finite speed for transmission of classical information (Theorem 9) does not apply to quantum information (Bell's theorem). Thus, it is not clear what the motivation is for the spin-1 particles to be entangled and then separated far away. Arguably, deterministic classical physics cannot support free will (Section 5.4) and utilizing theorems that pertain to classical information (such as Theorem 9) could hardly lead to any result relevant to free will. On the contrary, by noticing the nonlocal correlations of the outcomes produced by the individual spin-1 particles, one can argue that each of the two entangled particles could not have possibly had individual independent free will; rather the composite two-particle entangled system should have acted as if it had a single mind that imposed its own free will upon both of the component particles. Identifying entangled systems with a single mind is exactly the content of our Axiom 6.1.1.

Second, without an explicit definition of what a single mind is, nested *minds within minds* cannot be excluded, and consequently nested *free will within free will* contradictions are bound to occur: If the quantum particles within my brain do have their own free will and I also do have my own free will, who is then ultimately responsible for my actions? What would happen if I decide to do exactly the opposite of what the individual quantum particles in my brain choose to do? Will the universe be torn apart by such a contradiction or will the physical laws override someone's free will? Does a component quantum subsystem lose its free will as it enters into the composition of a quantum entangled state? Deciding whose free will is operating at a given instant of time cannot be meaningfully addressed in the framework used by Conway and Kochen, as is seen by their admission that ultimately free will should be attributed to the whole universe, rather than to physical objects inside the universe:

> our assertion that "the particles make a free decision" is merely a shorthand form of the more precise statement that "the Universe makes this free decision in the neighborhood of the particles." [99, p. 1456]

Theorem 25. (*Elaborate free will theorem*) *Only minds possess free will. Quantum particles do not always have their own free will, but will eventually be a component of some quantum entangled system (mind) that will exercise its own free will. Even though non-factorizable quantum entangled systems do have a single mind, this mind cannot exercise its free will unless a certain energy threshold $\mathcal{E}$ for objective reduction is reached. Thus, free will is exercised only by a mind that undergoes objective reduction thereby choosing one of the possible outcomes according to the Born rule.*

Proof. Axiom 6.1.3 defines the energy threshold $\mathcal{E}$ required for the indeterministic objective reduction to occur. From Axiom 6.1.1 it follows that the non-factorizable quantum entangled system that undergoes objective reduction is a single conscious mind. Then by Definition 1.3, this mind makes a choice among multiple available alternatives thereby exercising its own free will. Finally, from Axiom 6.1.2 it follows that it is possible to have a collection of minds without their own free will if none of these minds has reached the energy threshold $\mathcal{E}$. Because the time evolution leads to external entanglements and growth of the quantum entangled clusters, eventually every quantum entangled cluster will reach the energy threshold $\mathcal{E}$ and undergo an objective reduction thereby exercising its own free will. Overall, no free will within free will contradictions occur in the quantum information theory of consciousness. □

6.5.4 Schrödinger's cat and objective reduction

Quantum effects are often described as being minor fluctuations that could hardly affect the dynamics of macroscopic bodies. Slightly more pretentious is the assertion that quantum effects are negligible in the domain of large-scale objects. Such a claim is incorrect, however, because initial quantum superpositions of individual elementary particles could be amplified into macroscopic superpositions of large composite objects due to the linearity of time evolution of quantum systems according to the Schrödinger equation (4.157). In fact, Erwin Schrödinger himself showed how quantum theory is able to produce macroscopic superposition of a cat being both dead and alive [423].

Example 6.5. (*Schrödinger's cat*) *Suppose that we put a cat into a box. Let us also arrange two photon detectors such that if a photon arrives at detector D_1, it triggers a mechanism that releases toxic gas into the box with the cat, whereas if a photon arrives at detector D_2, no toxic gas is released. If the photon states arriving at each detector are $|\gamma_1\rangle$ and $|\gamma_2\rangle$, the initial state of the cat is $|cat\rangle$, and the two possible final cat states are $|dead\rangle$ and $|alive\rangle$, the action of the quantum time evolution operator $\hat{U}$ is*

$$\hat{U}|\gamma_1\rangle|cat\rangle = |D_1\rangle|dead\rangle \tag{6.29}$$
$$\hat{U}|\gamma_2\rangle|cat\rangle = |D_2\rangle|alive\rangle \tag{6.30}$$

The linearity of $\hat{U}$ implies that if we prepare a quantum superposition of the photon state $\frac{1}{\sqrt{2}}(|\gamma_1\rangle + |\gamma_2\rangle)$ with the use of a beam splitter, the cat will also evolve into a superposition of dead and alive cat states (Fig. 6.7), where each cat state is entangled with

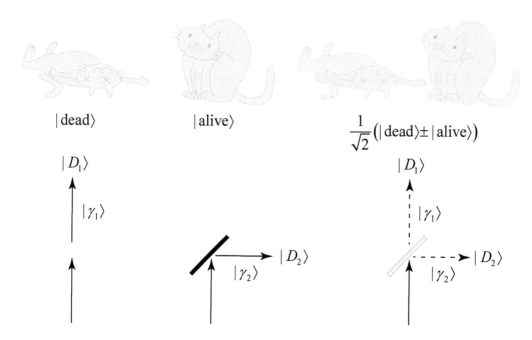

Figure 6.7   Linearity of the Schrödinger equation easily generates macroscopic quantum superposition of a cat being both dead and alive. All that is needed is a single photon detector $|D_1\rangle$ equipped with a mechanism that kills the cat, and another photon detector $|D_2\rangle$ that does no harm to the cat. Then, the input of a quantum superposed photon state $\frac{1}{\sqrt{2}}(|\gamma_1\rangle + |\gamma_2\rangle)$ produced by a beam splitter would inevitably lead to a macroscopic superposition resulting from the entanglement between the photon detections at D_1 or D_2 and the corresponding dead or alive cat states.

the firing of one of the two detectors $|D_1\rangle$ or $|D_2\rangle$

$$\hat{U}\frac{1}{\sqrt{2}}(|\gamma_1\rangle + |\gamma_2\rangle)|cat\rangle = \frac{1}{\sqrt{2}}(|D_1\rangle|dead\rangle + |D_2\rangle|alive\rangle) \qquad (6.31)$$

Before we open the box, the cat appears to be in a paradoxical quantum superposition. Because we never observe macroscopic superpositions of dead and alive cats, there should be some physical process that leads to objective reduction of the quantum state at a certain energy threshold $\mathcal{E}$. Most likely, the objective reduction would be associated with gravity due to the fact that macroscopic objects never seem to reduce in a basis where their localization in space is superposed [135, 136, 137, 138, 376, 377, 382].

The absence of Schrödinger's cats in our everyday life is the strongest indication that the objective reduction is a real physical process. Indeed, theoretical attempts to explain the absence of Schrödinger's cats without collapse, such as the existence of both dead and alive cat histories in decoherent parallel universes as asserted by *Everett's many worlds interpretation* [161, 162], do not really solve the problem but rather utilize word-jugglery to disguise it. For example, one could use the vector nature of quantum states (Section 6.4.1) and rewrite the superposition of dead and alive cat states as

$$|\Psi\rangle \quad = \quad \frac{1}{\sqrt{2}}(|D_1\rangle|\text{dead}\rangle + |D_2\rangle|\text{alive}\rangle) \tag{6.32}$$

$$= \quad \frac{1}{\sqrt{2}}(|D_+\rangle|\text{cat+}\rangle + |D_-\rangle|\text{cat-}\rangle) \tag{6.33}$$

where

$$|D_+\rangle \quad = \quad \frac{1}{\sqrt{2}}(|D_1\rangle + |D_2\rangle) \tag{6.34}$$

$$|D_-\rangle \quad = \quad \frac{1}{\sqrt{2}}(|D_1\rangle - |D_2\rangle) \tag{6.35}$$

$$|\text{cat+}\rangle \quad = \quad \frac{1}{\sqrt{2}}(|\text{dead}\rangle + |\text{alive}\rangle) \tag{6.36}$$

$$|\text{cat-}\rangle \quad = \quad \frac{1}{\sqrt{2}}(|\text{dead}\rangle - |\text{alive}\rangle) \tag{6.37}$$

Rewriting the entangled state $|\Psi\rangle$ using the quantum superposed states $|\text{cat+}\rangle$ and $|\text{cat-}\rangle$ shows that both of these quantum superposed states are also decoherent. Therefore, there is no reason that would force the parallel universes to split in the basis $|\text{dead}\rangle$, $|\text{alive}\rangle$, and it should be equally likely that the parallel universes split in the basis $|\text{cat+}\rangle$, $|\text{cat-}\rangle$. This is a well-known shortcoming of the decoherence process and is usually referred to as the *preferred basis problem* [537]. In other words, if the universe can never split in the basis $|\text{cat+}\rangle$, $|\text{cat-}\rangle$, and always splits in a certain preferred basis as predicted by theories with objective reduction, the process of universe splitting becomes just another disguised word for collapse.

Example 6.6. *(Quantum resurrection) Insisting that the universe splitting does not violate the Hilbert space formalism of quantum mechanics backfires because the $|dead\rangle$ and $|alive\rangle$ cat states can always be expressed as quantum superpositions in the $|cat+\rangle$, $|cat-\rangle$ basis as*

$$|dead\rangle \quad = \quad \frac{1}{\sqrt{2}}(|cat+\rangle + |cat-\rangle) \tag{6.38}$$

$$|alive\rangle \quad = \quad \frac{1}{\sqrt{2}}(|cat+\rangle - |cat-\rangle) \tag{6.39}$$

Therefore, quantum mechanically an experimenter can start with a cat in the $|dead\rangle$ state, then measure it into $|cat+\rangle$ or $|cat-\rangle$ state, and then re-measure one more time the obtained $|cat+\rangle$ or $|cat-\rangle$ state into $|alive\rangle$ state with probability of $\frac{1}{2}$. Since quantum resurrection seems to be impossible, it should be the case that objective reductions do indeed occur and eliminate all but one of the Everett's many worlds.

6.5.5 Debunking free-will skepticism

Free-will skepticism claims that free will is impossible regardless of whether determinism is true. Two of the most frequent claims made by the skeptics are that free will is incompatible with indeterminism or free will is incompatible with the causal potency of consciousness [386, pp. 80–90].

Example 6.7. *(The randomness problem) Consider an unstable state such as the ball on top of a potential hill (Fig. 5.11) and suppose that the indeterministic action upon the ball is decided by the rolling of an indeterministic dice. If the universe is rolling a dice in regard to your actions, then it is not up to you to decide what you are going to choose and free will is impossible. Intuitively, if you are standing on the edge of a bridge and a slight external push forces you out of balance, leading to your fall, then you will not consider yourself to have chosen freely to fall. Similarly, you should not consider yourself able to choose freely if your choices were the result of indeterministic jerks or spasms. Thus, it may seem that not only free will is incompatible with determinism, but free will is incompatible with indeterminism as well [386, p. 81].*

The apparent problem with indeterministic randomness disappears once it is pointed out that the external dice rolling is a misleading and altogether false analogy of what quantum indeterminism is. The quantum jump exhibited by a quantum system that has reached the energy threshold $\mathcal{E}$ for objective reduction (Fig. 6.6) is an inherent choice made by the system itself and is not decided by the universe. Thus, the indeterministic universal physical laws only force the system to make a choice but do not tell the system what the choice should be. In other words, the quantum system neither gives up nor loses its own free will. When the time for objective reduction comes, the system has to choose one way or another. This is not controversial at all, since it is not up to us to decide whether we have free will or not, or whether we will use our free will or not. What is up to us is only choosing what future course of action we are going to actualize among the set of physically available alternatives.

Example 6.8. *(The exercise problem) Our conscious experiences, feelings, desires or beliefs are assumed to be causally potent in the physical world. The causal potency, however, requires that our actions be determined by our mental states. On the other hand, the free will is incompatible with determinism and our freedom to make choices seems to require that our actions are not caused by our prior desires. Thus, it may seem that one is faced with a dilemma: give up on the libertanian understanding of free will or give up on the causal potency of our conscious experiences [386, p. 83].*

The causal potency of consciousness cannot be due to something less strong than a logical implication. In particular, consciousness cannot be said to just increase or decrease the likelihood for an action to happen. Because the quantum wave function ψ determines exactly the probability $\text{Prob}(x,t) = |\psi(x,t)|^2$ for the occurrence of an event at any point x at time t, the likelihood for the event could not be altered without violating the Born rule. If consciousness causally determines our actions in the sense of a logical implication, however, the origin of the exercise problem has to be found in a possible misunderstanding of the claim that our free choices are not caused by our prior desires. Indeed, correct description of the relationship between consciousness and free will requires proper understanding of the discontinuous nature of the physical process of objective reduction. When viewed retrospectively, the quantum wave function exhibits abrupt quantum jumps at time points where free decisions were taken. In the example shown in Figure 6.6, the quantum jump occurs at time t_1 when $|\psi(t_1)\rangle$ goes to $|\psi'(t_1)\rangle$. Because a mathematical function cannot have two values at a single point, we can set the value at t_1 to be $\psi'(t_1)$ by definition. The discontinuous wave function will exhibit $\psi(t_1)$ as a left-hand limit at t_1 and $\psi'(t_1)$ as a right-hand limit at t_1. Since the quantum wave function models the mind of the system and the discontinuous quantum jump models a free choice, it is true that the free choice made at t_1 is not determined by prior desires, namely, the value of $\psi'(t_1)$ cannot be determined by knowing any $\psi(t < t_1)$. On the other hand, the value of the wave function $\psi'(t_1)$ at the point t_1 exactly determines (logically implies) the choice made at t_1.

Discussing explicitly our past choices does not make them unfree. Because free will is exercised in time, the flow of time itself cannot threaten the existence of free will. What happens with the flow of time is that with each decision one of many possible choices gets actualized. Thus, it is possible to discuss free will from a block universe perspective where the time dimension is explicitly given. In the block perspective, free will is manifested in the fact that the value $\psi'(t_1)$ of the wave function at a given time t_1 is independent of and cannot be predicted from any knowledge of the wave function values $\psi(t < t_1)$ at arbitrary prior time $t < t_1$. In other words, the exercise problem is avoided provided that the cause and effect occur simultaneously at the same point in time. In a deterministic universe one can use uncritically the term "prior" because all future states of the universe are implied by the knowledge of the initial state of the universe. In a quantum indeterministic universe, however, there is no such universal implication, and from knowing the wave function value $\psi(t)$ at time t one can make deterministic predictions only up to the nearest future quantum jump forward in time, or up to the nearest past quantum jump backward in time. Consequently, the claim that the free decision is not determined by a prior desire, yet it is determined by the mind, should be understood as the statement that the decision at t_1 is not implied by any $\psi(t < t_1)$, but it is implied by $\psi'(t_1)$.

6.5.6 Quantum existentialism

The philosophy of existentialism promoted by Jean-Paul Sartre states that we are born free and each of us is able to choose what the meaning of his or her own life is [420]. Thus, man is free, man is freedom. The morality does not come from some external moral standard, but from the fact that we are beings endowed with free will. Being free, we are left with no excuses for our actions. The best that we can do is to develop our rational faculties to foresee the consequences of our own actions upon the others, and then use our rationality to consciously withhold ourselves from doing certain actions that are harmful to others in spite of possible temptations to act otherwise. Once humanity finds a moral principle that is good for the whole society of human beings, it is then better to ingrain that principle into the law and the education system. For example, slavery was considered perfectly acceptable for thousands of years in human history, but at present we consider slavery an abomination. We no longer take other human beings for slaves, not because we are unfree to do so, but because we are capable of rationally reasoning and understanding that slavery is not good for us as humans.

Existentialism clashes seriously with determinism. Classical deterministic physics allows no free will and no objective morality. In contrast, quantum indeterminism guarantees the fundamental free will exhibited by quantum systems and endorses quantum existentialism. Because the physical laws are immutable and do not evolve, it follows that free will cannot evolve too. Foreseeing the consequences of one's own actions, however, could evolve as it is dependent on accumulated knowledge. Thus, our morality could evolve in time as we learn how to use our free will to build up a just society [411] in which all people could live in peace and realize their dreams to the extent that it does not interfere with someone else's happiness or freedom. *Sets of moral values* in a society of free individuals could be then viewed as *objective*, if they maximize the quality of life for as many individuals in the society as possible. Thus, for agents with free will, probably the most important of all moral principles is not to do to others what you do not want to be done to you.

6.6   Inner privacy of consciousness in quantum physics

Conscious experiences are subjective, private and inaccessible for external observers. Indeed, suppose that you are eating a chocolate while undergoing open skull neurosurgery. What the surgeon would see is the pinkish-gray, walnut-shaped, jelly-like substance of your brain. If a microscope or other measuring devices are used, it would be possible for the surgeon to further zoom in on individual neurons and record various complicated physical processes. But he would not be able to observe the taste of chocolate, because your experiences are inside your mind with a *kind of insideness* that is different from the way in which your brain is inside your head [352]. Namely, conscious experiences are *unobservable*.

6.6.1   Observability and unobservability

In classical physics, everything is observable (Section 3.19). Consequently, there is no room left for unobservable consciousness. A fundamental no-go theorem in quantum information theory, however, establishes that the quantum state vector $|\psi\rangle$ of any physical system is not observable (Theorem 15). As a corollary, one can deduce that the density matrix $\hat{\rho}$ of any quantum physical system is not observable too. Thus, quantum mechanics contains unobservable fundamental physical entities such as the state vector $|\psi\rangle$ or the density matrix $\hat{\rho}$ that, if identified with conscious experiences of the system, will not contradict the characteristic subjective, first-person accessibility of conscious experiences and their external inaccessibility by third-person observers [207]. Because Theorem 15 provides more than one unobservable physical entity, deciding whether consciousness should be identified with the quantum state vector $|\psi\rangle$ or the quantum density matrix $\hat{\rho}$ needs to be done with the help of additional considerations. Two important constraints are the need to explain what determines the boundaries of individual minds (Section 1.1) and what binds conscious experiences together (Section 1.2).

In Section 6.2, we have shown that the minds within minds problem is avoided if the quantum state vector, rather than the density matrix, of non-factorizable quantum systems is identified with a conscious mental state. The rationale is that all quantum systems have their own density matrix $\hat{\rho}$ (Section 4.18), whereas only quantum systems that are not entangled externally have their own state vector $|\psi\rangle$. As a result, it becomes clear that quantum subsystems of an entangled state would not have their own minds, and quantum entanglement is the glue that binds conscious experiences together. Nevertheless, the above rationale does not imply that our criterion of setting the boundaries of individual minds cannot be reformulated in the language of density matrices. The mathematics of quantum mechanics allows direct translation of statements written in the language of state vectors into the language of density matrices at the price of introducing several new quantum information concepts.

6.6.2 Quantum purity

Quantum purity γ is defined as

$$\gamma = \mathrm{Tr}\left(\hat{\rho}\hat{\rho}\right) = \mathrm{Tr}\left(\hat{\rho}^2\right) \tag{6.40}$$

Because the density matrix $\hat{\rho}$ is a Hermitian matrix $\hat{\rho} = \hat{\rho}^\dagger$, if one chooses the eigenvectors $|i\rangle$ of $\hat{\rho}$ as a representation basis, the matrix representation of $\hat{\rho}$ will be diagonal and exhibit the corresponding eigenvalues λ_i on the main diagonal

$$\hat{\rho} = \sum_{i=1}^{n} \lambda_i |i\rangle\langle i| = \begin{pmatrix} \lambda_1 & 0 & \cdots & 0 \\ 0 & \lambda_2 & \cdots & 0 \\ \vdots & \vdots & \ddots & \vdots \\ 0 & 0 & \cdots & \lambda_n \end{pmatrix} \tag{6.41}$$

Due to normalization of probabilities, the density matrix has a unit trace

$$\mathrm{Tr}\left(\hat{\rho}\right) = \sum_{i=1}^{n} \lambda_i = 1 \tag{6.42}$$

Further, it is straightforward to calculate that

$$\hat{\rho}^2 = \sum_{i=1}^{n} \lambda_i^2 |i\rangle\langle i| = \begin{pmatrix} \lambda_1^2 & 0 & \cdots & 0 \\ 0 & \lambda_2^2 & \cdots & 0 \\ \vdots & \vdots & \ddots & \vdots \\ 0 & 0 & \cdots & \lambda_n^2 \end{pmatrix} \tag{6.43}$$

$$\text{Tr}\left(\hat{\rho}^2\right) = \sum_{i=1}^{n} \lambda_i^2 \leq 1 = 1^2 = \left(\sum_{i=1}^{n} \lambda_i\right)^2 \tag{6.44}$$

All matrix calculations, including the calculation of the trace of a matrix, are easier to perform in the eigenbasis of the density matrix. The calculated trace, however, will hold true in general, because the trace of a matrix is a quantity that is invariant with respect to a change of basis.

For a pure state $\hat{\rho} = \hat{\rho}^2 = |\psi\rangle\langle\psi|$, there is a one-to-one correspondence between the density matrix and the state vector, $\hat{\rho} \leftrightarrow |\psi\rangle$, and the quantum purity is maximal

$$\gamma = \text{Tr}\left(\hat{\rho}^2\right) = \text{Tr}\left(|\psi\rangle\langle\psi||\psi\rangle\langle\psi|\right) = \text{Tr}\left(\hat{\rho}\right) = \text{Tr}\left[\begin{pmatrix} 1 & 0 & \cdots & 0 \\ 0 & 0 & \cdots & 0 \\ \vdots & \vdots & \ddots & \vdots \\ 0 & 0 & \cdots & 0 \end{pmatrix}\right] = 1 \tag{6.45}$$

For a mixed state $\hat{\rho} = \sum_{i=1}^{n} \lambda_i |i\rangle\langle i|$, in which there is at least one eigenvalue $0 < \lambda_k < 1$, the state does not have a corresponding state vector $|\psi\rangle$, and the quantum purity is submaximal

$$\gamma = \text{Tr}\left(\hat{\rho}^2\right) = \text{Tr}\left(\sum_{i=1}^{n} \lambda_i^2 |i\rangle\langle i|\right) = \sum_{i=1}^{n} \lambda_i^2 < 1 = 1^2 = \left(\sum_{i=1}^{n} \lambda_i\right)^2 \tag{6.46}$$

The maximally mixed state $\hat{\rho} = \frac{1}{n}\hat{I}$ has the minimal purity of $\frac{1}{n}$

$$\gamma_{\min} = \text{Tr}\left(\frac{1}{n^2}\hat{I}\right) = \text{Tr}\left[\begin{pmatrix} \frac{1}{n^2} & 0 & \cdots & 0 \\ 0 & \frac{1}{n^2} & \cdots & 0 \\ \vdots & \vdots & \ddots & \vdots \\ 0 & 0 & \cdots & \frac{1}{n^2} \end{pmatrix}\right] = n\frac{1}{n^2} = \frac{1}{n} \tag{6.47}$$

6.6.3 Quantum entropy

Quantum entropy S, or von Neumann entropy, measured in bits is

$$S = -\text{Tr}\left(\hat{\rho}\log_2 \hat{\rho}\right) = -\sum_{i=1}^{n} \lambda_i \log_2 \lambda_i \tag{6.48}$$

Pure states $\hat{\rho} = \hat{\rho}^2 = |\psi\rangle\langle\psi|$ always have a single eigenvalue of 1, and $n-1$ eigenvalues that are 0. Because the function $f(x) = -x\log_2 x$ vanishes for $x = 0$ and $x = 1$ (Fig. 6.8), pure states are characterized with *zero* quantum entropy

$$S = -1\log_2 1 - (n-1)0\log_2 0 = 0 \tag{6.49}$$

The zero quantum entropy reflects the fact that the system is with certainty in a single quantum state, namely, $\hat{\rho} \leftrightarrow |\psi\rangle$.

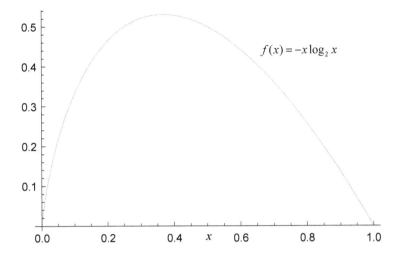

Figure 6.8 The function $f(x) = -x \log_2 x$ is concave within the unit interval $x \in [0,1]$ and has a maximum value at $x = \frac{1}{e} = 0.36787\ldots$

For a mixed state $\hat{\rho} = \sum_{i=1}^{n} \lambda_i |i\rangle\langle i|$, in which there is at least one eigenvalue $0 < \lambda_k < 1$, the quantum entropy is always positive (Fig. 6.8)

$$S = -\sum_{i=1}^{n} \lambda_i \log_2 \lambda_i > 0 \tag{6.50}$$

Theorem 26. *The quantum entropy S is a concave functional [208, 361, 515], meaning that if $p_1, p_2, \ldots, p_n \geq 0$ and $\sum_i p_i = 1$, then*

$$S\left(\sum_i p_i \hat{\rho}_i\right) \geq \sum_i p_i S(\hat{\rho}_i) \tag{6.51}$$

The maximally mixed state $\hat{\rho} = \frac{1}{n}\hat{I}$ has the maximal quantum entropy of $\log_2 n$

$$S_{\text{max}} = S\left[\left(\begin{matrix} \frac{1}{n} & 0 & \cdots & 0 \\ 0 & \frac{1}{n} & \cdots & 0 \\ \vdots & \vdots & \ddots & \vdots \\ 0 & 0 & \cdots & \frac{1}{n} \end{matrix}\right)\right] = -n\left(\frac{1}{n}\log_2 \frac{1}{n}\right) = \log_2 n \tag{6.52}$$

Quantum systems, for which $\hat{\rho} \leftrightarrow |\psi\rangle$ holds, have purity $\gamma = 1$ and quantum entropy $S = 0$. Every such system is in a pure quantum state $\hat{\rho} = \hat{\rho}^2$. Quantum systems that are part of an entangled state do not have their own state vector, but have mixed density matrices $\hat{\rho} \neq \hat{\rho}^2$ that are not pure $\gamma < 1$ and whose quantum entropy is positive $S > 0$. Thus, the postulate identifying the boundaries of individual minds (Axiom 6.1.1) equivalently states that only non-factorizable pure state density matrices $\hat{\rho} = \hat{\rho}^2$ with *zero* quantum entropy correspond to individual minds. In this form, the postulate becomes cumbersome to comprehend, unless the statement for a pure state with zero quantum entropy is viewed as a criterion for the existence of the state vector $|\psi\rangle$.

6.6.4 Quantum coherence

Quantum coherence and *decoherence* are frequently used in discussions on the feasibility of quantum approaches to consciousness [229, 233, 475]. Because quantum coherence is required for interference phenomena, it is an essential resource for quantum information processing [24, 33]. Quantum coherence can be quantified in several alternative ways including the *relative entropy of coherence* or the *ℓ_1 norm of coherence* [24, 33, 46, 529]. A distinctive feature of quantum coherence is its *basis dependence*.

Let $\hat{\rho}$ be the density matrix of a quantum system and $\{|\psi_i\rangle\}_{i=1}^n$ is a given orthonormal basis. The matrix representation of $\hat{\rho}$ in the given basis will be

$$\hat{\rho} = \begin{pmatrix} a_{11} & a_{12} & \cdots & a_{1n} \\ a_{21} & a_{22} & \cdots & a_{2n} \\ \vdots & \vdots & \ddots & \vdots \\ a_{n1} & a_{n2} & \cdots & a_{nn} \end{pmatrix} \tag{6.53}$$

where $a_{ij} = \langle\psi_i|\hat{\rho}|\psi_j\rangle$ are complex numbers. Since $\hat{\rho} = \hat{\rho}^\dagger$ is Hermitian, it also follows that $a_{ij} = a_{ji}^*$. In general, the quantum coherence in the given basis $\{|\psi_i\rangle\}_{i=1}^n$ will be dependent on the presence of non-zero off-diagonal entries.

The *relative entropy of coherence* C_{RE} in the basis $\{|\psi_i\rangle\}_{i=1}^n$ is defined as

$$\begin{aligned} C_{RE} &= S\left[\begin{pmatrix} a_{11} & 0 & \cdots & 0 \\ 0 & a_{22} & \cdots & 0 \\ \vdots & \vdots & \ddots & \vdots \\ 0 & 0 & \cdots & a_{nn} \end{pmatrix}\right] - S\left[\begin{pmatrix} a_{11} & a_{12} & \cdots & a_{1n} \\ a_{21} & a_{22} & \cdots & a_{2n} \\ \vdots & \vdots & \ddots & \vdots \\ a_{n1} & a_{n2} & \cdots & a_{nn} \end{pmatrix}\right] \\ &= S\left(\hat{\rho}_{\text{diagonal}}\right) - S\left(\hat{\rho}\right) \geq 0 \end{aligned} \tag{6.54}$$

The maximally coherent state $|\Psi\rangle$ in the basis $\{|\psi_i\rangle\}_{i=1}^n$ is a pure state given by

$$|\Psi\rangle = \frac{1}{\sqrt{n}} \sum_{i=1}^n |\psi_i\rangle \tag{6.55}$$

The maximally coherent density matrix $\hat{\rho}$ in the basis $\{|\psi_i\rangle\}_{i=1}^n$ is

$$\hat{\rho} = \frac{1}{n} \begin{pmatrix} 1 & 1 & \cdots & 1 \\ 1 & 1 & \cdots & 1 \\ \vdots & \vdots & \ddots & \vdots \\ 1 & 1 & \cdots & 1 \end{pmatrix} \tag{6.56}$$

The maximally coherent state has maximal relative entropy of coherence $C_{RE}^{\max} = \log_2 n$.

The maximally mixed quantum state $\hat{\rho} = \frac{1}{n}\hat{I}$ is invariant under change of basis, and as a consequence it has minimal relative entropy of coherence $C_{RE}^{\min} = 0$ in every basis.

The ℓ_1 *norm of coherence* C_{ℓ_1} is intuitively a clearer measure of quantum coherence defined with the use of the sum of all off-diagonal moduli

$$C_{\ell_1} = \frac{1}{n-1} \sum_{i \neq j} |\langle \psi_i | \hat{\rho} | \psi_j \rangle| = \frac{1}{n-1} \sum_{i \neq j} |a_{ij}| \tag{6.57}$$

The maximally coherent density matrix given by Eq. (6.56) has $(n^2 - n)$ off-diagonal entries of $\frac{1}{n}$, and consequently has the maximal ℓ_1 norm of coherence

$$C_{\ell_1}^{\max} = \frac{1}{n-1} \left(n^2 - n \right) \frac{1}{n} = \frac{n(n-1)}{(n-1)n} = 1 \tag{6.58}$$

Notice that if the maximally coherent density matrix $\hat{\rho}$ in the basis $\{|\psi_i\rangle\}_{i=1}^n$, is rewritten in its eigenbasis $\{|i\rangle\}_{i=1}^n$, it will become

$$\hat{\rho} = \sum_{i=1}^n \lambda_i |i\rangle\langle i| = \begin{pmatrix} 1 & 0 & \cdots & 0 \\ 0 & 0 & \cdots & 0 \\ \vdots & \vdots & \ddots & \vdots \\ 0 & 0 & \cdots & 0 \end{pmatrix} \tag{6.59}$$

Thus, in its eigenbasis the pure state $\hat{\rho}$ is not quantum coherent, since the ℓ_1 *norm of coherence* C_{ℓ_1} becomes *zero*.

Theorem 27. *For every pure state $|\Psi\rangle$ there is always an orthonormal basis in which the state is maximally coherent. Conversely, every maximally coherent state is pure.*

Theorem 28. *In the trivial orthonormal basis in which the pure state $|\Psi\rangle$ is one of the basis vectors, the quantum coherence is zero.*

Theorem 29. *Every maximally coherent composite (multiparticle) state is factorizable. Because quantum entangled states are non-factorizable, it follows that all quantum entangled states are not maximally coherent.*

Proof. Let the k-level quantum system A be in a maximally coherent state (in the basis $\{|A_i\rangle\}_{i=1}^k$) in k-dimensional Hilbert space $\mathcal{H}_A$ and the n-level quantum system B be in a maximally coherent state (in the basis $\{|B_j\rangle\}_{j=1}^n$) in n-dimensional Hilbert space $\mathcal{H}_B$. The state of the composite system $|\Psi_{AB}\rangle = |\Psi_A\rangle \otimes |\Psi_B\rangle$ is also maximally coherent (in the tensor product basis $\{|A_i\rangle|B_j\rangle\}_{i,j=1,1}^{k,n}$) in $k \times n$-dimensional Hilbert space $\mathcal{H}_A \otimes \mathcal{H}_B$

$$|\Psi_{AB}\rangle = \frac{1}{\sqrt{k}} (|A_1\rangle + |A_2\rangle + \ldots + |A_k\rangle) \otimes \frac{1}{\sqrt{n}} (|B_1\rangle + |B_2\rangle + \ldots + |B_n\rangle) \tag{6.60}$$

$$= \frac{1}{\sqrt{k}} \sum_{i=1}^k |A_i\rangle \otimes \frac{1}{\sqrt{n}} \sum_{j=1}^n |B_j\rangle = \frac{1}{\sqrt{kn}} \sum_{i=1}^k \sum_{j=1}^n |A_i\rangle|B_j\rangle \tag{6.61}$$

$$= \frac{1}{\sqrt{kn}} (|A_1\rangle|B_1\rangle + |A_1\rangle|B_2\rangle + |A_2\rangle|B_1\rangle + \ldots + |A_k\rangle|B_n\rangle) \tag{6.62}$$

Doing the above calculation in the opposite direction transforms the maximally coherent composite state given by Eq. (6.62) into the nicely factorized state given by Eq. (6.60). $\qquad \square$

Since quantum coherence is basis dependent, it is not very useful for the axiomatic formulation of a quantum theory of consciousness. Even worse, Theorem 29 implies that if every maximally coherent state corresponded to a conscious mind, then the collection of any two non-interacting minds would comprise another global mind since the tensor product of any two maximally coherent states is another maximally coherent state. Previous works have only considered that quantum mind theories need to be supported by *quantum coherent states* [233, 475]. Because maximally coherent quantum states are described by pure state density matrices $\hat{\rho} = \hat{\rho}^2 = |\psi\rangle\langle\psi|$, it follows that to every maximally coherent quantum system always there is a corresponding state vector $|\psi\rangle$ (Section 6.6.4). Conversely, the existence of the state vector $|\psi\rangle$ implies that the quantum system is maximally coherent in some basis. Quantum coherence, however, does not provide a plausible rule for outlining the boundaries of individual minds, because it suffers from the *minds within minds* problem as exemplified by Theorem 29 and the discussion following Eq. (6.2). To avoid the possibility of minds within minds, we have foresightedly postulated that only non-factorizable pure quantum states correspond to individual minds (Axiom 6.1.1). Noteworthy, from Axiom 6.1.2 and Theorem 29, it follows that every multiparticle maximally coherent state corresponds to a *collection of minds* in quantum information theory of consciousness.

6.6.5   Communicability and incommunicability

Conscious experiences are private, subjective and inaccessible to external observers. Theorem 20 together with Axiom 6.1.1 provide insight into the origin of the inner privacy of conscious experiences, namely, if consciousness is composed of quantum information then it cannot be converted into bits of classical information. Deriving only the inner privacy of conscious experiences, however, leaves certain important problems untouched. Confronted with unobservable consciousness, philosophers have had a hard time of explaining how is it possible that we can communicate anything about our conscious experiences to others. Indeed, classically if something is communicable then it is observable, or by modus tollens, if it is unobservable it is incommunicable. Ludwig Wittgenstein (1889–1951) used two related arguments, namely, the *private language argument* and the *beetle in the box argument*, in order to highlight the problem [224].

Example 6.9. (*Wittgenstein's private language argument*) *Conscious experiences including my believing, seeing, imagining and loving are inner, private and inaccessible to anyone else. That very claim, however, is expressed in words that we all understand: "believing," "seeing," "imagining" and "loving." We have learned these words from our parents or teachers through correcting incorrect uses and praising correct uses of the words [224]. Thus, if consciousness is inaccessible to anyone else, then we should not have learned these words. Yet, we have learned these words; therefore it appears that consciousness has to be accessible [523, §243–271].*

Example 6.10. (*Wittgenstein's beetle in the box argument*) *Suppose that each of us had a beetle in a box into which no one else could look. If I say "My beetle is fiddled-edee," you may answer "Mine is too" or "No, it is more flummadiddle than fiddlededee." Such conversation is nonsensical and the words like "fiddlededee" or "flummadiddle" can never acquire any meaning. Since mental terms possess meaning, it appears that consciousness cannot be private [523, §293].*

Qualia are not subject to exteriorization because we do not have a way to communicate in words or symbols what qualia are. Still, qualia can be introspectively compared and certain relationships between qualia can be encoded and communicated. For example, sounds can be loud or low, pleasant or unpleasant, etc., meaning that there is some order that can be captured in words and communicated, even though the phenomenal nature of each sound quale cannot be communicated. Classical understanding of information is very restrictive and incapable of reconciling the inner privacy of conscious experiences with the undeniable fact that we can talk about our experiences in a meaningful way. Quantum information theory, however, provides a deeper insight into the problem. Even though the quantum information carried by quantum systems is not observable (Theorem 15) and cannot be converted into classical information (Theorem 20), each quantum system can carry a certain amount of accessible classical information subject to Holevo's theorem. Indeed, non-orthogonal quantum states cannot be distinguished through measurement or observation, whereas orthogonal quantum states can. Thus, something meaningful can be communicated in the form of classical information about conscious experiences that are generated by orthogonal quantum states. In essence, the quantum information is not completely inaccessible; there is some accessible part that is bounded by a certain amount of bits of classical information (Holevo's theorem). Therefore, both the *private language argument* and the *beetle in the box argument* are deficient when viewed within the framework of quantum information theory. Only if we were able to say everything there is about our consciousness, it would have followed that consciousness is observable or accessible. From the fact that we can say something meaningful about our conscious experiences does not follow that we can say everything there is about these experiences. Hence, consciousness can be private insofar we can communicate only a limited number of meaningful things (classical bits) in regard to the content of our conscious experiences.

That certain aspects of our conscious experiences are not communicable can be shown by the inverted qualia thought experiment (Example 6.11). In particular, it can be established that when we talk about our sensations, we never actually communicate the phenomenal aspect of the qualia associated with those sensations; rather we communicate the objective circumstances under which our sensations occurred.

Example 6.11. *(Locke's inverted qualia thought experiment) Suppose that there is a person who subjectively experiences yellowness when he is looking at a violet and blueness when he is looking at a marigold. The conscious experiences of that person would be qualitatively inverted compared to what you may experience when looking at the flowers violet or marigold [320, p. 257]. Because we neither have direct access to someone else's mind, nor are we able to communicate to others what the blueness or the yellowness of our experiences are, we can never be sure that others do experience the same thing when put into an identical situation or can experience what we are capable of experiencing. Thus, two people can both agree that they see a yellow marigold, even though one of them may have inverted qualia compared with the other. Moreover, each person is entitled to consider his quale to be normal, because there is no objective test that can determine whose quale of yellowness is the normal one.*

Example 6.12. *(It is impossible to explain what the colors are to a color blind person) The incommunicability of conscious experiences can be better appreciated by considering real color blind people instead of hypothetical individuals with inverted qualia. Color blindness is a genetically inherited sex-linked condition due to mutations in the genes that produce retinal photopigments in the eye. Because color blind people do experience some color qualia when they look at certain colors, usually they are unaware of their condition before they get tested for color blindness in a medical check. What the medical tests can reveal is that the individual does not see a difference between two or more different colors that are distinguished by people who are not color blind. The tests cannot find, however, what color exactly is experienced by the color blind person. If you were able to explain in words what color exactly you are seeing, you could have been able to cure color blind people only with your words. Since you do not have the power to cure color blind people by words, it follows that conscious experiences are incommunicable. The best you can do is to compare your experiences in different experimental situations and then report whether the experiences you had were the same or not.*

The three philosophers from the Vienna circle, Rudolf Carnap (1891–1970), Hans Hahn (1879–1934) and Otto Neurath (1882–1945), have succinctly summarized the scientific attitude toward the *incommunicability* of the phenomenal nature of qualia as follows:

> A scientific description can contain only the *structure* (form of order) of objects, not their 'essence'. [...] Subjectively experienced qualities – redness, pleasure – are as such only experiences, not [communicable] knowledge; physical optics admits only what is in principle understandable by a blind man too. [71, pp. 309–310]

Wittgenstein further warns us that attempting to communicate the incommunicable is logically inconsistent:

> What we cannot speak about we must pass over in silence. [522, p. 111]

Asking how it is possible that incommunicable aspects of our conscious minds do exist, however, is a legitimate scientific question. The quantum information approach to consciousness, including Theorems 15, 20 and 21, explains the origin of the inner privacy of consciousness and the related communicability/incommunicability of conscious experiences. Furthermore, it provides comprehensible and consistent analysis of certain philosophical problems (such as those that arise from the private language argument or the beetle in the box argument) that appeared intractable within classical physics.

6.6.6 Quantum support of classical information

Quantum systems are able to encode classical information and execute classical algorithms. The converse, however, is not true. Classical systems are unable to support quantum information and are incapable of executing quantum algorithms.

To show that quantum systems can support classical information, we can choose a single representation basis $\{e_i\}_{i=1}^{n} = |e_1\rangle, |e_2\rangle, \ldots, |e_n\rangle$ in the Hilbert space $\mathcal{H} = \mathbb{C}^n$ of n-level quantum system Q and encode a string of $\log_2 n$ classical bits using the fact that the basis states are orthogonal and distinguishable from one another, namely, $\langle e_i | e_j \rangle = 0$ for $i \neq j$. To distinguish between the different orthogonal quantum states, we could perform a measurement using the operator $\hat{M} = \sum_i \lambda_i |e_i\rangle\langle e_i|$ such that all eigenvalues λ_i are distinct, namely, $\lambda_i \neq \lambda_j$ for $i \neq j$. Since from the measured eigenvalue λ_i we can uniquely determine the state $|e_i\rangle$, encoded strings of classical bits could always be retrieved with certainty from a quantum system Q provided that we keep the system Q undisturbed.

Example 6.13. *(Quantum encoding of classical information) For $n = 4$, we can encode a string of two classical bits as follows: $|e_1\rangle \leftrightarrow 00$, $|e_2\rangle \leftrightarrow 10$, $|e_3\rangle \leftrightarrow 01$, $|e_4\rangle \leftrightarrow 11$. Since we have all possible 2-bit messages encoded in orthogonal quantum states, we can send every single one of them to another receiver who knows the representation basis and uses the operator $\hat{M} = \sum_i \lambda_i |e_i\rangle\langle e_i|$ to read the message. In particular, if we want to send the string of bits 10, we prepare the state $|e_2\rangle$ and send it to the receiver. Since $\hat{M}|e_2\rangle = \lambda_2|e_2\rangle$, the receiver measures λ_2 with certainty and correctly recovers the message 10, as would be the case if we had used a classical channel for communication.*

6.6.7 Quantum versus classical computation

The physical reality of quantum information carried by quantum systems is exemplified by the fact that *classical computers* cannot perform quantum computation. In order to execute a quantum algorithm one needs a quantum computer that is physically real. In this sense, *quantum computers* tap directly into the fundamental fabric of physical reality. Typically, at the end of the quantum computation a quantum measurement is performed that collapses all of the qubits in the computational basis $|0\rangle, |1\rangle$. Due to the probabilistic nature of the collapse, the outcome of the quantum measurement may not necessarily give the correct answer to the problem whose solution is searched by the quantum algorithm. For example, it may be the case that the correct answer is obtained only half of the time, hence with probability of success that is $p = 0.5$. To check with absolute certainty whether the obtained answer is correct or not, one needs to run a classical algorithm on a classical computer. If the answer is correct, the problem is solved, but if the answer is incorrect, the quantum algorithm needs to be run again and again until the correct answer is obtained. For n runs, the probability of success will be $p = 1 - 2^{-n}$, which tends to 1 for $n \to \infty$. Thus, to use in practice any form of quantum computation, both a quantum and a classical computer will be needed.

One may naively suppose that classical computers are better than quantum computers, because classical algorithms output the correct answer with certainty. Such a viewpoint, however, fails to consider the speed of computation and the amount of physical time needed to complete the algorithm. Quantum computers are more powerful than classical computers due to their incredible speed gained from utilization of quantum superpositions and quantum entanglement.

Definition 6.4. *(Time complexity) The time complexity of an algorithm quantifies the amount of time t taken by the algorithm to run as a function of the length x of the input string of symbols. If all steps in the algorithm take a unit of time to perform, then the time complexity counts the number of steps N needed to complete the algorithm.*

The time complexity is expressed with the use of *big $\mathcal{O}$ notation* that describes the limiting behavior of a function $f(x)$ when the argument tends toward infinity, $x \to \infty$. More precisely, if $f(x)$ and $g(x)$ are two functions, one writes

$$f(x) = \mathcal{O}(g(x)) \tag{6.63}$$

if and only if there exists a positive real constant $C \in \mathbb{R}$ and a real argument value $x_0 \in \mathbb{R}$ such that $|f(x)| \le C|g(x)|$ for all $x \ge x_0$.

Definition 6.5. *(Polynomial time algorithm) An algorithm is solvable in polynomial time if the number of steps N required to complete the algorithm for a given input x is at worst $N(x) = \mathcal{O}(x^k)$ for some $k > 0$.*

Definition 6.6. *(Exponential time algorithm) An algorithm is solvable in exponential time if the number of steps N required to complete the algorithm for a given input x is at worst $N(x) = \mathcal{O}(2^x)$.*

Since exponential growth overwhelms polynomial growth for large x (Fig. 6.9), problems that can be solved by classical algorithms no faster than in exponential time are considered to be computationally hard. If you have guessed a solution for such a hard problem, however, it is usually easy to check whether the solution is correct or not in polynomial time. Quantum computers are capable of exploiting this asymmetry between finding a solution and checking the correctness of the solution. Currently, for finding the solution of a certain type of hard problem there are quantum algorithms that are solvable in polynomial time, but there are only exponential time classical algorithms [93, 283]. Given a polynomial time classical algorithm for checking the correctness of a guessed solution, it would be much faster to run the polynomial time quantum algorithm several times and check the

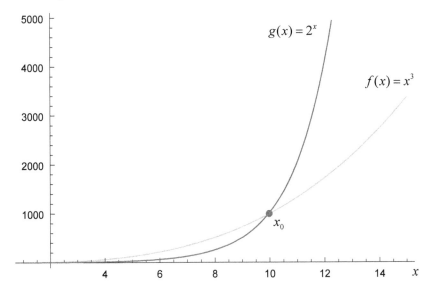

Figure 6.9   Comparison of polynomial and exponential growth. The exponential function $g(x) = 2^x$ exceeds a test polynomial function $f(x) = x^3$ for all x greater than $x_0 = 9.93954...$ For example, $g(10) = 2^{10} = 1024 > f(10) = 10^3 = 1000$.

correctness of the quantum answers, rather than solving the problem with any of the exponential time classical algorithms. Thus, the speed achieved by quantum computers is payed for by the uncertainty in the correctness of the quantum results obtained. Nonetheless, this is a fair and reasonable price since quantum algorithms can be run multiple times until the correct answer is found.

Remarkably, the work of human intuition for solving hard problems closely resembles the work of quantum computers. Introspectively, it appears to us that after a period of time spent on thinking about the problem, intuitive answers pop out as possible solutions and then we need to do classical checks to see whether the answers are correct or not. In most cases, our intuition does not succeed from the first attempt, yet after multiple trials, we may eventually find the correct solution. Thus, the inner workings of human intuition are hard to explain classically by a random guess-and-check mechanism, but they become explicable if the human mind is viewed as executing a quantum algorithm followed by classical checks by electrically active neural networks in the brain.

6.7   Mind–brain relationship in quantum physics

Through our senses we obtain information of how the surrounding world is, whereas with our conscious decisions we change the world toward our vision of what we would like the world to be. In order to transform the surrounding world so that it suits our needs, we need a continuous sensory feedback both for correcting execution errors and for optimizing our actions so that they achieve more efficiently the desired goals.

The classical evolutionary approach asserts that it is our *brain* that changes the surrounding world. The brain (Figs. 1.3 and 1.4) is connected through the nerves with the rest of the body (Fig. 1.2). The electric activity of sensory nerves brings sensory information from the surrounding world and from our body toward the brain. Conversely, the brain influences the surrounding world through the electric activity of motor nerves that control our body muscles whose contraction leads to body motion. Thus, neuroscience provides a comprehensible description of the interaction between the brain and the surrounding world through a bidirectional flow of classical information. Unfortunately, the *brain* cannot be identified with the *mind* because anesthetized brains, crushed brains or dead brains do not generate conscious experiences (Section 5.3.1). Yet, understanding how the mind and the brain relate to each other is all that is needed for a complete explanation of how our conscious minds are able to interact with the world we live in. If the mind can change the brain, then the mind will be able to also change the surrounding world through the electric activity of the motor nerves originating from the brain (Fig. 1.5). Conversely, if the brain can change the mind, then the surrounding world will be able to also change the mind through the sensory information delivered to the brain in the form of electric impulses coming from the sensory organs (Fig. 5.4).

6.7.1 Quantum interactionism

The axioms of quantum information theory of consciousness (Section 6.1) lead naturally to a form of *quantum interactionism* in which the mind and the brain are physically distinct but capable of interacting with each other (Fig. 5.13). The brain informs the mind by inputting classical sensory information, whereas the mind transforms the brain using its free will to make choices among multiple future courses of action. The cycles of mind–brain interaction occur repeatedly.

Each *conscious mind* is modeled with a single non-factorizable quantum state $|\psi\rangle$ of multiple quantum entangled physical components (Axiom 6.1.1). The wave function ψ represents the unobservable quantum information that makes up the fabric of the conscious mind. Because the quantum information is composed of quantum probability amplitudes for potential future events, the conscious mind is endowed with free will to actualize one of those events in a quantum jump (Fig. 6.6). Here, it is important to note that free will is manifested only by sufficiently large minds that reach a certain energy threshold $\mathcal{E}$ for objective reduction (Axiom 6.1.3). Smaller minds that have not attained the energy threshold $\mathcal{E}$ can only evolve unitarily according to the Schrödinger equation (4.157) and entangle with neighboring minds in order to bind their conscious experiences together into larger, more complex minds (Figs. 6.2 and 6.3).

In contrast to the unobservable mind, the *brain* is observable. Therefore, if the mind is composed of quantum information, the brain has to be the accessible n bits of classical information allowed by Holevo's theorem. Because the accessible information extracted from a quantum system is obtained in the process of measurement, and we require the measurements to produce a single outcome rather than a superposition of Schrödinger's cat states (Section 6.5.4), the brain should emerge out of the process of objective reduction given by $|\Psi'\rangle \rightarrow |\psi_1'\rangle \otimes |\psi_2'\rangle \otimes ... \otimes |\psi_n'\rangle$. In particular, the brain has to be modeled with the accessible classical information that characterizes the tensor product collection of individual minds. For example, if the objective reduction generates a tensor product of qubits $|1\rangle \otimes |0\rangle \otimes |0\rangle \otimes |1\rangle \otimes ... \otimes |0\rangle$, the brain state will be given by the string of classical bits $1001...0$, while the quantum state $|1\rangle \otimes |0\rangle \otimes |0\rangle \otimes |1\rangle \otimes ... \otimes |0\rangle$ will be a collection of n individual minds according to Axiom 6.1.2.

Example 6.14. *(The brain is the classical information that records the past mind choices) The accessible classical information that records the past mind choices has to be the brain. Just before the objective reduction takes place, the composite quantum state $|\Psi'\rangle$ of n qubits can be expressed in the basis in which the reduction takes place as*

$$
\begin{aligned}
|\Psi'\rangle &= \sum_{i=1}^{2^n} a_i |\psi_i'\rangle \\
&= a_{00...0}|0\rangle|0\rangle...|0\rangle + a_{10...0}|1\rangle|0\rangle...|0\rangle + a_{01...0}|0\rangle|1\rangle...|0\rangle \\
&\quad + a_{00...1}|0\rangle|0\rangle...|1\rangle + a_{11...0}|1\rangle|1\rangle...|0\rangle + a_{10...1}|1\rangle|0\rangle...|1\rangle \\
&\quad + a_{01...1}|0\rangle|1\rangle...|1\rangle + ... + a_{11...1}|1\rangle|1\rangle...|1\rangle
\end{aligned}
\tag{6.64}
$$

There are 2^n possible outcomes $|\psi_i'\rangle$, each of which can be actualized with the corresponding probability $Prob(i) = |a_i|^2$. Once a given outcome is chosen, say $|1\rangle|1\rangle...|1\rangle$, the mind $|\Psi'\rangle$ disentangles into n qubits whose pure states are consistent with the chosen outcome. For the outcome $|1\rangle|1\rangle...|1\rangle$, each qubit gets transformed into the state $|1\rangle$. Since the basis for objective reduction is fixed, the chosen outcome can be encoded with n bits of classical information as $11...1$. The brain state $11...1$ is the classical record of the mind choice $|1\rangle|1\rangle...|1\rangle$. Because the objective reduction is irreversible, the mind cannot undo past choices once they are actualized. Thus, the brain could be viewed as the accumulated classical record of past mind choices.

Example 6.15. *(The brain is the mental image constructed from classical information in the mind of the observer) The classical information that is referred to as the brain, is experienced by the individual minds that have witnessed the objective reduction. If the process of objective reduction is given by $|\Psi'\rangle \rightarrow |\psi_1'\rangle \otimes |\psi_2'\rangle \otimes ... \otimes |\psi_n'\rangle$, the individual minds $|\psi_i'\rangle$ that are the product of the reduction are also observers witnessing the reduction. If some of those $|\psi_i'\rangle$ are emitted as quantum particles into the environment they will deliver the witnessed information to other remote observers as well. For example, visual photons reflected from a living brain and focused through a microscope could provide an image of the neurons inside the brain, whereas electromagnetic fields recorded within or around the brain could indicate which neurons are electrically active and which are not. Suppose that you are observing the brain of your friend who is undergoing open skull surgery. Your friend could be conscious, but you will not observe his conscious experiences. What you will observe and experience in your mind will be the electric activities of neurons inside the brain of your friend. These electric activities will be the outcome of the decisions made by the conscious mind of your friend. Classically, it is assumed that the brain exists with all of its physical properties regardless* *of whether someone is observing it or not. In a quantum physical world, however, the outcomes of quantum measurements are contextual and could not have been preexisting before the measurement is done (Kochen–Specker theorem). Thus, the brain is the observable classical information that is witnessed by others when a large mind makes a choice. In other words, the brain does not exist before the mind creates it. On the other hand, because the brain is information that exists within the mind of the observer, it can affect the probability distribution of future choices made by the observer.*

Because the *brain* is accessible classical information (Section 3.19), it can be observed, shared, communicated, copied and analyzed by multiple observer minds. The observability of the brain makes neuroscience possible due to the fact that multiple scientists can have simultaneous access to multiple identical copies of information about the same brain. Furthermore, it is consistent to say that the brain is both a past record of someone's mind choices and a mental image in another observer mind. Classical information remains the same regardless of the medium in which it is encoded. A poem remains exactly the same poem without regard to being written on paper, carved in stone, or typeset on a personal computer. In a sense, the simulation of classical information is equivalent to a copy of the original classical information. In contrast to classical functional approaches to consciousness, which claim that consciousness can be uploaded to a computer chip (Section 5.3.1), here we have developed a quantum information theory (Section 6.1) according to which it is actually the brain that can be uploaded to a chip as classical information, whereas the conscious mind as quantum information is inseparable from the quantum physical system it is attached to.

6.7.2   Quantum panpsychism

Because the Hilbert space of the universe $\mathcal{H}_U$ hosts a *collection of minds*, it could be said that the physical reality is fundamentally built up from mental stuff. Quantum entanglement of individual mental units leads to binding of their conscious experiences into larger and larger conscious minds (Section 6.3), until the energy threshold $\mathcal{E}$ for objective reduction is reached and the overgrown mind is forced by physical laws to make a decision (Section 6.5). In the particular case, when the *human mind* makes a decision, the accessible classical information is the *human brain* (Figs. 1.3, 1.4 and 6.3). Conceptually, the distinction between the *mind* and the *brain* is equivalent with the distinction between what a physical system *is* and what a physical systems *appears to be*. From this, however, does not follow that all physical systems should look like human brains. The accessible information from the water looks like water, from the rocks looks like rocks, and so on.

Due to the *chemical stability* of the biomolecules and their specific organization in the living neurons, the objective reduction in the human brain does not lead to a tensor product of quantum particles that all fly apart. Instead, only a minor part of the product is thermal radiation that is emitted, while the remaining biomolecules such as phospholipid membranes, protein voltage-gated ion channels, etc. (Fig. 1.8) are able to quantum entangle with water molecules and dissolved ions to form again a complex human mind that is only slightly different from the one before the reduction. Because the objective reductions occur at a very fast rate of over 100 GHz [207], the human mind makes as many choices, and with each choice the available future courses of action become irreversibly changed. Since the classical information for the past mind choices is the brain, it could be said that the brain sets the boundary conditions for solving the Schrödinger equation of the mind and directly affects the probability distribution for future mind courses of action. Thus, quantum interactionism endorses a picture in which the mind is engaged in making choices, whereas the brain is manifested as the causal influence of past mind choices on the probability distribution for future mind choices. The continuous existence of a complex mind such as the human mind requires that the majority of the disentangled component subsystems following the objective reduction stay together for multiple cycles of entanglement-disentanglement. In the case when the majority of disentangled components stay together, it is meaningful to say that there is a mind that interacts with its brain. The small portion of emitted disentangled components that inform other observer minds about someone's mind decisions could be viewed as a means of communication of someone's mind through its brain to other observer minds.

Example 6.16. *(Fleeting existence of collections of minds in inanimate objects) The picture of a single mind interacting with its brain becomes inapplicable if after the objective reduction all disentangled component subsystems fly apart. Consider the water in lakes, rivers or oceans. The quantum entanglement will produce minds whose existence will be just the time for a single objective reduction (Fig. 6.2). Because the water molecules are in fluid motion, once the objective reduction occurs the individual water molecules will fly apart and mix with other water molecules from nearby minds. Such a fleeting existence of minds in the inanimate matter gives the impression that the lakes, rivers or oceans are unconscious. Indeed, inanimate objects do not possess a single mind, because they are a collection of ephemeral minds popping in and out of existence. The apparent determinism in inanimate objects comes from the predictability of the probability distribution for large numbers according to the Born rule. For example, sending a single photon toward a beam splitter will result in an unpredictable outcome since either the photon will be transmitted or it will be reflected. The outcome of the same experiment performed with a beam of billions of photons, however, is easier to predict since about half of the photons will be reflected and about half of the photons will be transmitted.*

The fleeting existence of collections of minds in inanimate objects explains why quantum panpsychism does not have the classical problem with turning the consciousness on or off. Even though all physical systems possess mental properties, we do not see evidence of conscious activity in dead brains because these do not support a single conscious mind, but rather a stochastic collection of short-lived minds. When the human mind dies, the consciousness is not just turned off, but a process of mind disbinding and disintegration occurs that is opposite to the process of mind binding and integration. As the postmortem brain decays, the elementary conscious minds obtained through disbinding and disintegration get dispersed and the original complex human mind is never revived again.

Example 6.17. *(The quantum brain) Classically, the word "brain" stands both for the anatomical organ that exists inside the skull and for what can be observed when someone is looking at that organ. This is because in classical physics everything that exists is observable. In quantum theory, however, what exists is not what can be observed. Consequently, in the quantum context one needs to explicitly make it clear whether the word "brain" refers to the "quantum brain" that is the physically existing anatomical organ inside the skull or the "observable brain" that is the classical information about that anatomical organ obtained through observation. Within the quantum information theory of consciousness, the physically existing quantum brain that is described by a density matrix $\hat{\rho}$ may correspond to one of the following three cases: First, if the brain density matrix $\hat{\rho}$ is mixed, then it describes a fictitious collection of mind parts that belong to different minds. This is because mixed density matrices provide an incomplete description of the physical reality and represent fictitious physical systems created by arbitrary assembly of physical particles disregarding any quantum correlations with other external physical particles that remain outside of this arbitrary assembly. Second, if the brain density matrix $\hat{\rho}$ is pure and factorizable, then it describes a collection of minds each of which is a whole mind. Because the density matrix of the universe is always pure and factorizable due to ongoing objective reductions, the universe is always a collection of whole minds. Third, if the brain density matrix $\hat{\rho}$ is pure and non-factorizable, then it corresponds to a single mind. Thus, the quantum information theory of consciousness in which each mind is identified with a single non-factorizable state vector $|\psi\rangle$ is a form of identity theory, but one that is severely restrained in such a way to both provide a well-defined mind boundary and avoid the minds within minds problem.*

Example 6.18. *(Quantum brain in a vat) The brain communicates with the rest of the body through peripheral nerves (Fig. 1.2) that conduct classical information in the form of neural electric impulses. Afferent sensory nerves deliver sensory information from the body to the brain cortex (Fig. 5.4), whereas efferent motor nerves deliver motor information from the brain cortex to the body (Fig. 1.5). Because the brain cortex hosts the human mind, it is possible to disembody the human mind by dissecting the brain with its nerves from the body, immersing the brain into a vat filled with an electrolyte solution, and then reconnecting the nerves to electrodes that are controlled by a silicon computer chip. If the computer chip sends the appropriate electric impulses to the brain, the brain could be tricked into believing that it has a body while actually it has none. Classically, if the brain in a vat can be tricked into falsely believing that it has a body, then it can also be tricked into believing that the physical reality is different from what it is. In the quantum information theory of consciousness, however, there are multiple constraints that make the trickery hardly possible. First, a classical computer cannot perform quantum tasks. Therefore, in order to create virtually the quantum world that we see, the classical computer should either have access to a quantum reality to be interrogated on our behalf, or else the computer should already have a classical recording of the outcomes of a quantum physical system interrogated in the past. Second, a classical computer cannot support conscious experiences because classical information is observable, yet we have a privileged, direct introspective access to our own conscious experiences that are unobservable from a third person point of view. Thus, we cannot* *be a simulation inside a classical computer. Third, because conscious minds have a genuine free will, in order to create virtually the minds of other people with whom we interact, the classical computer should either have access to other human minds to be interrogated on our behalf, or else the computer should already have a classical recording of the replies from other human minds interrogated in the past. Fourth, the classical computer cannot deterministically predict in advance what our questions would be, but has to continuously interact with us. Since potentially we could ask any question, we could also ask questions about our place in the cosmos, our origins and our evolutionary history. Because the classical computer has to provide us with the answers obtained through interrogation of physically existing quantum systems at some point in time, present or past, we could use those answers to reconstruct what the true quantum physical reality looks like. Remarkably, a classical computer can trick us only with false facts that can be classically simulated such as how many moons there are in orbit around Mars; yet, because quantum effects cannot be classically simulated, if quantum physical systems never existed it would follow that we could not be brains in a vat tricked by a classical computer that we live in a quantum world. In essence, the classical computer can feed us with classical data produced by a quantum system, but cannot itself produce that data. For example, quantum computers can act as genuine random number generators, whereas classical computers can only be pseudorandom number generators [508]. Hence, once the algorithm behind the pseudorandom numbers is discovered, the trickery by the classical computer will be unmasked.*

6.7.3   Comparison with other quantum theories of mind

The merits of the axiomatic quantum information theory of consciousness could be better appreciated through comparison with other proposals for quantum theories of mind such as the Hameroff–Penrose Orch OR model [231, 232, 233] and Stapp's quantum Zeno model [451, 452]. Both of these models have been criticized extensively in previous works [206, 208]. Here, we will highlight the important conceptual differences in regard to the mind–brain relationship.

The *Hameroff–Penrose Orch OR model* [231, 233] postulates that conscious experiences occur at a frequency of 40 Hz in the form of discrete conscious *events* referred to as conscious *flashes*, conscious *bings*, or conscious *nows*, which emerge from orchestrated objective reduction (Orch OR) events inside stable neuronal microtubules. These emergent conscious Orch OR events are separated by prolonged 25 ms periods of subconscious quantum computation performed by tubulins in microtubules. To avoid decoherence, neuronal microtubules are further assumed to be shielded from the disturbing electric activity of neurons by a surrounding Debye layer of counterions, or through putative cycles of *actin* gel-sol transitions [229, 527]. Still, in order to be able to input the sensory information carried by the neuronal electric signals, dendritic microtubules have to be orchestrated by the attachment or detachment of *microtubule-associated proteins* (MAP2) [229, 233]. At the end, the output from the quantum computation in microtubules has to be delivered by a yet unknown mechanism to the voltage-gated ion channels in the axonal hillock where neuronal action potentials are generated [527].

The computational basis $|0\rangle$, $|1\rangle$ of each tubulin is assumed to consist of two macroscopically distinguishable conformations or dipole orientations. Based on Einstein's theory of general relativity, Hameroff and Penrose claimed that the tubulin states $|0\rangle$ and $|1\rangle$ correspond to clearly defined energy distributions and to well defined *space-time geometries*, whereas quantum superpositions of the form

$$a_1|0\rangle + a_2|1\rangle, \qquad \sum_i |a_i|^2 = 1 \tag{6.65}$$

do not correspond to clearly defined energy distributions and would lead to superpositions of different space-time geometries, which is a particularly awkward situation from the physical point of view [233]. Consequently, such macroscopic quantum superpositions were regarded as unstable even without environmental entanglement, and prone to decay to either $|0\rangle$ or $|1\rangle$, with relative probabilities $|a_1|^2 : |a_2|^2$ in a certain time scale T taken from the uncertainty principle to be

$$E = \frac{\hbar}{T} \tag{6.66}$$

where E is the gravitational self-energy of the displaced tubulins. During the period of quantum computation, the tubulins exhibit a certain pattern of quantum coherence in the computational basis, which is then reduced by the Orch OR event. In Section 6.6.4, we have shown that quantum coherence is not suitable for defining what the conscious mind is and where the mind boundaries are. The conscious experiences in the Orch OR model, however, are emergent events not directly related to the quantum wave function ψ. The mind is claimed to emerge from the objection reduction, whereas the objective reduction is supposed to be preceded by prolonged subconscious quantum activity given by the unitary Schrödinger evolution of the quantum wave function ψ. As a result, since the quantum coherence of ψ is related to subconscious activities, the minds within minds paradox does not affect the emergent consciousness in the Orch OR model. Nonetheless, multiple problems in regard to the mind–brain relationship remain:

(1) The emergence of conscious experiences from repeated objective reductions is as miraculous as in the classical theory of emergent consciousness.

(2) If consciousness is generated by the objective reduction, it is not possible for the conscious mind to exhibit free will. Indeed, if the indeterministic event is not under conscious control, then the external dice throwing argument by free-will skeptics holds (Section 6.5.5). Because the objective reduction is the only indeterministic event in quantum physics, it is necessary for consciousness to be already present in order to causally affect the outcome of the reduction (Section 6.5.1).

(3) The outcome of the objective reduction is accessible as classical information, hence it is in principle observable. Because consciousness is inner, private and unobservable, it should be the case that the Orch OR event has to create extra unobservable mental stuff besides the accessible classical information.

(4) It is incomprehensible how the brain and the mind interact. If the brain is identified with the physical substrate that supports both the Orch OR event and the subconscious quantum physical activity before the Orch OR event, it is hard

to explain where the conscious experiences affect causally the brain dynamics. Apparently consciousness cannot affect the brain dynamics through the Orch OR outcomes, because it is exactly the Orch OR outcomes that generate the emergent conscious experiences [231, Fig. 12].

Stapp's quantum Zeno model is an alternative proposal for a quantum theory of mind that puts certain aspects of the process of objective reduction under direct conscious control [451, 452]. Stapp describes the interaction between the mind and the brain with the use of three basic processes 1, 2 and 3, attributed to John von Neumann [451, 452]. In modern terminology, these processes can be referred to as (1) projective measurement, (2) unitary evolution and (3) objective reduction [208]. Stapp usually discusses these processes in the order 2, 1, 3 as they appear in his model of mind–brain interaction.

Process 2. The brain is considered to be an n-level quantum system whose states belong to the Hilbert space $\mathcal{H}$. Unless the brain interacts with the mind or the surrounding environment, the brain density matrix $\hat{\rho}$ evolves unitarily according to the Schrödinger equation

$$i\hbar \frac{\partial}{\partial t}\hat{\rho} = \left[\hat{H}, \hat{\rho}\right] \tag{6.67}$$

where the brackets denote a commutator (Eq. 4.76). If the Hamiltonian $\hat{H}$ is time-independent, the solution of the Schrödinger equation is given by

$$\hat{\rho}(t) = e^{-i\hat{H}t/\hbar}\hat{\rho}(0)e^{i\hat{H}t/\hbar} \tag{6.68}$$

According to Stapp, Process 2 generates a smear of classically alternative possibilities or a cloud of possible worlds, instead of the one world we actually experience.

Process 1. The mind is able to perform repeated projective measurements upon the brain using a freely chosen set of projection operators $\{\hat{P}_1, \hat{P}_2, \ldots, \hat{P}_n\}$, which are mutually orthogonal $\hat{P}_i \hat{P}_j = \delta_{ij}\hat{P}_j$ and complete to identity $\sum_j \hat{P}_j = \hat{I}$. After each projective measurement the brain density matrix undergoes non-unitary transition

$$\hat{\rho}(t) \rightarrow \sum_j \hat{P}_j \hat{\rho}(t) \hat{P}_j \tag{6.69}$$

According to Stapp, Process 1 extracts from the smear of possibilities generated by Process 2, a particular set of alternative possibilities among which only one is going to be actualized by Nature.

Process 3. The actualization of only one possibility, from the set of available possibilities, is done by Nature through objective reduction. Within the density matrix formalism, Process 3 is described by a non-unitary transition that converts the unconditional density matrix into conditional one

$$\sum_j \hat{P}_j \hat{\rho}(t) \hat{P}_j \rightarrow \frac{\hat{P}_k \hat{\rho}(t) \hat{P}_k}{\text{Tr}\left[\hat{P}_k \hat{\rho}(t)\right]} \tag{6.70}$$

where $\hat{P}_k$ is a particular projector from the set $\{\hat{P}_1, \hat{P}_2, \ldots, \hat{P}_n\}$ selected by Nature and $\text{Tr}\left[\hat{P}_k \hat{\rho}(t)\right]$ is the probability for the state to collapse to that particular state.

In essence, Stapp postulates that the conscious mind is able to choose freely both the timing and the basis in which objective reductions occur, even though the mind cannot choose the outcomes of those objective reductions. Then, he argues that the mind is able to exert a quantum Zeno effect upon the quantum brain by choosing to perform projective measurements in a basis of interest with time intervals Δt between the measurements being vanishingly small $\Delta t \to 0$. The quantum Zeno effect is supposed to be manifested as frozen time evolution of the quantum brain, namely, the brain stays with high probability in its initial quantum state at which the mind effort was initiated through the series of repeated measurements.

Despite the apparent interactionism, Stapp's model faces multiple problems when addressing the mind–brain relationship:

(1) The causal influence goes in only one direction from the mind to the brain. Since the mind is not related by any physical law to the quantum state $\hat{\rho}$ of the brain, it is hard to explain how the brain inputs sensory information to the mind.

(2) The conscious mind does not have its own quantum state, yet it can perform projective measurements upon the quantum brain. In quantum theory, any interaction between subsystems has to be included in the Hamiltonian $\hat{H}$ of the composite system that evolves by the Schrödinger equation. In Stapp's model, however, the mind–brain interaction does not appear in the Hamiltonian $\hat{H}$, thereby hiding the fact that if the mind action were physical, it would have delivered an infinite amount of energy to the brain according to the Planck–Einstein relation (7.2) at the infinitely fast frequency when $\Delta t \to 0$. Since the mind action is not restrained by a Hamiltonian, the mind in Stapp's model appears to behave like a *spirit*, a *soul* or a *ghost* attached to the brain without any physical law specifying that one mind cannot freely float around and act upon other nearby brains. In contrast, our Axiom 6.1.1 identifies each individual mind with the quantum information of a single non-factorizable wave function ψ, and by doing so explicitly rules out the possibility of one's consciousness floating around and affecting other brains.

(3) The power of the mind to choose freely the timing and the basis in which the projective brain measurements are performed introduces almost complete lawlessness in the time evolution of the quantum brain. Even without the quantum Zeno effect, with the use of repeated alternation of measurements in two mutually unbiased bases, the mind in Stapp's model is able to post-select the brain in any desired quantum state as follows: Consider a two level quantum brain with computational basis $|0\rangle$, $|1\rangle$, and a mind attempting to force the brain into the state $|1\rangle$. For any initial quantum brain state $|\psi\rangle$, the mind can first measure the brain in the basis $\frac{1}{\sqrt{2}}(|0\rangle + |1\rangle)$, $\frac{1}{\sqrt{2}}(|0\rangle - |1\rangle)$, followed by another measurement in the basis $|0\rangle$, $|1\rangle$. Such a cycle will return the state $|1\rangle$ with a probability of $\frac{1}{2}$. For n cycles, the probability for obtaining the brain state $|1\rangle$ at least once is $\text{Prob}(|1\rangle) = 1 - 2^{-n}$. Letting $\Delta t \to 0$, the mind can perform in an arbitrarily short period of time as many cycles of alternative measurements as needed in order to post-select the brain state $|1\rangle$. Direct generalization of the argument to any k-level quantum brain shows that the brain can be post-selected in any state desired by the mind, provided that the state is a solution of the Schrödinger equation.

(4) The mind's ability to choose the basis in which the quantum brain is measured appears to be at odds with the fact that we do not consciously experience what the physical state of our brain is. If we do not know which neuron in our brain is electrically firing and which is electrically silent, it is not clear how the mind in Stapp's model can choose projective measurements upon the brain in different bases without knowing what the basis states of those different bases are.

The quantum information theory of consciousness (Section 6.1) neither leads to miraculously emergent conscious experiences as in the Hameroff–Penrose Orch OR model, nor introduces ghostly minds that may or may not float around acting upon brains as in Stapp's model. Instead, Axioms 6.1.1 and 6.1.2 establish a physical correspondence between non-factorizable quantum states and conscious minds, while Axiom 6.1.3 attributes free will to some, sufficiently large minds. The brain then is the classically accessible result of the past mind choices.

6.7.4 Intertwining consciousness and quantum mechanics

The linearity of quantum mechanics works well in the realm of microscopic quantum systems. Straightforward amplification of quantum superpositions at the macroscopic scale, however, appears to somehow fail, as we do not see Schrödinger's cats around us (Section 6.5.4). Remarkably, if we enter ourselves into Schrödinger's cat experiment, we will obtain bizarre predictions in regard to our own conscious experiences. Indeed, let us experience happiness $|\odot\rangle$ when we see the living cat, sadness $|\ominus\rangle$ when we see the dead cat, and anxiety $|\ominus\rangle$ at the start of the experiment when we do not yet know what the cat outcome is. The vector nature of quantum states (Section 6.4.1) allows us to define two other orthogonal quantum superposed states of ourselves such that

$$|^{8}+\rangle \;\; = \;\; \frac{1}{\sqrt{2}}(|\odot\rangle + |\ominus\rangle) \tag{6.71}$$

$$|^{8}-\rangle \;\; = \;\; \frac{1}{\sqrt{2}}(|\odot\rangle - |\ominus\rangle) \tag{6.72}$$

At first glance, these states appear difficult to interpret and one may wonder what they could possibly mean. The answer can be obtained after algebraic rewriting. At the end of the modified Schrödinger's cat experiment, the quantum state includes ourselves together with the cat

$$\hat{U}\frac{1}{\sqrt{2}}(|\gamma_1\rangle + |\gamma_2\rangle)|\text{cat}\rangle|\ominus\rangle = \frac{1}{\sqrt{2}}(|D_1\rangle|\text{dead}\rangle|\ominus\rangle + |D_2\rangle|\text{alive}\rangle|\odot\rangle) \tag{6.73}$$

$$= \frac{1}{2}(|D_+\rangle|\text{cat}+\rangle|^{8}+\rangle + |D_-\rangle|\text{cat}-\rangle|^{8}+\rangle - |D_+\rangle|\text{cat}-\rangle|^{8}-\rangle - |D_-\rangle|\text{cat}+\rangle|^{8}-\rangle) \tag{6.74}$$

$$= \frac{1}{2}(|D_1\rangle|\text{dead}\rangle|^{8}+\rangle + |D_2\rangle|\text{alive}\rangle|^{8}+\rangle - |D_1\rangle|\text{dead}\rangle|^{8}-\rangle + |D_2\rangle|\text{alive}\rangle|^{8}-\rangle) \tag{6.75}$$

Thus, the state $|^{8}+\rangle$ represents conscious observation of the quantum superposed state $\frac{1}{\sqrt{2}}(|D_1\rangle|\text{dead}\rangle + |D_2\rangle|\text{alive}\rangle)$, whereas the state $|^{8}-\rangle$ represents conscious observation of the quantum superposed state $\frac{1}{\sqrt{2}}(|D_1\rangle|\text{dead}\rangle - |D_2\rangle|\text{alive}\rangle)$. At this point, a serious problem arises. Physically, the quantum states given by Eqs. (6.73) and (6.75) are identical. The interpretation of the states, however, seems to be dependent on the representation basis. If $|☺⟩$ and $|☹⟩$ are chosen as basis states, it appears that the happy state observes the living cat, whereas the sad state observes the dead cat. On the other hand, if the states $|♀+⟩$ and $|♀-⟩$ are chosen as basis states, it would be the case that conscious observations of various quantum superposed dead and alive cat states have to occur. Because we never find ourselves in states $|♀+⟩$ and $|♀-⟩$, it either follows that a physical process of objective reduction destroys states of the form given by Eq. (6.73), or that the quantum theory should be complemented with additional physical laws that involve conscious experiences as a fundamental ingredient of reality, thereby introducing asymmetry in the admissible ways for vector decomposition of the quantum states.

Example 6.19. *(Wigner's friend experiment) In 1961, Eugene Wigner (1902–1995) introduced a twist to the Schrödinger's cat experiment [521]. Instead of putting a cat in the box, he locked his friend inside a dark room, where he could either observe an incoming flash produced by a photon $|☺⟩$ or stay in the dark observing no flash $|☹⟩$. Similarly to the quantum superposed cat states, we can define superposed states of Wigner's friend*

$$|♀+⟩ = \frac{1}{\sqrt{2}}(|☺⟩ + |☹⟩) \tag{6.76}$$

$$|♀-⟩ = \frac{1}{\sqrt{2}}(|☺⟩ - |☹⟩) \tag{6.77}$$

Wigner then wondered what would happen when he opened the door, and asked his friend whether he saw the flash. If Wigner's conscious states are $|☺⟩$ when hearing that his friend did see the flash and $|☹⟩$ when hearing that his friend did not see the flash, we can also define corresponding quantum superposed states given by Eqs. (6.71) and (6.72). The linearity of quantum mechanics predicts that at the end of the experiment the combined quantum state of Wigner and his friend will be

$$\hat{U}\frac{1}{\sqrt{2}}(|\gamma_1⟩ + |\gamma_2⟩)|☹⟩|☹⟩ = \frac{1}{\sqrt{2}}(|D_1⟩|☹⟩|☺⟩ + |D_2⟩|☺⟩|☹⟩) \tag{6.78}$$

$$= \frac{1}{2}(|D_+⟩|♀+⟩|♀+⟩ - |D_-⟩|♀-⟩|♀+⟩ + |D_+⟩|♀-⟩|♀-⟩ - |D_-⟩|♀+⟩|♀-⟩) \tag{6.79}$$

Because the friend always answers that either he did see or he did not see the flash, Wigner concluded that the friend can never be in states $|♀+⟩$ or $|♀-⟩$, hence quantum superpositions should reduce as soon as they enter someone's consciousness. Such a speculation, however, brings us back to the preferred basis problem.

Introducing consciousness into quantum physics requires additional physical laws defining what consciousness is. Noteworthy, only saying that consciousness is a fundamental ingredient of quantum reality is not sufficient to keep the quantum theory consistent: If states such as $|♀+⟩$ and $|♀-⟩$ are invalid, whereas states such as $|☺⟩$ and $|☹⟩$ are valid conscious states, it follows that the set of all possible conscious states does not form a Hilbert space $\mathcal{H}$, hence conscious states cannot be solutions to the Schrödinger equation. But if the conscious states are not vectors, the very usage of bra-ket notation for writing conscious states becomes nonsense. Moreover, if equations of the form (6.78) are meaningless, it is hard to tell in what respect such a theory of consciousness is quantum.

Proponents of *consciousness causes collapse* interpretation of quantum mechanics have claimed that the theory is quantum in the sense that consciousness can exert paranormal influences upon quantum systems, altering the observed interference pattern of double slit experiments or the output of quantum random number generators [51, 276, 404, 405]. Experimental tests, however, invariably refute the existence of paranormal effects, while sporadic positive reports of paranormal phenomena do typically exhibit poor science, including inadequate statistical analysis, biased data sampling or data fabrication [3, 230, 280, 509, 510]. Thus, it seems that one's consciousness, whatever it is, cannot collapse the wave functions of other physical objects outside of one's own brain.

Everett's many worlds interpretation of quantum mechanics is an alternative attempt to introduce consciousness into the physical world in the form of *many minds* that share a common history up to a certain point in time [130, 161, 162, 474, 476]. In 1955, Hugh Everett III (1930–1982) described his idea of many minds using the metaphor of an *intelligent amoeba* that splits in parallel universes:

> As an analogy one can imagine an intelligent amoeba with a good memory. As time progresses the amoeba is constantly splitting, each time the resulting amoebas having the same memories as the parent. Our amoeba hence does not have a life line, but a life tree. The question of the identity or non identity of two amoebas at a later time is somewhat vague. At any time we can consider two of them, and they will possess common memories up to a point (common parent) after which they will diverge according to their separate lives thereafter. [162, p. 69]

The motivation behind Everett's interpretation is to use the Schrödinger equation as the sole physical law governing the time evolution of the universe and dispense with the need for objective reductions. Introducing consciousness into the picture of splitting universes, however, is much harder than originally envisioned.

Example 6.20. *(The preferred basis problem in Everett's interpretation) If the Schrödinger equation (4.157) is valid for the universe as a whole, it has to be the case that the space of conscious states satisfies the axioms of a Hilbert space $\mathcal{H}$. Only if the conscious states behave as vectors, would it be permissible to represent them using ket vectors such as $|\smiley\rangle$ or $|\frownie\rangle$. Since from the Schrödinger's cat and the Wigner's friend experiments we know that macroscopic superpositions of the form (6.78) are predicted by the Schrödinger equation, conscious states such as $|\text{\Neutral}+\rangle$ or $|\text{\Neutral}-\rangle$ could theoretically split into parallel universes and be experienced by us. Because experimentally we always experience the states $|\smiley\rangle$ or $|\frownie\rangle$ instead of the states $|\text{\Neutral}+\rangle$ or $|\text{\Neutral}-\rangle$, the conscious states do not appear to satisfy the axioms of Hilbert space $\mathcal{H}$. But if the conscious states cannot be written as ket vectors and cannot form tensor products with the physical states of observed objects, discussing multiple minds in parallel universes becomes impossible.*

Patching Everett's interpretation with a preferred basis hardly solves any of the main problems of consciousness. Indeed, suppose that only brain states enter into the Schrödinger equation and let only a certain basis of the brain states generate conscious experiences. The conscious experiences would not be ket vectors, but a new kind of mind states that are subjective in character such as ☺ and ☹. To these mental states would correspond unique quantum brain states such as |☺⟩ and |☹⟩ that are just shorthand ways of writing very complicated quantum states composed of a zillion elementary particles in the brain. The one-to-one correspondence between the conscious states and some of the brain states would allow one to write down a ket vector brain state every time a conscious state is mentioned

$$|☺⟩ \leftrightarrow ☺, \qquad |☹⟩ \leftrightarrow ☹, \qquad \ldots \tag{6.80}$$

On the other hand, the difference between the brain states and the mind states would explain away the experimental inadequacy observed in the Schrödinger's cat and the Wigner's friend experiments as follows: The brain states can evolve unitarily according to the Schrödinger equation and thus satisfy the axioms of a Hilbert space $\mathcal{H}$. In particular, every linear combination of brain states

$$|⚡⟩ = a|☺⟩ + b|☹⟩ \tag{6.81}$$

where $|a|^2 + |b|^2 = 1$ with $a, b \in \mathbb{C}$ is another valid brain state. To the brain state $|⚡⟩$, however, will not correspond a single conscious state, but a large number N of universes such that in $N|a|^2$ universes is experienced the conscious state ☺, whereas in $N|b|^2$ universes the conscious state ☹. The latter distribution of minds reproduces correctly the quantum probabilities obtained by the Born rule. Thus, the preferred basis could lead to universe splitting in a way that populates the universes with *many minds* that share a common memory up to a point in time, as envisioned by Everett. Unfortunately, the price of this construction is intolerably high due to the conversion of consciousness into a causally ineffective epiphenomenon (Fig. 5.13). Indeed, all of the quantum physical interference effects that causally affect the events inside the Everett universe will be due to the time evolution of the quantum amplitudes a and b and will have nothing to do with the conscious states ☺ and ☹. If consciousness is an epiphenomenon, it will not have a survival value and cannot be selected for by natural selection in the process of evolution. In addition, many of the main questions related to consciousness will be either ignored or swept under the carpet. For example, free will will be impossible for epiphenomenal minds, the inner privacy and binding of consciousness will be unexplained, it will not be clear how the mind boundaries are defined in terms of how many brain particles are required to produce any form of conscious experience, and there will be no hope of understanding what exactly generates the conscious experiences in the living brain and turns off those experiences in the dead brain.

The existence of living conscious brains and dead unconscious brains shows that the mind is not the brain. One might be wondering, however, why our axiomatic quantum information theory of consciousness (Section 6.1) is able to identify non-factorizable quantum state vectors $|\psi⟩$ with conscious mind states, whereas the Everett's interpretation is unable to identify quantum brain states with conscious mind states. The answer is to be found in the physical process that leads to objective reductions in the form of quantum jumps in the time evolution of the quantum states (Axiom 6.1.3). First, the objective reduction leads to disentanglement and creates localized in space component subsystems. This avoids the problematic entanglement of the whole universe into a single cosmic mind and explains why we experience ourselves as minds that are localized in space rather than delocalized on a cosmic scale. Second, the objective reductions account for the free will that is impossible only with deterministic time evolution according to the Schrödinger equation. Third, conscious states such as $|\text{\textbf{?}}\rangle$ are not experienced in experiments not because conscious states are different from ket vectors, but because such superposed conscious states are well above the energy threshold $\mathcal{E}$ and are physically eliminated by the process of objective reduction.

6.8   The hard problem of consciousness in quantum physics

The hard problem of consciousness is a hallmark of *functional theories* of consciousness, in which conscious experiences are assumed to be generated by the brain in the process of performing a certain kind of function [83, 84]. Classically, if you crash or scramble a living conscious brain, the result will be a dead unconscious brain jelly. Because the content of chemical atoms in both the living and the scrambled brain is the same, it appears that consciousness has to be a product of the brain functional organization. And here is where the difficulty arises, namely, why should it be the case that performing a certain kind of function would produce any conscious experience at all? Why the brain does not do what it does but in a mindless brain mode (Fig. 1.9) without any conscious experiences whatsoever?

The hard problem does not occur for *reductive theories* of consciousness, in which conscious experiences are identified with physical states. Indeed, if the logical identity relation makes the physical state and the mind equivalent, it becomes impossible to have the same physical state without being a mind.

Classical reductionism identifies consciousness either with a known physical field, such as the electromagnetic field (Section 5.3.2), or with a novel kind of mental field composed of yet undiscovered mental particles. Even though such an identification avoids the hard problem, it creates serious difficulties due to the fact that classical information (Section 3.19) is not well tailored to support subjective, private, and incommunicable conscious experiences or free will.

Quantum reductionism, however, is capable of successfully identifying conscious states with quantum information (Section 4.20) carried by non-factorizable pure quantum states $|\psi\rangle$ (Section 6.1). By doing so, the Hilbert space of the universe $\mathcal{H}_U$ becomes populated with minds only. The main difference between quantum reductionism and Berkeley's idealism is that the interaction between minds is not mediated by a divine creator, but is due to the quantum physical laws. The quantum physical laws govern how component minds bind into larger and more complex minds through quantum entanglement, and how those lager minds make choices and disbind through objective reduction into disentangled localized simpler minds (Fig. 6.3). The mind choices of each disentangled mind get observed and recorded in the form of classical information by the remaining observer minds who interpret that information as the brain of the disentangled mind. Consequently, the hard problem cannot be even meaningfully posed within the quantum reductive approach. If it is the mind that creates the observable reality of the brain through its past mind choices, then it is logically inconsistent to speculate that the brain is able to operate as it does in a mindless brain mode uncaused by the mind. Noteworthy, the fleeting existence of minds within inanimate objects produces accessible classical information such as the waves in the ocean or the shape of the rocks that we usually refer to as an observable reality rather than a brain. Thus, it is true that the mind creates reality, but it is not always our mind that does the job. Our minds can create only the observable reality of our own brain. Everything else in the inanimate world outside us is created by minds with fleeting existence whose experiences we cannot observe but can infer to exist from the axioms in Section 6.1.

Alternative quantum theories of mind, such as the Hameroff–Penrose Orch OR model or Stapp's quantum Zeno model, are clueless about the utility of quantum information theorems for explaining some of the characteristic features of conscious experiences. Consequently, each of these models introduces consciousness in a bizarre way that only exacerbates the hard problem.

In the Hameroff–Penrose Orch OR model, the quantum processes are claimed to be subconscious, whereas the flashes of conscious experiences emerge out of the orchestrated objective reduction (Orch OR) events. However, if consciousness is not a fundamental ingredient of reality but an emergent phenomenon, there is little hope for solving the hard problem. Claiming that the moral values and the potentiality for conscious experiences are embedded in the fundamental space-time geometry of the universe or in spin networks at a subconscious quantum level [231, 232] is incomprehensible and does not solve any problem of consciousness. Because the emergent consciousness would be a causally ineffective epiphenomenon as the indeterministic Orch OR event is outside of conscious control, the hard problem is exacerbated: We have to explain not only why we do have conscious experiences, but also why these experiences are evolutionary useless.

On the other extreme, Stapp's model introduces consciousness as a fundamental ingredient of the physical world but in an unphysical way. The mind does not have its own state vector or a density matrix, but nonetheless is able to act upon the quantum brain with any complete set of orthogonal projection operators at any time. Without a sufficient number of physical laws that could be used to answer questions related to consciousness, the theory becomes powerless. For example, if there is only a law that postulates that consciousness exists and that it can do things upon the brain, but there is no law that explains how the brain affects consciousness, one is free to speculate that consciousness is immortal and reincarnates from one brain to another once the old brain is dead. Such a speculation could hardly pass the rigorous criteria for a scientific theory.

CHAPTER 7

Toward a quantum neuroscience

The quantum description of the natural world in terms of quantum probability amplitudes that comprise individual conscious minds (Axiom 6.1.1) highlights the qualitative similarity between the physical processes that occur in inanimate and animate objects. As a result of the fundamentally mental nature of those processes, the natural evolution of complex conscious minds on Earth out of primitive ones in the course of ≈ 3.5 billion years of evolution does not appear to be miraculous at all, but a logical consequence of the gradual accumulation of traits that helped the individual minds adapt to their environment and produce a larger number of surviving offspring. Thus, according to the quantum information theory of consciousness (Section 6.1), what distinguishes humans from other animal species is not consciousness, as these animals are sentient too, but rather the developed ability of humans to discover facts (in the form of classical information) about the surrounding world, to memorize those facts, and then to teach them to their offspring, propagating the accumulated knowledge into the future.

In the rest of this chapter, we will discuss several previously published results on possible applications of quantum information theory to neuroscience [207, 210, 211], and will present a number of open questions that deserve further study.

7.1 Protein engines of life

Cells are the basic living units of organisms. Each cell contains water, inorganic ions and organic biomolecules. Water accounts for over 70% of the cell mass and its interaction with biomolecules directs many of the biological processes that sustain life. Water is a polar solvent that readily dissolves electrically charged ions and other polar molecules, but repels nonpolar ones. Amphipatic molecules containing both polar and nonpolar parts, such as cholesterol and phospholipids, interact with water differently at different ends of the molecule thereby forming double layered membranes with a polar (hydrophilic) surface and a nonpolar (hydrophobic) interior. Membranes enclose the cell, protecting it against unbeneficial factors in the extracellular space and organize the cell by dividing it into a multitude of specialized compartments called *organelles*.

The nucleus is the main cellular organelle that contains a complete copy of the organism's genome in the form of double-stranded deoxyribonucleic acid (DNA) molecules. The genes, if active, are used as templates for production of proteins, which are the molecular engines of life. For example, proteins can catalyze biochemical reactions that produce the required concentration of biomolecule ingredients, can assemble into cellular structures or organelles, can integrate into phospholipid membranes and act as ion channels or biomolecule transporters, can

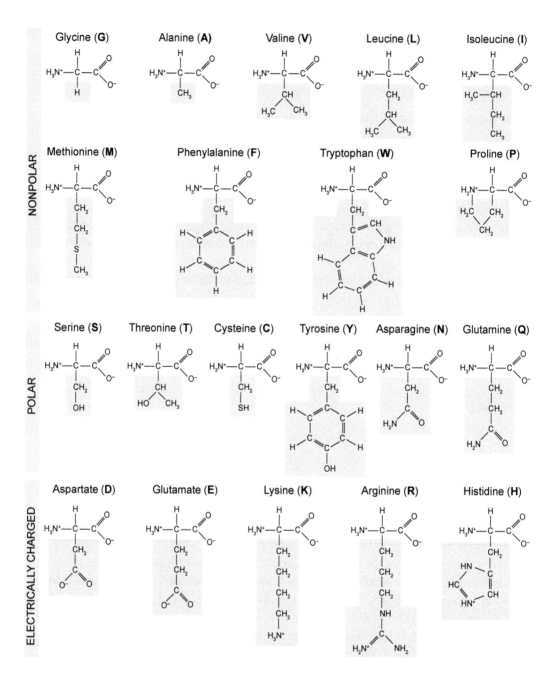

Figure 7.1   The amino acid alphabet. Amino acids differ from each other by the chemical structure of their functional R groups (highlighted in blue). Nonpolar R groups are hydrophobic and have a low propensity to be in contact with water, whereas polar or electrically charged R groups are hydrophilic as their contact with water is energetically favorable.

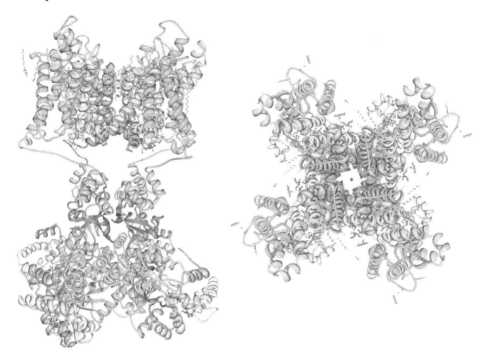

Figure 7.2 Side and top views of Kv1.2–2.1 chimera voltage-gated potassium ion channel visualized with RCSB Protein Workshop from Protein Data Bank entry 4JTA. doi:10.2210/pdb4jta/pdb

contract and expand, thereby changing the shape of the cell, can act as signaling molecules or can synthesize other messenger molecules with signaling functions, can regulate the gene expression by turning on or off different genes, etc. This diverse portfolio of protein functions is made possible by the great variety of conformations between which a single protein can switch.

The primary structure of proteins is determined by the linear assembly of amino acids within the protein polypeptide chain. In biological systems, there are 20 different amino acids, each of which possesses a distinct amino acid R group (Fig. 7.1). Twisting or turning of the protein polypeptide chain allows the formation of *hydrogen bonds* between different non-adjacent amino acids, thereby stabilizing two main types of secondary structure, the α-helix and the β-strand [371, 370]. Further organization of the protein α-helices and β-strands generates tertiary structure domains and motifs that determine the geometric shape of the protein. Multiple protein subunits, each possessing its own geometric shape, may further bind into protein complexes with a quaternary structure. Voltage-gated ion channels provide a typical example of protein complexes that are composed of multiple protein subunits (Fig. 7.2).

7.2 Neuronal ion channels and electric excitability

The nervous system of multicellular organisms allows them to convert different environmental stimuli, such as light, sound waves, temperature, mechanical stretching, or chemical concentration, into electric signals that propagate along the plasma membranes of their neurons (Fig. 1.8). The physical process by which a neuron converts one kind of signal (stimulus) into another (most commonly electric or chemical signal) is called *signal transduction*. The transduction ensures that all information about the surrounding world is encoded in the same neural language and memories for different in nature, but important, events can be recorded and retrieved using the same kind of biological memory traces. For example, short-term memories can be encoded in reversible changes of the neuronal excitability due to alteration of the number of protein ion channels incorporated in the plasma membrane or modulation of the channel conductivities through biochemical modifications such as phosphorylation or dephosphorylation. Long-term memories in neurons can be stored in morphological changes of the neurite diameter, length, or branching (Fig. 1.8) and different patterns of synaptic connectivity due to the differential expression, transport and assembly of scaffold proteins (see Fig. 7.7).

All sensory information from the five senses (taste, sight, touch, smell and hearing) is encoded and delivered to the brain cortex in the form of electric impulses [266, 285]. Clinical experiments by Wilder Penfield and William Dobelle, discussed in Section 1.1, further show that direct electric stimulation of the brain cortex elicits conscious experiences in human subjects [141, 374]. Because the electric activity of neurons is due to opening or closing of ion channels incorporated in the plasma membrane, as demonstrated in large squid axons whose internal cytosolic content has been replaced with artificial electrolyte solutions [25, 26], it is plausible that human consciousness is generated by quantum entangled states of the membrane-bound ion channels (Fig. 1.8) underlying electric excitation or inhibition of cortical neurons. The quantum behavior of individual ion channels is clearly manifested in the discrete changes observed in their channel conductances. Single-channel recordings have revealed that the electrical conductance of each channel can take only two discrete values: in the closed channel conformation the conductance is *zero*, whereas in the open conformation the channel has a characteristic *non-zero* single channel conductance in the range 0.1–100 pS. At a given transmembrane voltage of the neuronal plasma membrane, each voltage-gated ion channel undergoes stochastic transitions between open and closed states characterized by a certain probability $p(O)$ for the given channel type to be in the open conformation [419]. The transitions between open and closed states are due to quantum tunneling of electrons inside the voltage-sensing α-helices of the ion channels [86]. If the ion channel type is abundantly expressed, the macroscopic electric current generated by the neuron would appear to be stable and predictable due to the fact that on average the fraction of open channels will be close to $p(O)$, even though some channels open and others close all the time. Small quantum deviations from $p(O)$, however, could still be decisive for nonlinear threshold phenomena such as the firing of electric spikes [210].

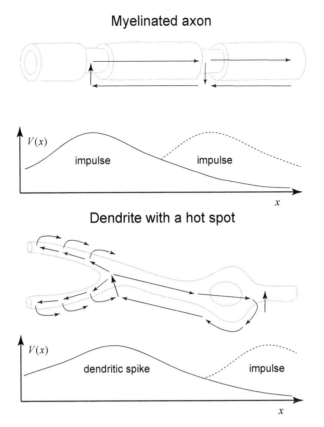

Figure 7.3 Spatial spread of the electric voltage $V(x)$ in neuronal axons or dendrites. In the myelinated axon, the electric impulse jumps from one node of Ranvier to another adjacent node. Similarly, the electric impulse can jump from a hot spot in a dendrite down to the axonal hillock where axonal spikes are generated. The impulse amplitudes (in millivolts) are shown in their spatial extent along each neurite at an instant of time. The extent of the current spread is governed by the cable properties of the neurite.

Quantum effects inside the pores of ion channels confer selectivity for passage of a certain type of ion [63, 64, 286, 287, 289, 288]. Ion selectivity divides channels into excitatory or inhibitory. Sodium and calcium channels are excitatory because they let positively charged Na^+ and Ca^{2+} ions, respectively, enter into the neuronal cytosol. Conversely, potassium channels are inhibitory, as they let positively charged K^+ ions escape from the neuronal cytosol toward the extracellular space [209]. Chloride channels are also inhibitory as they let negatively charged Cl^- ions enter into the neuron. At places where positively charged ions enter the neuron, the electric voltage $V(x)$ across the membrane increases and the membrane depolarizes (Fig. 7.3). Alternatively, if negatively charged ions enter the neuron, the electric voltage decreases and the membrane hyperpolarizes. Both depolarizations and hyperpolarizations spread along the neuronal projections and summate with each other, thereby performing various forms of computation that subserve the cognitive processes [1, 89, 282, 530].

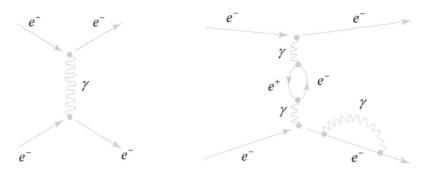

Figure 7.4   Feynman diagrams showing the interaction between two electrons e^- through emission and absorption of photons γ. The photons can create virtual electron–positron pairs $e^- + e^+$ that exist for only a brief moment of time before they annihilate each other. For each such diagram there is a quantum probability amplitude. Different quantum histories represented by different Feynman diagrams can exist in a quantum superposition. The time on the diagrams flows from left to right.

The covalent bonds that hold the atoms together within individual biomolecules are fundamentally quantum in nature, as they involve quantum superpositions of electron pairs that are shared between different atoms. Thus, moving from the biomolecule level down to quantum probability amplitudes is a relatively straightforward task that involves structural quantum chemistry. The really interesting problem, however, is to explain how different biomolecules entangle with each other predominantly along the membranes of individual neurons (Fig. 6.3), rather than proceeding with a stochastic bubble like entanglement as in the case of inanimate matter (Fig. 6.2).

At this point, neuronal electrical activity comes to the rescue. The electric field across the plasma membrane reaches 10^7 V m^{-1} due to the fact that the phospholipid bilayer is only 10 nm thick. In quantum field theory, the electric forces exerted by the electric field onto charged physical particles such as electrons are due to emission and absorption of photons in a multitude of superposed quantum histories [173, 175, 177, 178] that can be illustrated with the use of *Feynman diagrams* (Fig. 7.4). Because the voltage sensors of the neuronal ion channels are electrically charged (Fig. 1.8), the open or closed channel conformations will rapidly entangle with each other through the emission and absorption of photons. The process will continue until millions of cortical neurons are involved in a macroscopic superposition of firing patterns for which the energy threshold $\mathcal{E}$ for objective reduction is reached (Section 6.3, Fig. 6.3). The ensemble of cortical neurons will then make a decision utilizing its inherent free will (Section 6.5, Fig. 6.6), and will disentangle into a single pattern of electric firing that gets observed by the rest of the brain, thereby outputting the mind action toward the skeletal muscles and the surrounding world (Fig. 1.5). Applying the quantum information theory of consciousness to modeling the electric processes that occur in neuronal networks in the brain is an open scientific problem that deserves further study.

7.3   Dynamic timescale of individual conscious steps

Individual conscious steps in the quantum information theory of consciousness are defined by the time interval elapsed between two consecutive mind decisions that execute the conscious free will. Each decision is accomplished by a cycle including quantum entanglement of individual quantum systems until the energy threshold $\mathcal{E}$ is reached, followed by objective reduction and disentanglement of the component quantum subsystems (Section 6.3, Fig. 6.3). The cycles of entanglement-disentanglement in the brain cortex would be very fast and likely to exceed the frequency of 100 GHz [207]. This is an important prediction of quantum information theory of consciousness that avoids known *decoherence arguments* against other quantum mind theories [475] and eliminates possible psychophysical paradoxes in regard to our subjective experience of time [207]. An argument for such extremely fast elementary conscious steps, based on clinical observations of human subjects with time agnosia and psychophysical measurements of the shortest time interval for which two consecutive events in time are nonetheless experienced as simultaneous, will be presented below.

7.3.1   Conscious perception of time and time agnosia

Clinical data from human patients shows that the physical flow of time in the brain does not itself generate subjective feeling for a passage of time. Cerebrovascular incidents, traumatic brain injuries or psychiatric diseases are all capable of causing *time agnosia*, that is, a loss of the ability to subjectively experience the flow of time or to comprehend the succession and duration of events [104, 108, 207, 217]. Because subjects with time agnosia continue to have conscious experiences, it follows that consecutive individual conscious steps could occur without an associated experience of a time flow.

   In contrast to subjects suffering from time agnosia, healthy human subjects do experience subjectively the flow of time. The subjective time flow is constructed by the brain cortex via reading of packets of classical information called *time labels*, produced by the electrical activity of the neurons in subcortical brain regions, such as the right basal ganglia, that act as an internal brain timekeeper [410, 480, 524]. The brain cortex observes the ticking of the internal brain clock in the basal ganglia and labels the subjective conscious experiences with subjective time labels $T_1, T_2, \ldots, T_n$ (Fig. 7.5). Because the minimal time interval that is sufficient for the conscious mind to subjectively discern whether two consecutive events in time are nonsimultaneous is $\approx$ 30 ms [254, 284, 391, 392, 393], it follows that the time interval between any two time labels T_n and T_{n+1} is also $\Delta T = T_{n+1} - T_n \approx$ 30 ms. However, there could be a large number of individual conscious events that share the same time label T_n and are subjectively reported as having occurred simultaneously even though they did not occur simultaneously. For example, due to the fact that the three conscious events ♣, ♦ and ♡ in Fig. 7.5 share the same time label T_1, our mind is unable to tell whether these events occurred simultaneously or not in objective physical time $t_1, t_2, t_3, \ldots$. The conscious event ♣, however, has another time label T_2, which makes it subjectively possible to discern that this event

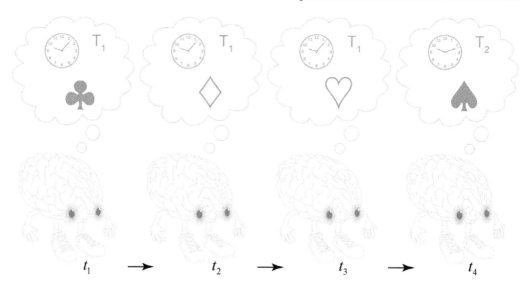

Figure 7.5   Objective versus subjective time flow. The subjective experience of time flow is constructed by the brain cortex from classical information obtained through observation of an internal brain clock that is located subcortically in the right basal ganglia whose neurons produce electric signals called time labels. Between any two consecutive time labels T_1 and T_2 spaced by an objective time interval of ≈ 30 ms, the brain cortex generates multiple conscious events sharing the same time label. Our conscious mind is unable to tell whether the three conscious events ♣, ◊ and ♡ occurred simultaneously or not in objective physical time $t_1, t_2, t_3, \ldots$ because these events share the same time label T_1. The conscious event ♠, however, has another time label T_2, which makes it subjectively possible to discern that this event actually occurred later in time than the events ♣, ◊ and ♡.

occurred at a later time compared with the events ♣, ◊ and ♡. If the individual conscious step is defined as the objective physical time $\Delta t = t_{n+1} - t_n$ between any two consecutive experiential events t_n and t_{n+1}, it can be logically argued that the duration of an individual conscious step is no longer than the minimal time interval that allows for two conscious events to be discerned as not simultaneous, namely, $\Delta t \leq \Delta T \approx 30$ ms.

Both points above show that it is possible for a human mind to have multiple consecutive conscious events that are nonetheless subjectively experienced as occurring simultaneously. This finding, based on clinical observations and psychophysical measurements, is one of the most important discoveries that allows for successful construction of a *quantum theory of consciousness* [207]. In particular, the individual conscious steps could be extremely fast, as required by quantum physical calculations, yet since we do not experience the flow of time for such brief intervals there will be no psychophysical paradoxes.

7.3.2   Reaction times and inner monologue

The quantum physicist Max Tegmark has estimated the decoherence time for neu-ronal electric firing due to ion-ion collisions to be

$$\tau_{\text{dec}} \approx \frac{(4\pi\varepsilon_0)^2 \sqrt{m(kT)^3}}{Nq^4 n} \tag{7.1}$$

where $\varepsilon_0 = 8.85 \times 10^{-12}$ F m^{-1} is the electric permittivity of the vacuum, $m = 3.81 \times 10^{-26}$ kg is the mass of the sodium ion, $N = 10^6$ is the total number of sodium ions that enter into the axon of a single neuron per single action poten-tial, $n = 8.73 \times 10^{25}$ m^{-3} is the intracellular volume number density of positive ions corresponding to physiological osmolarity of 290 mOsm, $q = 1.6 \times 10^{-19}$ C is the elementary electric charge, $k = 1.38 \times 10^{-23}$ J K^{-1} is the Boltzmann constant, and $T = 310$ K is the physiological temperature [475]. For a single neuron, substitu-tion of all numeric values gives $\tau_{\text{dec}} \approx 10^{-20}$ s. Considering a classical prediction for the duration of individual conscious steps that is ≈ 25 ms based on Hameroff's identification of a single conscious event with a single cycle of thalamo-cortical reverberation at 40 Hz [229, 231, 232, 233], Tegmark concluded that the fast brain decoherence time rules out quantum effects as being relevant to consciousness [475]. The whole argument, however, only shows that Hameroff's classical pre-dictions cannot be extrapolated to the quantum domain [207]. Classical neuro-science provides physiological times of the order of 1 ms for the synaptic trans-mission of information between individual neurons or for the width of the electric spikes produced by cortical pyramidal neurons [285]. Our internal monologue, speech, or muscle reaction times are also performed at a millisecond timescale [73, 75, 74, 77, 76, 78], but all these classical processes involving the transfer of classical information do not imply that the duration of individual conscious events should be of the order of milliseconds. Because individual conscious steps cannot be directly observed and measured, their duration should be calculated from the physical theory of consciousness at hand. Classical theory will predict millisec-ond steps, whereas quantum theory will predict at most picosecond steps. Thus, extrapolation of classical predictions to the quantum theory is unwarranted.

Example 7.1. *(Human reaction times are much longer than the duration of individ-ual conscious events) To show that the measurement of human reaction times provides only an upper bound for the duration of individual conscious steps, consider the op-eration of a personal computer. The personal computer has a processor that is used to perform computational tasks, and a human interface to get the input (keyboard and mouse) and provide the output (monitor) (Fig. 7.6). The monitor operates at a millisec-ond timescale so that the displayed image is refreshed every 10 ms, corresponding to a frequency of 100 Hz. The processor, however, operates at a much faster picosecond timescale, so that the computational state is refreshed every 100 ps, corresponding to a frequency of 10 GHz. Clearly, measuring the refresh rate of the monitor does not really help in determining how fast the processor of the computer is. Such a measurement can establish only an upper bound according to which the processor could not operate at a*

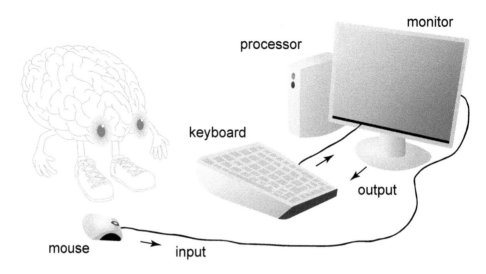

Figure 7.6 Communication with a personal computer takes place at a millisecond timescale (10 ms steps at a frequency of 100 Hz) whereas the processor computes at a picosecond timescale (100 ps steps at a frequency of 10 GHz).

timescale longer than 10 ms (or equivalently, at a frequency lower than 100 Hz). Essentially the same argument applies to physiological experiments measuring the reaction times of human volunteers [207]. While it is true that the brain cortex communicates at a millisecond timescale with the sensory organs to obtain information (Fig. 5.4) and with the muscles to output information (Fig. 1.5) using electric impulses propagating along the nerve fibers (axons), this does not prove that the duration of individual conscious events within the brain cortex also occurs at a millisecond timescale. If we can perform a cognitive task and react in a millisecond timescale, it would be irrational to construct a theory in which each conscious step lasts seconds or minutes. Yet, it is quite rational to expect that individual conscious steps are much faster than our reaction times exactly as the processor of a personal computer is much faster than the monitor.

Example 7.2. *(The inner monologue is a classical process) Because the word content of our inner thinking can be written on a sheet of paper, it follows that the inner monologue has to be expressible in the form of communicable classical information. Conscious experiences, however, could occur without inner monologue. For example, while silently thinking "Oh, I have to go to the library to return that book" we also perform short mental pauses between the individual words. We continue to be conscious during the time intervals for which we mentally pause between the words, so there is no contradiction in stating that the individual conscious steps are much faster and due to dynamics of quantum information. Quantum processes that are responsible for our intuition are not communicable, and indeed we are unable to explain what exactly we do before we put the next word in our inner monologue. On the other hand, once we insert a word in our inner monologue it becomes communicable classical information, exactly as it should be due to the fact that the brain is the classical record of past mind choices.*

Classically, one does not have a good reason to predict conscious steps faster than a millisecond, because classical physics does not have the temporal resolution to explain in what respect the brain state during one extremely short interval of time differs from the next such short time interval. In quantum theory, however, the temporal resolution is drastically increased due to the Planck–Einstein relation

$$E = h\nu = \frac{h\omega}{2\pi} = \hbar\omega \tag{7.2}$$

The angular frequency ω of the temporal oscillations exhibited by the quantum amplitudes ψ in the solutions of the Schrödinger equation (4.157), is linearly dependent on the energy of the quantum system $\omega = E/\hbar$. Because the reduced Planck constant $\hbar = 1.0545718 \times 10^{-34}$ J s is an extremely small number, the typical angular frequencies for quantum systems are quite large. Due to the possibility of quantum interference effects in biological systems at a picosecond or even a subpicosecond timescale [207, 210, 211], it is no longer rational to expect that the brain cortex will generate conscious experiences with 25 ms gaps from one conscious event to another. Instead, the time interval Δt between two consecutive objective reductions should be calculated from the energy threshold $\mathcal{E}$ as

$$\Delta t = \frac{\hbar}{\mathcal{E}} \tag{7.3}$$

The Diósi–Penrose model for objective reductions estimates the energy threshold $\mathcal{E}$ in terms of the gravitational interaction energy between two displaced macroscopic superpositions with equal mass [135, 136, 137, 138, 376, 377, 382]

$$\mathcal{E} = G\frac{m^2}{r} \tag{7.4}$$

where $G = 6.67408 \times 10^{-11}$ m^3 kg^{-1} s^{-2} is the gravitational constant and r is a short distance cutoff that for biological systems could be taken to be the diameter of a single carbon atom $r \approx 2.7 \times 10^{-15}$ m.

The gray matter of the brain cortex of humans weighs ≈ 316.3 g and contains $N \approx 6.18 \times 10^9$ neurons [21]. Pyramidal neurons comprise $\approx 70\%$ of all cortical neurons [359]. On average, the volume of a single pyramidal neuron is $V \approx 5 \times 10^4$ μm^3 and its mass density is close to that of water $\rho \approx 1$ g cm^{-3}. For the total mass of all cortical pyramidal neurons, we get $m = 0.7NV\rho \approx 216.3$ g. The time needed for the objective reduction to occur is

$$\Delta t = \frac{\hbar r}{Gm^2} \tag{7.5}$$

which would give $\Delta t \approx 10^{-37}$ s for each conscious step if all cortical pyramidal neurons are involved. The calculated answer is feasible because it is not smaller than the Planck time [91, p. 151] that is the shortest physically meaningful time step given by

$$t_P \equiv \sqrt{\frac{\hbar G}{c^5}} \approx 5.39106 \times 10^{-44} \text{ s} \tag{7.6}$$

The above estimate provides a very fast rate for individual conscious steps because we assumed that the total neuronal mass enters into the objective reduction. Yet, it is more likely that only proteins such as voltage-gated ion channels, electrolyte solution and photons near the neuronal plasma membranes are involved in the generation of human consciousness. Therefore, the junk mass contributed by DNA in the neuronal nuclei together with various end products of the neuronal metabolism stored in different intraneuronal organelles may be ignored. Taking the decoherence time for the firing of one million cortical neurons as the time needed for the objective reduction, namely, $\Delta t \approx 10^{-26}$ s based on Tegmark's formula (7.1), we can conversely calculate the mass of the neuronal membrane components involved as

$$m = \sqrt{\frac{\hbar r}{G\Delta t}} \qquad\qquad (7.7)$$

which would give $m \approx 6.5 \times 10^{-4}$ g as the lower bound on the amount of neuronal matter involved in the generation of a single conscious event. Here, we should point out that Tegmark's formula is used only to set an upper bound on how fast objective reductions should be in order to avoid environmental decoherence and the associated quantum entanglement between the brain cortex and its environment. Better estimates of the time for individual conscious steps should be available when a complete physical theory of quantum gravity and more detailed models for the quantum entanglement of protein voltage-gated ion channels along the neuronal projections are constructed.

7.4   Quantum tunneling in synaptic communication

Neurons communicate with each other at synaptic contacts [238] that permit the transmission of electric or chemical signals from one neuron to another (Fig. 7.7). At the *synapse*, the plasma membranes of the two neurons come to close opposition to each other and a specialized molecular machinery allows the transmission of the signal from the presynaptic (output) neuron to the postsynaptic (input) neuron. Functionally, there are two types of synapses: electrical and chemical.

Electrical synapses provide a direct electrical coupling between the two neurons through gap junctions that are nanopores constructed from connexin proteins [383]. The electrical synapses are rapid as there is no synaptic delay, but they transmit the electric potential passively, which means that the amplitude of the signal decays exponentially with the distance. Because the gap junctions are pores, the transmission is bidirectional and signals can pass from each of the two neurons to the other one. Besides the passage of ions that carry electric current, the gap junctions allow diffusion of small chemical molecules with signaling action, such as inositol 1,4,5-trisphosphate (IP_3) or cyclic adenosine monophosphate (cAMP). During brain development, electrical synapses provide a blueprint for the formation of neuronal networks, but once the chemical synapses are formed the electric coupling in most neuronal cell types is eliminated [383].

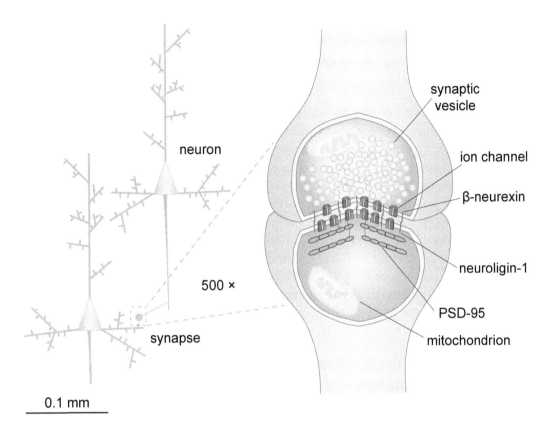

neuron

500 ×

synapse

0.1 mm

synaptic
vesicle

ion channel

β-neurexin

neuroligin-1

PSD-95

mitochondrion

Figure 7.7   Communication between neurons takes place at synaptic contacts. Excitatory synapses are typically formed between presynaptic axonal boutons and postsynaptic
dendritic spines. The presynaptic terminal has a large number of synaptic vesicles that
contain neurotransmitter molecules. When the axonal bouton is electrically excited, the
Ca^{2+} entry through presynaptic voltage-gated calcium channels leads to the fusion of a
single synaptic vesicle with the plasma membrane at the active zone, thereby releasing
its neurotransmitter content into the synaptic cleft. The neurotransmitter molecules then
bind to postsynaptic ligand-gated ion channels and generate a postsynaptic electric potential in the target neuron. The synapse is held together by protein-protein bridges such
as the β-neurexin-neuroligin-1 complexes whose anchoring in the presynaptic and postsynaptic plasma membranes sets the width of the synaptic cleft in the range of 20–40 nm.
Presynaptically, the intracellular part of β-neurexin also participates in the docking of
synaptic vesicles. Postsynaptically, the intracellular part of neuroligin-1 interacts with
scaffolding proteins such as postsynaptic density protein 95 (PSD-95) that organize the
postsynaptic protein machinery and anchor the ligand-gated ion channels. The presence
of mitochondria in both the axonal boutons and the dendritic spines ensures that the high
energy demands of the active synapses are effectively met.

Chemical synapses are the most abundant type of synaptic contact in the brain. The main feature of the chemical synapse is that the electric signal from the presynaptic neuron is converted into a chemical signal in the form of *neurotransmitter* molecules released from synaptic vesicles into the synaptic cleft (Fig. 7.7). The neurotransmitter molecules then bind to ligand-gated ion channels whose opening leads to the generation of postsynaptic electric potential. The action of the chemical synapses is slow (there is ≈ 1 ms delay) due to the conversion of the signal from electrical to chemical and back to electrical form; however, what is achieved is that the postsynaptic potentials can now be either positive or negative in sign, depending on the type of neurotransmitter released [149]. In the brain cortex, $\approx 84\%$ of the synapses release an excitatory neurotransmitter such as glutamate that binds to postsynaptic excitatory ligand-gated ion channels generating a depolarizing, positive postsynaptic electric potential, whereas $\approx 16\%$ of the synapses release an inhibitory neurotransmitter such as γ-aminobutyric acid (GABA) that binds to postsynaptic inhibitory ligand-gated ion channels generating a hyperpolarizing, negative postsynaptic electric potential [35]. The transmission at chemical synapses is predominantly unidirectional from the presynaptic axonal boutons toward the postsynaptic dendrite, soma or axon of the target neuron. Still, protein-protein bridges such as the β-neurexin-neuroligin-1 complexes traversing the synaptic cleft are also able to transmit information retrogradely from the postsynaptic toward the presynaptic neuron [119, 210, 459].

The fusion of synaptic vesicles with the plasma membrane allows for extrusion of the vesicle content out of the cell, an event that is referred to as *exocytosis* [365]. The release of neurotransmitter due to synaptic vesicle exocytosis is a probabilistic event that occurs with a frequency of 0.35 ± 0.23 on all axonal spikes [142]. On average, the neurons in the human cortex have about 7000 synapses, each for intracortical reception and exchange of information [363]. A conservative estimate for a single neuron with only $n = 1000$ axon terminals predicts that on average $k = 350$ of them will release neurotransmitter molecules upon each action potential. The number N of possible combinations of active synapses is then given by the *binomial coefficient* indexed by n and k as follows

$$N = \binom{n}{k} = \frac{n!}{k!(n-k)!} = \frac{1000!}{350!\,650!} = 4 \times 10^{279} \tag{7.8}$$

where ! is the *factorial function* such that $n! = 1 \times 2 \times 3 \times \ldots \times n$. If the synaptic vesicle release were a completely random event, then there would be sheer disorganization in the function of cortical neuronal networks within seconds [211]. Fortunately, accumulated biomolecular evidence shows that an elaborate protein machinery regulates the fusion process between the synaptic vesicle and the plasma membrane [153, 458, 464]. Furthermore, the triggering of that protein machinery could be theoretically modeled with a quantum effect as would be expected from a quantum information theory of consciousness.

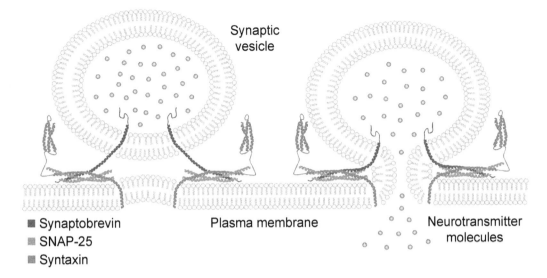

- ■ Synaptobrevin
- ▦ SNAP-25
- ▨ Syntaxin

Plasma membrane

Synaptic vesicle

Neurotransmitter molecules

Figure 7.8   SNARE zipping in neurotransmitter release. Synaptic vesicles are docked at the active zone of the synapse through hemi-zipped SNARE complexes (left). The docking interaction between the SNARE proteins leads to a close proximity of the synaptic vesicle and the membrane. The fusion pore is to be opened between the transmembrane domains of 5–8 circularly arranged syntaxin molecules. Full zipping of the core SNARE complex is due to formation of a four-α-helix bundle as synaptobrevin fits into the syntaxin/SNAP-25 groove (right). SNARE zipping leads either to a transient opening and closure of the fusion pore, which is known as the kiss-and-run mode of neurotransmitter release, or to a complete fusion of the synaptic vesicle with the presynaptic membrane. In both cases, the released neurotransmitter molecules traverse the synaptic cleft and bind to postsynaptic ligand-gated receptors that electrically excite or inhibit the postsynaptic neuron.

7.4.1   SNARE proteins and synaptic vesicle exocytosis

Synaptobrevin, syntaxin and *SNAP-25* are so-called *SNARE proteins* that dock the synaptic vesicles to the presynaptic active zone [461] and sustain the fusion pore while letting the neurotransmitter molecules pour into the synaptic cleft (Fig. 7.8). Synaptobrevin is anchored in the membrane of the synaptic vesicle, whereas syntaxin and SNAP-25 are anchored in the plasma membrane [90]. All three SNARE proteins zip together into a four-α-helix bundle whose twisting applies a traction force on the opposing phospholipid bilayers of the synaptic vesicle and the plasma membrane until they merge with each other [415, 535]. The three SNARE proteins synaptobrevin, syntaxin and SNAP-25 form the minimal molecular machinery sufficient to complete the fusion of liposomes in vitro at a physiological temperature of 37°C [512]. The three SNARE proteins are also able to assemble and tether different liposomes at a lower temperature of 4°C; however, the fusion of liposomes does not occur in this case [512]. Another important fact is that although 5–8 SNARE protein complexes line the fusion pore of each synaptic vesicle [234], only one SNARE complex is sufficient to drive the fusion reaction [496].

In neurons, the neurotransmitter release is synchronized with the axonal firing and requires Ca^{2+} influx in the presynaptic bouton [2, 460]. In the resting condition, the hemi-zipped SNARE complex of docked vesicles is clamped by the Ca^{2+} sensor protein called *synaptotagmin-1* that represses the complete SNARE zipping and the spontaneous release of synaptic vesicles [23, 87, 92, 319, 335]. Under axonal firing, however, opening of voltage-gated calcium channels leads to Ca^{2+} entry upon which the synaptotagmin-1 molecule binds at least four Ca^{2+} ions [143, 87, 455, 462] and detaches from the SNARE complex thereby both deinhibiting the SNARE function and actively assisting in the membrane fusion through phospholipid interaction [88, 462, 463].

7.4.2 Protein α-helix structure and conformational distortions

The SNARE complex is zipped by four protein α-helices: synaptobrevin and syntaxin contribute one α-helix each, while SNAP-25 contributes two α-helices [416, 494]. Because we will focus on modeling the zippering of the four-α-helix, next we will briefly describe the secondary structure of protein α-helices.

Geometrically, the protein α-helix is a right-handed spiral with 3.6 amino acid residues per turn [371], where the N–H group of an amino acid forms a hydrogen bond with the C=O group of the amino acid four residues earlier in the polypeptide chain (Fig. 7.9). Three longitudinal chains of hydrogen bonds referred to as *α-helix spines* that run parallel to the helical axis stabilize the α-helix structure.

At the quantum level, the interaction of the amide I excitation (due to C=O bond stretching) with the vibrations of the hydrogen bonds along the α-helix spines leads to localization of the amide I excitation within a region spanning a few amino acids [428]. The composite quantum state constituted by an amide I excitation and its associated hydrogen bond distortions could be viewed as a quantum quasiparticle called the *Davydov soliton* [109, 110, 111, 322, 428]. Since the propagation of the Davydov soliton along the protein α-helix spines is capable of twisting the α-helix structure [60, 322], Davydov solitons may be instrumental in zipping the SNARE complexes in docked synaptic vesicles and thus triggering the process of neurotransmitter release [211].

The original Davydov model is a mathematical idealization that does not take into account amide II excitations (due to N–H bond stretching), amino acid R-side chains, or surrounding water molecules; however, it allows for analytic solution of the resulting equations [109, 110, 111]. Because adding extra terms to Davydov's Hamiltonian does not change qualitatively the nature of the soliton, throughout we will use the term *Davydov soliton* to refer to any quantum quiasiparticle or waveform that arises from the Davydov model together with its possible extensions [327, 403, 465, 466]. Noteworthy, in spite of controversy on the lifetime of Davydov solitons [103, 112, 311, 321, 426, 511], extensive computer simulations by Wolfgang Förner have verified the stability of Davydov solitons at physiological temperature $T = 300$ K for up to 30 ps [182, 183, 184, 185, 186, 187, 188, 189, 190], a time sufficient for the soliton to traverse 3 times the length of any of the α-helices in the core SNARE complex at velocity of 1260 m s^{-1} [271].

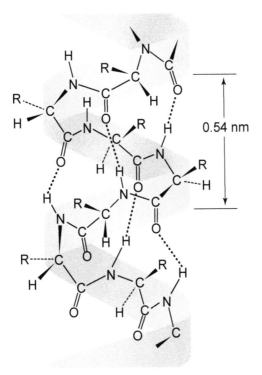

Figure 7.9   Structure of the protein α-helix with 3.6 amino acid residues per turn. Three α-helix spines consisting of longitudinal chains of hydrogen bonds ($\cdots$) within vertically aligned –C=O$\cdots$H–N–complexes stabilize the helical structure. Quantum quasiparticles called Davydov solitons could form, propagate along, and conformationally twist the α-helix due to the interaction of the amide I (C=O) vibrations with the hydrogen bonds in the α-helix spines. The vertical distance between consecutive turns of the helix is 0.54 nm.

7.4.3   Quantum tunneling through rectangular potential barrier

The quantum nature of Davydov soliton-assisted zipping of the SNARE complexes is manifested in the fact that the physical quantities involved are quantum probability amplitudes $\Psi(x,t)$ that dynamically evolve according to the Schrödinger equation (4.157). Considering only the longitudinal direction along the four-α-helix bundle of the SNARE complex, we can write the Hamiltonian $\hat{H}$ governing the dynamics of the quantum quasiparticle as

$$\hat{H} = \left[\frac{\hat{p}^2}{2m} + V(x) \right] \tag{7.9}$$

where $m = 1.1 \times 10^{-28}$ kg is the effective mass of the Davydov soliton whose energy is $E_0 = 3.2 \times 10^{-20}$ J [211], $\hat{p}$ is the quasiparticle momentum operator, and $V(x)$ is the potential energy. Expressing the momentum operator in position basis as $\hat{p} = -i\hbar \frac{\partial}{\partial x}$, we can rewrite the Hamiltonian as

$$\hat{H} = \left[-\frac{\hbar^2}{2m} \frac{\partial^2}{\partial x^2} + V(x) \right] \tag{7.10}$$

The one-dimensional time-dependent Schrödinger equation then becomes

$$i\hbar\frac{\partial}{\partial t}\Psi(x,t) = \left[-\frac{\hbar^2}{2m}\frac{\partial^2}{\partial x^2} + V(x)\right]\Psi(x,t) \tag{7.11}$$

Solving the Schrödinger equation for an arbitrary potential $V(x)$ is an extremely hard problem that requires advanced perturbation theory [40, 41], but we could solve the problem exactly for a rectangular potential barrier. Because the zipping of the SNARE complex occurs in two stages, hemi-zipped and fully-zipped, we could model the potential energy in the form of a rectangular potential barrier with *width a* (extending from $x = 0$ to $x = a$) and *height* $V_0 > E_0$ (if the barrier height were $V_0 < E_0$ then the long-lived hemi-zipped SNARE complex would not have been possible)

$$V(x) = V_0[\Theta(x) - \Theta(x-a)] \tag{7.12}$$

where $\Theta(x)$ is the *Heaviside step function* given by

$$\Theta(x) = \begin{cases} 0, & x < 0 \\ 1, & x \geq 0 \end{cases} \tag{7.13}$$

The assumption that the quantum quasiparticle has a definite energy E_0 allows us to factor the wave function $\Psi(x,t)$ into a product of two functions: $\psi(x)$ depending only on the spatial variable and $\varphi(t)$ depending only on time

$$\Psi(x,t) = \psi(x)\varphi(t) \tag{7.14}$$

Substitution into the time-dependent Schrödinger equation (7.11) gives

$$i\hbar\frac{\partial}{\partial t}[\psi(x)\varphi(t)] = \left[-\frac{\hbar^2}{2m}\frac{\partial^2}{\partial x^2} + V(x)\right][\psi(x)\varphi(t)] \tag{7.15}$$

Dividing both sides by $\psi(x)\varphi(t)$ separates the variables

$$\frac{i\hbar\frac{\partial}{\partial t}\varphi(t)}{\varphi(t)} = \frac{\left[-\frac{\hbar^2}{2m}\frac{\partial^2}{\partial x^2} + V(x)\right]\psi(x)}{\psi(x)} \tag{7.16}$$

The left-hand side of Eq. (7.16) is a function only of t, not of x, whereas the right-hand side is a function only of x, not of t. Therefore, Eq. (7.16) can hold only if the two sides are equal to a constant, which is exactly the energy E_0 of the quantum quasiparticle [325, p. 168]. In essence, we have split the time-dependent Schrödinger equation (7.11) into a system of two equations

$$i\hbar\frac{\partial}{\partial t}\varphi(t) = E_0\,\varphi(t) \tag{7.17}$$

$$\left[-\frac{\hbar^2}{2m}\frac{\partial^2}{\partial x^2} + V(x)\right]\psi(x) = E_0\,\psi(x) \tag{7.18}$$

The solution of Eq. (7.17) gives the phase time dependence

$$\varphi(t) = \varphi_0 e^{-\frac{i}{\hbar}E_0 t} = \varphi_0 e^{-i\omega_0 t} \tag{7.19}$$

where $\omega_0 = E_0/\hbar$.

To solve the remaining time-independent Schrödinger equation (7.18), we first write the general solutions $\psi_I(x)$ in the region I on the left side of the barrier, $\psi_{II}(x)$ in the region II inside the barrier, and $\psi_{III}(x)$ in the region III on the right side of the barrier (Fig. 7.10) using coefficients that need to be determined by the boundary conditions of the problem [309, p. 78]

$$\psi_I(x) = Ae^{ik_1 x} + Be^{-ik_1 x}, \quad x \le 0 \tag{7.20}$$
$$\psi_{II}(x) = Ce^{k_0 x} + De^{-k_0 x}, \quad 0 \le x \le a \tag{7.21}$$
$$\psi_{III}(x) = Fe^{ik_1 x} + Ge^{-ik_1 x}, \quad x \ge a \tag{7.22}$$

where

$$k_1 = \frac{\sqrt{2mE_0}}{\hbar} \tag{7.23}$$

$$k_0 = \frac{\sqrt{2m(V_0 - E_0)}}{\hbar} \tag{7.24}$$

Here, we note that $Ae^{ik_1 x}$ and $Fe^{ik_1 x}$ represent waves traveling along the x-axis in the positive direction, whereas $Be^{-ik_1 x}$ and $Ge^{-ik_1 x}$ represent waves traveling in the negative direction. Normalization of the quantum amplitude incident on the barrier sets $A = 1$ [309, p. 76]. Also, because reflection from the barrier could occur only for $x \le a$, but not for $x > a$, we are able to set $G = 0$ [309, p. 78].

After differentiation, we further obtain

$$\frac{\partial}{\partial x}\psi_I(x) = ik_1\left(e^{ik_1 x} - Be^{-ik_1 x}\right) \tag{7.25}$$

$$\frac{\partial}{\partial x}\psi_{II}(x) = k_0\left(Ce^{k_0 x} - De^{-k_0 x}\right) \tag{7.26}$$

$$\frac{\partial}{\partial x}\psi_{III}(x) = ik_1 Fe^{ik_1 x} \tag{7.27}$$

The boundary conditions require that both the wave function $\psi(x)$ and its first spatial derivative $\frac{\partial\psi(x)}{\partial x}$ are continuous at the barrier boundaries $x = 0$ and $x = a$ in order to ensure that the second spatial derivative $\frac{\partial^2\psi(x)}{\partial x^2}$ exists. Therefore, the following four equations should hold [309, 491]

$$\psi_I(0) = \psi_{II}(0) \tag{7.28}$$

$$\left.\frac{\partial\psi_I(x)}{\partial x}\right|_{x=0} = \left.\frac{\partial\psi_{II}(x)}{\partial x}\right|_{x=0} \tag{7.29}$$

$$\psi_{II}(a) = \psi_{III}(a) \tag{7.30}$$

$$\left.\frac{\partial\psi_{II}(x)}{\partial x}\right|_{x=a} = \left.\frac{\partial\psi_{III}(x)}{\partial x}\right|_{x=a} \tag{7.31}$$

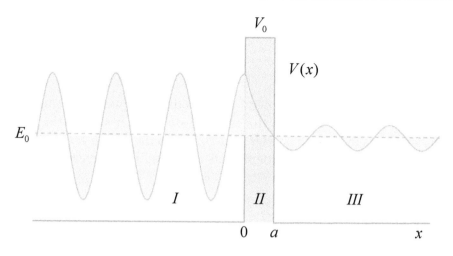

Figure 7.10 Quantum tunneling through a potential barrier. According to classical physics, an incoming particle of energy E_0 less than the barrier height V_0 cannot penetrate into the classically forbidden region II inside the barrier. In quantum physics, however, because the particle quantum wave function $\Psi(x,t)$ must be continuous in all regions I, II and III, it penetrates inside the barrier, undergoing an exponential decay until it exits the barrier. In general, the wave function $\Psi(x,t)$ on the other side of the barrier will not be *zero*, so there will be a finite probability that the particle will tunnel through the barrier and emerge on the other side. Importantly, the energy of the particle E_0 is the same on both sides of the barrier; what changes is the quantum probability amplitude of the wave.

For $x = 0$ or $x = a$, further substitution of Eqs. (7.20,7.21,7.22) and (7.25,7.26,7.27) into Eqs. (7.28,7.29,7.30,7.31) gives [491, p. 141]

$$1 + B = C + D \tag{7.32}$$
$$\imath k_1(1 - B) = k_0(C - D) \tag{7.33}$$
$$Ce^{ak_0} + De^{-ak_0} = Fe^{\imath ak_1} \tag{7.34}$$
$$k_0(Ce^{ak_0} - De^{-ak_0}) = \imath k_1 Fe^{\imath ak_1} \tag{7.35}$$

Multiplying Eq. (7.32) by $\imath k_1$ on both sides and adding or subtracting Eq. (7.33) gives

$$2\imath k_1 = (\imath k_1 + k_0)C + (\imath k_1 - k_0)D \tag{7.36}$$
$$2\imath k_1 B = (\imath k_1 - k_0)C + (\imath k_1 + k_0)D \tag{7.37}$$

Similarly, multiplying Eq. (7.34) by k_0 on both sides and adding or subtracting Eq. (7.35) gives

$$2k_0 Ce^{ak_0} = (k_0 + \imath k_1)Fe^{\imath ak_1} \tag{7.38}$$
$$2k_0 De^{-ak_0} = (k_0 - \imath k_1)Fe^{\imath ak_1} \tag{7.39}$$

That can be further manipulated in a form ready for substitution

$$C = \frac{(ik_1 + k_0)}{2k_0} F e^{a(ik_1 - k_0)} \tag{7.40}$$

$$D = -\frac{(ik_1 - k_0)}{2k_0} F e^{a(ik_1 + k_0)} \tag{7.41}$$

Now, we are ready to find the transmitted quantum amplitude F. Substitute Eqs. (7.40) and (7.41) into Eq. (7.36) to get

$$2ik_1 = \frac{(ik_1 + k_0)^2}{2k_0} F e^{a(ik_1 - k_0)} - \frac{(ik_1 - k_0)^2}{2k_0} F e^{a(ik_1 + k_0)} \tag{7.42}$$

$$4ik_1 k_0 e^{-iak_1} = \left[-(k_1^2 - k_0^2) + 2ik_1 k_0 \right] F e^{-ak_0} + (k_1^2 - k_0^2 + 2ik_1 k_0) F e^{ak_0} \tag{7.43}$$

Taking into account the formulas for hyperbolic functions

$$2\sinh x = e^x - e^{-x} \tag{7.44}$$

$$2\cosh x = e^x + e^{-x} \tag{7.45}$$

we can further simplify

$$4ik_1 k_0 e^{-iak_1} = 2\left[(k_1^2 - k_0^2)\sinh(ak_0) + 2ik_1 k_0 \cosh(ak_0) \right] F \tag{7.46}$$

$$F = \frac{2ik_1 k_0 e^{-iak_1}}{(k_1^2 - k_0^2)\sinh(ak_0) + 2ik_1 k_0 \cosh(ak_0)} \tag{7.47}$$

Similarly, we can find the reflected quantum amplitude B. Substitute Eqs. (7.40) and (7.41) into Eq. (7.37) to get

$$2ik_1 B = \frac{(k_1^2 + k_0^2)}{2k_0} F e^{a(ik_1 + k_0)} - \frac{(k_1^2 + k_0^2)}{2k_0} F e^{a(ik_1 - k_0)} \tag{7.48}$$

$$4ik_1 k_0 B e^{-iak_1} = (k_1^2 + k_0^2)(F e^{ak_0} - F e^{-ak_0}) \tag{7.49}$$

$$4ik_1 k_0 B e^{-iak_1} = (k_1^2 + k_0^2)\sinh(ak_0) F \tag{7.50}$$

Further substitution of Eq. (7.47) in Eq. (7.50) followed by rearranging gives

$$B = \frac{(k_1^2 + k_0^2)\sinh(ak_0)}{(k_1^2 - k_0^2)\sinh(ak_0) + 2ik_1 k_0 \cosh(ak_0)} \tag{7.51}$$

The probabilities for the particle to be reflected from the barrier R or to be transmitted to the other side of the barrier T are given by the Born rule [309, p. 76]

$$R = BB^* = |B|^2 \tag{7.52}$$

$$T = FF^* = |F|^2 \tag{7.53}$$

From Eq. (7.47) together with its complex conjugated equation, we can calculate the transmission coefficient T as

$$T = \frac{4k_1^2 k_0^2}{(k_1^2 - k_0^2)^2 \sinh^2(ak_0) + 4k_1^2 k_0^2 \cosh^2(ak_0)} \tag{7.54}$$

Taking into account the hyperbolic function identity

$$\cosh^2 x = 1 + \sinh^2 x \tag{7.55}$$

we can further simplify

$$T \; = \; \frac{4k_1^2 k_0^2}{(k_1^2 + k_0^2)^2 \sinh^2(ak_0) + 4k_1^2 k_0^2} \tag{7.56}$$

The reflection coefficient R is calculated analogously from Eq. (7.51) together with its complex conjugated equation as

$$R \; = \; \frac{(k_1^2 + k_0^2)^2 \sinh^2(ak_0)}{(k_1^2 - k_0^2)^2 \sinh^2(ak_0) + 4k_1^2 k_0^2 \cosh^2(ak_0)} \tag{7.57}$$

$$R \; = \; \frac{(k_1^2 + k_0^2)^2 \sinh^2(ak_0)}{(k_1^2 + k_0^2)^2 \sinh^2(ak_0) + 4k_1^2 k_0^2} \tag{7.58}$$

Adding together Eqs. (7.56) and (7.58) shows that the probability is conserved

$$T + R = 1 \tag{7.59}$$

Classically, an incoming particle of energy E_0 less than the barrier height V_0 always gets reflected. Quantum mechanically, however, the particle can tunnel through the barrier and appear on the other side (Fig. 7.10). Thus, the existence of quantum tunneling effects in brain function could be viewed as a direct evidence that classical physics is fundamentally inadequate for addressing the mind–brain problem. Quantum tunneling has already been confirmed in enzymes that catalyze biochemical reactions in the brain [30, 31, 429, 469]. The same biochemical method, based on the detection of the so-called *kinetic isotope effect*, could also be adapted for confirming quantum tunneling in SNARE complex zipping [211].

John Eccles (1903–1997), who shared the 1963 Nobel Prize in Physiology or Medicine for his work on the ionic mechanisms involved in neuronal excitation and inhibition, and the quantum physicist Friedrich Beck (1927–2008) calculated the probability p for synaptic vesicle exocytosis in terms of the quantum tunneling transmission coefficient T as

$$p = \omega_0 \Delta t \, T \tag{7.60}$$

where the angular frequency of the quantum particle $\omega_0 = E_0/\hbar$ is interpreted as the number of attempts that the particle undertakes to cross the barrier, and Δt is the duration of the quasistable situation in the presynaptic axonal bouton, after which the metastable state changes into a stable one where no exocytosis is possible [37, 36, 38, 152]. The model by Beck and Eccles correctly derived the observed probability for neurotransmitter release using a set of biologically feasible parameters, but importantly lacked concrete biomolecular implementation.

The SNARE zipping model not only uses quantum tunneling similarly to the Beck and Eccles model, but also provides an insight into the protein machinery involved in exocytosis. The main stages of neurotransmitter release are as follows:

(1) In the resting axonal bouton, the free Ca^{2+} levels are very low. The synaptic vesicle is docked in close proximity to presynaptic voltage-gated calcium ion channels that are inactive and remain closed. The SNARE complex is hemi-zipped and reliably clamped in this conformation by the Ca^{2+}-sensor synaptotagmin-1.

(2) When the neuronal axon fires, the electric signal depolarizes the presynaptic bouton thereby activating the voltage-gated calcium channels. The free Ca^{2+} levels in the bouton rise up in the form of a cloud-like microdomain that persists near the docked vesicle for a time period $\Delta t = 0.3$ ms [433, 434].

(3) Synaptotagmin-1 binds Ca^{2+} ions and rapidly releases the clamp on the SNARE complex. The energy released by synaptotagmin-1 detachment induces a Davydov soliton with energy E_0 that propagates along the unzipped portion of the synaptobrevin α-helix attempting to zip completely the four-α-helix bundle of the SNARE complex. The Davydov soliton has to cross a potential energy barrier parameterized by width a and height $V_0 > E_0$.

(4) If the Davydov soliton is reflected by the potential barrier V_0 it propagates back to the point where synaptobrevin is anchored in the synaptic vesicle membrane, undergoes another reflection and aims again toward the potential barrier V_0. The frequency n_0 of the attempts to cross the barrier could be much lower than the angular frequency ω_0 of the Davydov soliton, $n_0 \ll \omega_0$.

(5) If the Davydov soliton tunnels through the potential barrier V_0 the full zipping of the SNARE complex opens the fusion pore and exerts a traction force that merges the synaptic vesicle membrane with the plasma membrane. Through the open fusion pore, neurotransmitter molecules pour into the synaptic cleft.

At this point, we can perform some quantitative analysis. Quantum tunneling can occur only while the Ca^{2+} levels are elevated in the microdomain near the SNARE complex, so $\Delta t = 0.3$ ms [433, 434]. The soliton velocity in the protein chain is $v = 1260$ m s^{-1} [271]. Taking two turns of the α-helix as the distance traveled by the reflected Davydov soliton forth and back, we obtain $n_0 = 6 \times 10^{11}$ Hz. Setting the width of the potential barrier also to two turns of the α-helix, namely, $a = 1.08$ nm, allows plotting the probability p for neurotransmitter release as a function of V_0 using the equation

$$p = n_0 \Delta t \, T \tag{7.61}$$

The results show that barrier heights near the value of $V_0 = 3.7 \times 10^{-20}$ J give feasible probabilities for neurotransmitter release that are in the biological range 0.35 ± 0.23 (Fig. 7.11). For potential barriers with different shapes or subject to thermal oscillations, the height V_0 could be even higher [211]. Indeed, the observed temperature dependence of exocytosis suggests that vibrationally assisted tunneling could be involved in the process of SNARE zipping [211], similarly to the vibrationally assisted tunneling observed in the action of enzymes, a class of proteins with catalytic action [5, 30, 429, 469].

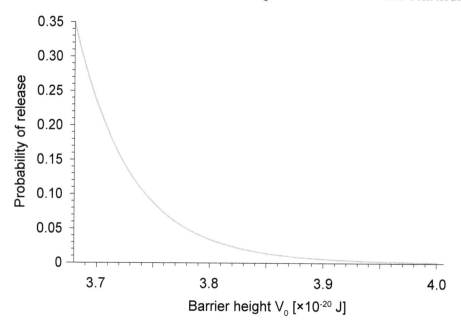

Figure 7.11   Probability for neurotransmitter release as a function of the potential barrier height V_0 in a quantum tunneling model utilizing the Davydov soliton to zip the SNARE complex. The width of the potential barrier is taken to be two turns of the protein α-helix.

7.4.4   SNARE proteins and volatile anesthesia

Multiple studies aimed at delineating the mechanism of action of volatile anesthetics that are routinely used in clinical practice for administration of general anesthesia provide strong experimental support for the direct involvement of SNARE proteins in consciousness [211]. First, general anesthesia with clinical concentrations of volatile anesthetics inhibits neurotransmitter release [329] and selectively erases consciousness but not all cortical responses; in particular, evoked potentials can be recorded from the visual cortex of anesthetized animals [272, 307, 308]. Second, volatile anesthetics bind with high affinity into the hydrophobic core of the SNARE four-α-helix bundle [281, 353]. Third, the only known mutation that confers resistance to volatile anesthetics is located in the syntaxin gene and leads to the production of a truncated form of syntaxin [497, 498]. The truncated form of syntaxin interferes with the binding of volatile anesthetics to the core SNARE complex, resulting in resistance to anesthesia manifested as higher doses of volatile anesthetic required to cause unconsciousness.

In addition to providing insight into the origin of probabilities in synaptic vesicle exocytosis, the quantum theory is capable of explaining the observed long-range correlation between synaptic vesicles in the axonal bouton expressed in the fact that the synaptic boutons always release a single synaptic vesicle. Classically, if two events are independent of each other, they can jointly occur with probability $p(A \cap B) = p(A)p(B)$, where $p(A)$ and $p(B)$ are the corresponding probabilities for each of the two events A and B. Therefore, if the fusions of synaptic vesicles were independent events each occurring with $p = 0.35$, then the release of two synaptic vesicles should have been observed with probability $p^2 = 0.12$, instead of *zero*. To solve that discrepancy, Beck and Eccles resorted to quantum mechanics and used the nonlocal physical correlation between spatially separated systems provided by *quantum entanglement* in order to explain the systematic release of only a single synaptic vesicle upon depolarization of the axon terminal [37, 152]. Their argument goes as follows: Attribute to each of N vesicles in the presynaptic bouton two states, $|\psi_0\rangle$ and $|\psi_1\rangle$, where $|\psi_0\rangle$ is the state with closed fusion pore and $|\psi_1\rangle$ the state with open fusion pore. In the resting state, the composite wave function of all vesicles (each indexed by superscript) can be expressed in a product form

$$|\Psi_0\rangle = |\psi_0^1\rangle|\psi_0^2\rangle\ldots|\psi_0^N\rangle \tag{7.62}$$

Under axonal depolarization, the synaptic vesicles get entangled in the form

$$
\begin{aligned}
|\Psi_1\rangle \quad = \quad & \sqrt{1-p}\left[|\psi_0^1\rangle|\psi_0^2\rangle|\psi_0^3\rangle\ldots|\psi_0^N\rangle\right] \\
& + \sqrt{\frac{p}{N}}\left[|\psi_1^1\rangle|\psi_0^2\rangle|\psi_0^3\rangle\ldots|\psi_0^N\rangle + |\psi_0^1\rangle|\psi_1^2\rangle|\psi_0^3\rangle\ldots|\psi_0^N\rangle\right. \\
& \left. + |\psi_0^1\rangle|\psi_0^2\rangle\ldots|\psi_1^{N-1}\rangle|\psi_0^N\rangle + |\psi_0^1\rangle|\psi_0^2\rangle\ldots|\psi_0^{N-1}\rangle|\psi_1^N\rangle\right]
\end{aligned}
\tag{7.63}
$$

Calculating the probability for exocytosis from the N-body wave functions $|\Psi_0\rangle$ and $|\Psi_1\rangle$, one obtains the same result as for the barrier penetration problem of one vesicle, namely, without an axonal spike the probability for release is *zero*, whereas after the occurrence of an axonal spike the probability for release of exactly one vesicle is p, for two or more vesicles is *zero*, and for no vesicles is $1-p$. This leads to the observable consequence that the probability for exocytosis in an axonal bouton does not depend on the number of docked synaptic vesicles [37, 152].

7.4.5 Comparison with interactionism proposed by John Eccles

Remarkable as it may be, the mind–brain interactionism at the quantum level proposed by John Eccles [37, 148, 150, 151, 152, 398] lacks physical rigor. The quantum probabilities for possible brain states are indeed calculated with the use of the Schrödinger equation but are then supposed to be somehow momentarily affected by the non-material mind [150, 152] without physical justification of why this is possible. Similarly to the objections raised to Stapp's model in Section 6.7.3, one may ask what prevents other minds from acting upon your brain as well?

In the quantum information theory of consciousness (Section 6.1), we have implemented some of the ideas championed by John Eccles, but in a physically meaningful way that is consistent with the modern developments in quantum information theory. In Chapter 6, we have shown how the main questions related to consciousness could be satisfactorily addressed and have established the conditions required for the theory to produce logically consistent and empirically adequate results. Several features of quantum information theory of consciousness are of particular importance, as we shall discuss in detail next.

First, the mind is not just analogous to the probability fields in quantum mechanics. Rather, the quantum wave function ψ provides the only physical way to refer to different mental states, hence ψ stands for the mind within the physical theory (Axiom 6.1.1). If we use the map versus territory distinction [299, p. 58], we could say that the mind is the territory, the physical theory of mind is the map, and ψ as a theoretical concept is a location onto the map. Minds have certain aspects such as the phenomenal nature of qualia that cannot be communicated in words. Therefore, the theory can only infer that two different wave functions ψ_1 and ψ_2 refer to two different mental states, but not communicate the incommunicable aspects of those mental states.

Second, the mind is not an external entity acting upon the probability fields in quantum mechanics. Instead, the internal propensity of a complex conscious mind to make choices is modeled as an internal propensity of the wave function ψ to undergo objective reductions whenever it reaches a certain energy threshold $\mathcal{E}$ (Axiom 6.1.3). Thus, using its free will, the mind changes itself at the moment a conscious choice is made. Because the brain is the outcome of the mind choice and the mind cannot undo choices that are already done, it follows that the mind creates the brain but it cannot change past brain states. The resulting model of mind–brain interaction makes it clear that it is the mind that possesses the free will, not the brain. It also explains why the brain is needed in a theory of mind; namely, if the past mind choices are to be of any future importance for the mind who made the choices, then those past mind choices need to be physically recorded in the universe and that physical record is exactly what the brain is.

Third, the quantum entanglement between different physical systems not only leads to nonlocal correlations between observable physical quantities, but also binds the component conscious experiences into a single composite mind (Axiom 6.1.2). The binding of conscious experiences through quantum entanglement is implied by the fact that no quantum wave function ψ can be assigned to a physical system that is entangled with other physical systems, hence not every physical system can have its own mind, but every physical system can be an integrated part of a larger composite mind. Thus, the theory avoids paradoxical existence of minds within minds, and imposes an important constraint on the outcome of the objective reductions of the wave function ψ. In order to avoid universal quantum entanglement and creation of a single cosmic mind, the objective reductions should also lead to quantum disentanglement of the component physical subsystems (Section 6.3) that are then able to undergo subsequent cycles of entanglement-disentanglement.

Fourth, the consciousness is not emergent since it is identified with quantum information. Instead, a form of quantum panpsychism is endorsed according to which the fleeting existence of minds is predicted both in inanimate matter and in other parts of the living bodies outside of the brain cortex. What makes our conscious mind in the brain cortex really special is that it repeatedly binds a multitude of conscious experiences coming from different sensory organs, including vision, hearing, taste, smell and touch, and then stores memories of our conscious lives in the form of a narrative history about our own conscious "I."

In Section 1.2, we have shown that split-brain patients whose corpus callosum is surgically cut behave as if they host two different conscious minds, each of which is located in one of the two cerebral hemispheres. Because the corpus callosum contains the axons of cortical pyramidal neurons that deliver electrical signals from one hemisphere to the opposite hemisphere, it follows that the unity of consciousness depends on the physiological activity of intact axons. This clinical observation is consistent with the quantum mechanism for binding of conscious experiences described in Section 7.2, namely, the strong electric fields across the neuronal membranes lead to predominant quantum entanglement of charged amino acids in the neuronal ion channels along the neuronal projections or across the synaptic cleft of anatomically connected cortical neurons until the energy threshold $\mathcal{E}$ for objective reduction is reached. Then the quantum entangled neuronal cluster disentangles and another cycle of entanglement begins.

In certain psychiatric diseases such as *multiple personality disorder*, impaired functional connectivity between cortical areas could lead to a situation in which multiple human-like minds or personalities are hosted inside a single brain. These different personalities may be unaware of each other and may have different memories. At any instant of time, one of the personalities may be controlling the behavior of a person, but different personalities may switch in taking control over behavior at different times. Thus, the quantum information theory of consciousness is able to predict the existence of multiple minds with their own identity and personality inside a single brain, and could explain how this is physically possible through formation of multiple disconnected quantum entagled clusters.

7.5 Memory storage and retrieval

The electric excitability conferred by voltage-gated ion channels is essential for the normal physiological function of neurons. The sensory information is delivered to the brain cortex in the form of electric signals that propagate along the axons of afferent neurons (Fig. 5.4). The motor information that controls muscle contraction leaves the brain cortex and propagates along the axons of efferent neurons also in the form of electric signals (Fig. 1.5). Therefore, it is essential for the quantum information theory of consciousness to provide an input/output interface to the electric activity of cortical neurons through quantum entangled states of voltage-gated ion channels and other membrane-bound proteins (such as SNARE proteins involved in synaptic communication). To be able to store and retrieve memories, however, our conscious minds need to have access to biological processes that last for extended periods of time. Thus, intraneuronal or extraneuronal molecular components could also entangle with the plasma membrane components thereby contributing to consciousness with their memory content. The stored memories can only be in the form of classical information, because only classical information can be read and retrieved (Section 3.19). When memories are retrieved, however, they need to be converted into quantum entangled states of cortical neurons in order to be consciously experienced and relived once again (Section 6.1).

Depending on their strength, memories are classified into three types: short-term memories (seconds to hours), long-term memories (hours to months) and long-lasting memories (months to lifetime) [338]. Long-term memories are stored in pronounced anatomical changes of neuronal morphology and connectivity determined by the localization and number of synapses onto the dendrites of cortical neurons [45, 201, 262]. Synapses that are frequently stimulated by electric activity exhibit growth and remodeling, including synapse duplication [45], whereas inactive synapses degenerate and are eliminated by microglial cells [22, 165, 440]. In contrast, short-term memories are stored in subtle subcellular changes due to reversible biochemical processes. Intraneuronal structural proteins forming the neuronal cytoskeleton, such as actin, intermediate filaments and microtubules, may undergo reversible conformational transitions that store memories of the recent neuronal electric activity for short periods of time. For example, neuronal contraction of actin-myosin complexes changes the shape of electrically active dendritic spines thereby altering the generation of subsequent excitatory postsynaptic electric potentials [294], intracellular transport of cargo vesicles by kinesin motors delivers messenger ribonucleic acid (mRNA) in the vicinity of electrically active spines for local synthesis of proteins [297], and protein enzymes catalyzing phosphorylation/dephosphorylation of voltage-gated ion channels or neurotransmitter receptors alter reversibly the electric conductivities for different ion types across the neuronal plasma membrane [285]. Changes in the phosphorylation status of ion channels and quantum entanglement through the nuclear spin of phosphate ions [180, 181, 516] could lead to quantum superpositions of different firing patterns of cortical neurons. Diffusion of bare phosphate ions or Posner molecules, $Ca_9(PO_4)_6$, could further extend the quantum entanglement along the neuronal projections or across the plasma membrane through endocytosis [180, 181]. While such biochemical processes are much slower than the subpicosecond timescales calculated in Section 7.3, they may supply the quantum information theory of consciousness with molecular mechanisms for short-term memory storage and retrieval.

CHAPTER 8

Research programs and conscious experiences

Scientific theories are communicable knowledge (Section 2.2.2) and as such could be transmitted between scholars, could be taught by lecturers in universities, and could be understood and learned by students. Defining scientific concepts in an explicit manner and postulating the physical laws in the form of a formal mathematical system of axioms enhance communicability. The formalization of science also reduces the amount of intersubjective misunderstanding and allows scientists to study objectively the implications of scientific theories by proving theorems, deriving experimental predictions and empirically testing these predictions in the laboratory.

8.1   Verificationism and falsificationism

Because scientific theories are supposed to represent knowledge about the real world (Section 2.5), experiments are needed in order to provide singular factual statements about the real world upon which the scientific theories could be based. Nevertheless, singular statements provided by experimental observations alone are hardly capable of either proving or disproving any of the current scientific theories [301, 302]. Instead, scientific theories are accepted or rejected only after the accumulation of a sufficient number of novel theoretical results in the form theorems that allow the scientist to look upon the world through new eyes. New theories are not just different, they are always better than the old theories in explaining why the physical world is the way it is.

Example 8.1. *(The problem of induction) David Hume (1711–1776) argued that from singular statements based upon experimental observations one is unable to prove inductively universal statements that will hold true as fundamental physical laws with absolute certainty. For example, from observing a large number of white swans in Europe it cannot be inductively concluded that "All swans are white" as there is no guarantee that the next observation will not be a black swan. In order to be able to generalize from a finite number of observations to a universal statement that potentially involves an infinite number of cases, one needs to assume that the future will resemble the past [267, p. 37]. The assumption that the future experimental observations will resemble past ones, however, is generally incorrect as exemplified by the discovery of black swans in Australia. Thus, all scientific theories containing universal statements should be viewed as* conjectures *that cannot be verified by any finite amount of experimental data [395, 397]. The experimental testing of the theory will never end if the theory is correct, yet it could end in a finite amount of time if a critical falsification occurs exposing the theory as incorrect.*

In Section 2.2.3, we have shown that there is an asymmetry between universal and existential statements. Experiments cannot *verify* universal statements, but they can *falsify* them. For example, observing a large number of white swans does not verify the statement "All swans are white," whereas observing even a single black swan falsifies it. In contrast, experiments cannot *falsify* singular existential statements, but they can *verify* them. For example, observing a large number of white swans cannot falsify the statement "Exists a black swan," whereas observing a single black swan verifies it.

Since all scientific theories contain fundamental physical laws that are in the form of universal statements, Karl Popper (1902–1994) argued that the growth of scientific knowledge is not due to *verification* of scientific theories, but due to *falsification* of those theories whose experimental predictions did not match the experimental data [397]. For Popper, all scientific theories are just bold *conjectures* that one day may be falsified and the goal of science is not to prove theories, but to disprove theories. Thus, intellectual honesty requires a precise specification of the conditions under which one is willing to give up one's position [395]. Unfortunately, falsificationism fails to consider that there is no natural demarcation between observational and theoretical propositions, hence none of the modern scientific theories is objectively falsifiable [301].

Example 8.2. *(The problem of naive falsificationism) Imre Lakatos (1922–1974) argued that no experimental result alone is able to falsify a scientific theory [301]. Indeed, no scientific theory T composed of* physical laws, $L_1, L_2, ..., L_k$, *is able to predict alone any experimental result E. What is also needed for making a prediction is a set of initial* conditions, $C_1, C_2, ..., C_n$ *under which the experiment is performed (Eq. 2.5). Therefore, a negative experimental result $\neg E$ can falsify a theory T only given the assumption that the initial conditions were exactly $C_1, C_2, ..., C_n$ and there was no other disturbing factor C_{n+1} that could have affected the outcome of the experiment. The history of science shows that in many cases scientists do not use the negative result $\neg E$ as a falsifier of the theory T, but rather conclude that the initial conditions could not have possibly been $C_1, C_2, ..., C_n$ as assumed. For example, Newton's theory of universal gravitation was successfully used to calculate the orbits of the planets before the discovery of the planet Uranus. Then, the French astronomer Alexis Bouvard (1767–1843) observed irregularities in the motion of Uranus that appeared to falsify the mathematical prediction of Newton's theory under the assumption that there are seven planets orbiting the Sun. Rather than concluding that Newton's theory is falsified, Bouvard inferred the existence of an eighth planet in the solar system. In 1846, the French mathematician Urbain Le Verrier (1811–1877) calculated the position of the eighth planet Neptune based on the assumption that Newton's theory is correct, and sent a letter urging the astronomer Johann Gottfried Galle (1812–1910) in the Berlin Observatory to look for the new planet. On the evening of the day Galle received the letter, he discovered Neptune within 1° of where Le Verrier had predicted it to be. The discovery of Neptune was hailed as a triumph of Newton's theory. In 1859, Le Verrier found irregularities in the motion of the planet Mercury and hypothesized that they were the result of another hypothetical planet Vulcan orbiting between Mercury and the Sun. In 1860, the discovery* *of Vulcan was reported by the amateur astronomer Edmond Lescarbault (1814–1894), who was then awarded the Legion of Honour by the Academy of Sciences in Paris. Now we know that Vulcan does not exist and the orbit of Mercury indeed falsifies Newton's theory of universal gravitation.*

8.2   Theory-laden observations and shared knowledge

One sees only what one knows. Thus, all experimental facts are *theory-laden*. Because experimental facts are necessarily interpreted in some background knowledge, there is always uncertainty whether the experimental fact actually is what it is claimed to be. Consequently, if a scientist is interested in refuting a given scientific theory T, the refutation has to be done by experimental facts that are interpreted and accepted as such by the very same scientific theory T. Indeed, experimental evidence E produced by a new theory T_2 cannot be used to falsify an old theory T_1 provided that T_1 does not prove E as well. The latter point is best illustrated by the historical fact that some of the most important of Galileo's discoveries supporting the heliocentric model of the solar system were met with disbelief by his learned contemporaries.

Example 8.3. *(Discovery of the moons of Jupiter by Galileo) The geocentric model of the universe postulated that all celestial bodies circled around the Earth. Discovering objects that orbit another planet such as the moons of Jupiter seems to us to be a definitive refutation of the geocentric model that should have been clearly seen and understood by Galileo's learned contemporaries as well, but this was not the case. Galileo observed the moons of Jupiter using a telescope that he had constructed himself. Galileo's telescope worked fine on Earth for magnifying objects up to 30 times and making those objects appear as if they are brought closer to the observer. When used in the heavens, however, the telescope appeared to deceive the senses [171]. First, some observers whose eyes were not sharp enough were unable to see the moons of Jupiter at all. Second, the images produced by the telescope appeared to be inconsistent. Galileo wrote himself that the fixed stars when seen through the telescope by no means appear to be increased in magnitude in the same proportion as other objects. The moon and Jupiter were enlarged and brought nearer, whereas the apparent diameter of the fixed stars decreased and the stars appeared to be pushed away. Third, through the telescope some stars were seen double [171, pp. 123–128]. As a result, it is not surprising that Galileo's opponents maintained that what appeared in the telescope was just an optical illusion and the spots seen through the telescope near Jupiter served the sole purpose of satisfying Galileo's lust for money [264].*

Two people can see the same thing only if they have a common *shared knowledge* [236]. A toddler may see a wonderful toy in the wooden tube that is covered with leather and equipped with two pieces of glass at both tube ends. Galileo will see a telescope. The common knowledge shared between the toddler and Galileo would allow them both to see the object as a cylinder covered with leather. Seeing a telescope, however, is heavily theory-laden and could be achieved by the toddler only after many years of studying and understanding the necessary scientific concepts that are taught in geometry, optics or astronomy classes [236]. Thus, for the proponents of a scientific theory T, the refutation of T cannot come from external knowledge that is not shared by T, but rather from the accumulation of logical inconsistencies or empirical inadequacies generated by the theory T itself.

8.3   Bayesian inference and assessment of theories

The majority of scientific theories are neither verifiable nor falsifiable in absolute terms. Assessment of the probability for a given scientific theory to be true given the available experimental evidence can be done only in the framework of some background knowledge that is shared, undisputed and widely accepted as true. The axioms of logic (Section 2.3), the logical rules of inference (Section 2.4) and Kolmogorov's axioms of probability (Section 3.10) are good candidates for such a background knowledge that would allow scientists to assess objectively the merit of different scientific theories.

From Kolmogorov's definition of conditional probability given by Eqs. (3.57) and (3.58) follows Eq. (3.59), which is Bayes' theorem in a disguised form.

Theorem 30. *(Bayes' theorem) The posterior probability $p(T|E)$ for the theory T being true given that evidence E is observed could be expressed as*

$$p(T|E) = \frac{p(E|T)}{p(E)}p(T), \qquad p(E) \neq 0 \tag{8.1}$$

where $p(T)$ and $p(E)$ are, respectively, the prior probabilities of T and E without regard to each other, and $p(E|T)$ is the probability for observing the evidence E given that T is true. The prior probability $p(T)$ of a theory T is given by Eq. (2.15).

Bayesian inference is a statistical method that uses Bayes' theorem to update the probability for a theory (or a hypothesis) being true as more evidence becomes available [204, 213]. One starts with the prior probability $p(T)$ that is independent of any evidence and updates it to posterior probability $p(T|E)$ that reflects the confidence that we may put in the claim that T is true in the face of the collected evidence E. If E is collected in the form of mutually independent chunks of data $E_1, E_2, \ldots, E_n$ Bayes' formula takes a product form

$$p(T|E) = \left[\prod_{i=1}^{n} \frac{p(E_i|T)}{p(E_i)}\right]p(T) = \left[\frac{p(E_1|T)}{p(E_1)} \times \frac{p(E_2|T)}{p(E_2)} \times \ldots \times \frac{p(E_n|T)}{p(E_n)}\right]p(T) \tag{8.2}$$

Thus, Bayesian inference could increase or decrease our confidence in a given theory T as we collect more empirical data [205, 417].

Bayesian inference is meaningful only for consistent theories T. If a theory is logically inconsistent, it can prove every possible statement A together with its negation $\neg A$. If the probabilities of both A and its negation $\neg A$ are equal to 1, then such an inconsistent theory does not satisfy Kolmogorov's axioms of probability (Section 3.10) as $p(A) + p(\neg A) = 2$. Inconsistent theories, however, are rejected as nonsense on logical grounds alone, and one does not really need to further perform empirical testing.

If a theory is logically consistent, then the probability $p(A)$ of A being true is equal to $1 - p(\neg A)$, where $p(\neg A)$ is the probability of A being false, or equivalently $p(\neg A)$ is the probability of $\neg A$ (the negation of A) to be true. The total sum of probabilities for A and $\neg A$ is equal to unity

$$p(A) + p(\neg A) = 1 \tag{8.3}$$

Logical consistency allows calculation of the odds $O(A)$ for A being true using the probability $p(A)$ for A being true.

Definition 8.1. *(Odds) The odds $O(A)$ for A being true are given by*

$$O(A) = \frac{p(A)}{p(\neg A)} = \frac{p(A)}{1 - p(A)} \tag{8.4}$$

Conversely, the probability $p(A)$ for A being true can be expressed through the odds $O(A)$ as

$$p(A) = \frac{O(A)}{1 + O(A)} \tag{8.5}$$

Bayes' theorem could be rewritten in a form that uses the odds for a theory T as

$$O(T|E) = B(E, T)O(T) \tag{8.6}$$

where $O(T)$ are the prior odds for T being true, $O(T|E)$ are the posterior odds for T being true given the evidence E, and $B(E, T)$ is the *Bayes factor* given by

$$B(E, T) = \frac{p(E|T)}{p(E|\neg T)} \tag{8.7}$$

Proof. In order to derive Eq. (8.6), we can start from the conditional probabilities

$$p(E) = \frac{p(E|T)}{p(T|E)}p(T) \tag{8.8}$$

$$p(E) = \frac{p(E|\neg T)}{p(\neg T|E)}p(\neg T) \tag{8.9}$$

Summing the above two equations gives

$$2p(E) = \frac{p(E|T)}{p(T|E)}p(T) + \frac{p(E|\neg T)}{p(\neg T|E)}p(\neg T) \tag{8.10}$$

Substitution into Eq. (8.1) results in

$$p(T|E) = \frac{p(E|T)}{\frac{1}{2}\left[\frac{p(E|T)}{p(T|E)}p(T) + \frac{p(E|\neg T)}{p(\neg T|E)}p(\neg T)\right]}p(T)$$

$$p(T|E)\left[\frac{p(E|T)}{p(T|E)}p(T) + \frac{p(E|\neg T)}{p(\neg T|E)}p(\neg T)\right] = 2p(E|T)p(T)$$

$$\frac{p(T|E)}{p(\neg T|E)} = \frac{p(E|T)}{p(E|\neg T)}\frac{p(T)}{p(\neg T)} \tag{8.11}$$

After taking into consideration Eq. (8.3), we get

$$\frac{p(T|E)}{1 - p(T|E)} = \frac{p(E|T)}{p(E|\neg T)}\left[\frac{p(T)}{1 - p(T)}\right] \tag{8.12}$$

which in view of Eq. (8.4) is seen to be just an explicit way of writing Eq. (8.6).   □

Converting the posterior odds for T given evidence E into posterior probability for T given evidence E could be done with Eq. (8.5) as follows

$$p(T|E) = \frac{B(E,T)\left[\frac{p(T)}{1-p(T)}\right]}{1 + B(E,T)\left[\frac{p(T)}{1-p(T)}\right]} \tag{8.13}$$

In the limit when the Bayes factor tends to infinity, $B(E,T) \to \infty$, we can conclude that the theory T is true with certainty

$$\lim_{B(E,T)\to\infty} p(T|E) = 1 \tag{8.14}$$

Conversely, in the limit when the Bayes factor tends to *zero*, $B(E,T) \to 0$, we can conclude that the theory T is false with certainty

$$\lim_{B(E,T)\to 0} p(T|E) = 0 \tag{8.15}$$

Noticing that the Bayes factor given by Eq. (8.7) is theory dependent, we have

$$B(E,\neg T) = \frac{1}{B(E,T)} \tag{8.16}$$

hence from $B(E,T) \to \infty$ it follows that $B(E,\neg T) \to 0$ and from $B(E,T) \to 0$ it follows that $B(E,\neg T) \to \infty$.

If the theory T predicts that the evidence E is probable, $p(E|T) > 0$, while the negation of the theory T predicts that E is improbable, $p(E|\neg T) = 0$, the Bayes factor $B(E,T)$ is *undefined* due to impossible division by zero. Bayesian inference, however, allows us to conclude from $B(E,\neg T) = 0$ that $p(\neg T|E) = 0$, and then calculate $p(T|E) = 1 - p(\neg T|E) = 1$. In other words, we are able to conclude that $p(T|E) = 1$ exactly as we would have done in the limit $B(E,T) \to \infty$. Both Bayes factors $B(E,T)$ and $B(E,\neg T)$ are undefined only if $p(E|T) = 0$ and $p(E|\neg T) = 0$, but the latter condition implies improbable evidence $p(E) = 0$. Considering improbable evidence has been explicitly denied by the requirement for $p(E) \neq 0$ in Bayes' theorem.

8.4 Comparison of scientific research programs

Bayesian inference allows us to compare different theories by assessing their posterior probabilities given collected evidence. Very simple theories may appear to be provable or disprovable with certainty by collected evidence. For example, the theory T composed of a single axiom $\exists x P(x)$ stating that "Swans exist" is proved with certainty by the experimental observation E that "A pair of white swans is swimming in the lake." Of course, the theory T is incomplete and to make further predictions about the world, one needs to add more axioms. In principle, converting all of the collected experimental data into a list of axioms $T = \{E_1, E_2, \ldots, E_n\}$ makes the theory trivially provable by experiment since $E_1, E_2, \ldots, E_n \vdash E_1, E_2, \ldots, E_n$. A list of singular statements, however, is not a good scientific theory because it does not compress any information and is devoid of explanatory power (Section 2.2.3). To achieve significant compression of observable information, scientists have to resort to universal statements that may accommodate an infinite number of singular instances. If those universal statements hold for all places in the universe at all times, they are referred to as fundamental physical laws, $L_1, L_2, \ldots, L_k$ and could be used to explain the occurrence of an experimental phenomenon E given the appropriate set of initial conditions $C_1, C_2, \ldots, C_n$. The price for the huge compression of information, however, is that universal physical laws can no longer be proved by any finite amount of experimental data, as noted by Hume (Example 8.1).

The negation $\neg T$ of the theory T that "Swans exist," also has a single axiom $\neg \exists x P(x)$ stating that "Swans do not exist." The evidence E due to observation of a pair of white swans falsifies with certainty $\neg T$. Here, it may seem that the Bayesian inference has conjured a trick in falsifying a universal theory with certainty given some evidence E, something that was debated by Lakatos (Example 8.2). Lakatos, however, did not criticize the possibility of logically falsifying a theory given that E is true (in the sense that E is collected under certain specified conditions C that are true, hence E is indeed what it is interpreted to be), but rather questioned the possibility of ever being certain whether E is true. To falsify $\neg T$, the observation that "A pair of white swans is swimming in the lake" has to be interpreted as being collected under condition C_1 in which the observer is healthy and his visual senses can be trusted in providing correct information about the real world, hence $C_1 \vdash E$. It is possible, however, to interpret the observation as being collected under condition C_2 in which the observer has been intoxicated with a hallucinogenic drug. If the pair of swans is a hallucination, then no real swan has been observed, hence $C_2 \nvdash E$, and the theory $\neg T$ that "Swans do not exist" is not falsified. Another example is Galileo's observation of light spots near Jupiter (Example 8.3). Both Galileo and his contemporary opponents were well aware of the fact that the experimental evidence E of "moons orbiting Jupiter" falsifies with certainty the geocentric model of the universe. Accordingly, in dispute was not the fact that light spots can be seen near Jupiter through the telescope, but the interpretation of their meaning. Galileo interpreted the spots near Jupiter as being observed under condition C_1 in which the telescope provides a magnified and trustworthy image of both terrestrial and celestial objects thereby implying that if those spots moved like moons they are in fact moons of Jupiter, hence $C_1 \vdash E$. In contrast, Galileo's opponents interpreted the spots near Jupiter as being observed under condition C_2 in which the telescope provides deceiving images of the heavens that say nothing useful about the real celestial objects, hence $C_2 \vdash E$. Thus, the challenge faced by scientists is not in deriving conclusions once they know what the evidence E is, but in not knowing certainly under what conditions E has been collected, hence having no guarantee to have interpreted correctly what the evidence E really is.

Scientific theories cannot be falsified by experimental data alone, because no theory can be tested without making assumptions about the initial conditions under which the experiment is performed. Scientists are able, however, to incorporate each scientific theory T into a *research program* consisting of the following:

(1) Listing the initial conditions $C_1, C_2, \ldots, C_n$ that were controlled by the experimenter and believed to be true before the experimental result is obtained.

(2) Calculating the predicted experimental result E' based on the theory T together with the initial conditions $C_1, C_2, \ldots, C_n$.

(3) Recording the experimental result E as it is interpreted by the theory T.

(4) Calculating the error $\Delta E = E' - E$ between predicted E' and observed E experimental data.

(5) Retrodicting the uncontrolled factors $C_{n+1}, C_{n+2}, \ldots$ that should have acted alongside with the controlled factors $C_1, C_2, \ldots, C_n$ in order to generate outcome E rather than E'.

A scientist involved in a research program based on theory T, can always perform a *problem shift* and claim that the theory T has not been falsified since uncontrolled factors $C_{n+1}, C_{n+2}, \ldots$ have acted. If we have two competing research programs, however, Bayesian inference can compare the research programs by the size of the errors in their experimental predictions $\Delta E = E' - E$ and by the complexity of the problem shifts $C_{n+1}, C_{n+2}, \ldots$ needed to protect each theory against falsification. Keeping the best theory results in growth of scientific knowledge.

Due to the theory-laden nature of experimental facts, it is hardly possible to rationally compare different theories that disagree on what the actual evidence is. Therefore, growth of scientific knowledge is possible only if Bayesian inference is used to assess a collection of theories $T_1, T_2, \ldots, T_n$ that have some set of experimental data whose interpretation can be agreed upon. For example, both Newton's theory of universal gravitation [357] and Einstein's theory of general relativity [155] agreed on the fact that the observed orbit of Mercury with the available telescopes represents faithfully the motion of the planet in space. What the two theories disagreed upon is why Mercury orbits in the way it does: Newton's theory predicted that the initial conditions included another planet Vulcan orbiting between the sun and Mercury, whereas Einstein's theory predicted the observed orbit of Mercury assuming that the sun alone attracts Mercury. The equations of Einstein's theory of general relativity also predicted that Newton's theory will always encounter unexplainable anomalies in the orbits of other planets that are close to their stars due to the effects of space-time warping caused by the mass of the stars. Because we now know that Vulcan does not exist, it can be said that the Bayes factor for Newton's theory is *zero* and the theory has been falsified.

Since no experiment or observational report alone can lead to falsification of a theory, there is no falsification before the emergence of a better theory that can explain the experimental facts already known, pinpoint the sources of error in the old theory, and predict new results hitherto unknown [301]. Thus, the overly negative character of naive falsificationism has to be replaced with constructive criticism. If the falsification depends on the emergence of better theories, however, it is no longer a simple relation between a theory and the empirical basis, but a multiple relation between competing theories with their corresponding research programs [300, 304, 306]. Consequently, the comparative study of research programs based on different theories becomes the main tool to grow scientific knowledge as it provides insights into the origins of the failure of those theories that happen to lose the competition and highlights the merits of the theory that wins.

Example 8.4. *(Progressive and degenerative problem shifts) A dominant scientific theory T_1 that is widely adopted by the scientific community cannot be refuted by experiments alone, unless there is another rival theory T_2 that provides a better explanation of the observed phenomena. Without a rival theory, the dominant theory T_1 can always perform a problem shift and claim that the problematic experimental outcome E is only an apparent contradiction due to the fact that the initial conditions C_1 of the experiment have been interfered with by another factor C_2. Thus, the dominant theory T_1 can discard a large number of problematic experiments as failed experiments, stating that the reading of the apparatus is invalid due to technical malfunction, due to improper calibration of the apparatus, or due to unacceptably large levels of background noise. Such problem shifts are done by experimentalists every time the experimental outcome is an outlier, well outside the expected range of outcomes. If a source of apparatus malfunction is indeed identified for some of those problematic experimental outcomes, the problem shift performed by the dominant theory T_1 is viewed as a* progressive problem shift *and the confidence in the theory T_1 grows due to the large Bayes factor obtained from predicting an unlikely event. The discovery of Neptune with the use of Newton's theory of universal gravitation is an example of a progressive problem shift. If the interfering factor is not identified, however, it does not follow that we should start losing confidence in T_1. Indeed, nothing can be concluded from one's ignorance of why the experiment failed, otherwise scientists would have been able to refute a theory by merely choosing to stay ignorant about it. Only if there is a rival theory T_2 that explains some of the problematic experimental results without the need of interfering factors, it is possible for scientists to revisit their decision in dismissing the problematic evidence. From the viewpoint of the rival theory T_2, given some initial condition C_k the probability for the existence of an extra interfering factor C_{k+1} could be zero. Then, if T_2 is correct, by knowing the condition C_k and the predicted experimental outcome E, one is able to set a crucial experiment for rejecting T_1. For example, Einstein's theory of general relativity T_2 predicts correctly the observed orbit of Mercury and explains why the predictions given by Newton's theory T_1 fail near massive objects such as stars. Then, by collecting observational data from many planets orbiting close to their stars like Mercury, strong Bayesian evidence against Newton's theory T_1 can be obtained. Since T_1 predicts the existence of other unseen interfering planets, and these planets can have different masses,* *velocities and orbits, it would be extremely unlikely that all those unseen planets are part of a cosmic conspiracy such that the orbital motion of the observable planets close to their stars is exactly as T_2 predicts. Yet, if the predictions of T_2 are correct and it requires no unseen planets, it would be rational to conclude that we do not see these extra planets exactly because they do not exist. In essence, T_1 becomes vulnerable only when among the allegedly failed experiments a pattern is discovered that remains un-explainable under T_1, but is nicely explained by another theory T_2. Another example is Galileo's discovery of the moons of Jupiter. We have shown that the interpretation of Galileo's observations is theory-laden and his contemporary opponents accepted only that light spots near Jupiter are seen, but claimed that these were artifacts produced by the telescope. Because the two theories actually agree on the existence of light spots, Galileo was able to use his theory T_2 stating that those spots are moons of Jupiter, cal-culate the orbital periods of each moon and predict exactly how those light spots would behave in the next weeks or months to come. If Galileo's predictions are then confirmed by his opponents, he will have strong Bayesian evidence against their theory T_1 which states that those light spots are artifacts. Indeed, it would be extremely unlikely that his telescope produces deceiving image artifacts night after night with such a conspirative dynamics that resembles the motion of orbiting moons. Crucial experiments that lead to rejection of any theory T_1 always rely on revealing some of the problem shifts per-formed by supporters of T_1* as degenerative problem shifts *[303, 304]. The extent of degeneration of T_1, however, can only be appreciated in the light of a better theory T_2.*

8.5   Conscious experiences and protocol sentences

If two alternative theories T_1 and T_2 do not appear to have a common interpre-tation of the experimental data, at least they could try to agree on how the phe-nomena appear to us in our *conscious experiences*. For example, the brightness and the shape of visual images as they appear to our senses *can be agreed upon* even though it may not be known exactly what these images correspond to in reality. Similarly, the direction in which the needle of an apparatus is pointing could be agreed upon even though it may not be known exactly what quantity is measured.

Definition 8.2. *(Protocol sentences) Sentences that describe communicable regulari-ties in one's own conscious experiences are referred to as* protocol sentences. *Protocol sentences can describe what we think, how we feel, or how the surrounding world ap-pears to us as perceived through our senses [70, 356].*

Protocol sentences are able to provide true facts about the universe, because we are part of the universe (Axiom 2.5.2) and our conscious experiences are a thing that we can be absolutely sure of (Chapter 1). In other words, we can be wrong in what we think the surrounding world is, but we cannot be wrong on how this surrounding world appears to us through our senses. Because protocol sentences are generated through introspection, scientific communication between different scientists requires that protocol sentences are honestly reported without deliber-ate lies, distortion or fabrication. Noteworthy, protocol sentences do not need to communicate the phenomenal nature of qualia. Two people with inverted qualia (Example 6.11) may both state "I see a yellow marigold," even though one may be in fact experiencing yellowness and the other blueness. For testing a scientific theory is needed agreement on the verbal content on the protocol sentences, not on the actual phenomenological character of the qualia (Section 6.6.5). For example, Galileo and his opponents agreed upon the protocol statement that "Four spots of light are seen near Jupiter through the telescope." Having agreed upon how the world appears to them through their senses, Galileo and his contemporary opponents could have further used Bayesian inference based on the time dynamics of these light spots as predicted by the two alternative theories, one stating that the spots are moons and the other that the spots are telescope artifacts, in order to assess who is right.

Because we can only access the world through our subjective consciousness, ultimately every scientific theory should be decomposable down to protocol sentences describing the kind of conscious experiences that we are supposed to get under conditions well prescribed by the theory. If we happen to see unicorns under circumstances where nobody else can, then no amount of scientific testing is capable of convincing us that "Unicorns do not exist" is a better theory than "Everyone else is not gifted enough to see the unicorn" or "Everyone else sees the unicorn, but is conspiring to lie about it." Alternatively, if we happen to be unable to see unicorns when everyone else can, no amount of protocol sentences such as "I saw the unicorn" uttered by others would help to convince us that "Unicorns exist" is a good theory. Thus, objectivity in science is only possible to the extent that we happen to hear from other people reports similar to our own. Fortunately, in real life we do hear from other people confirmations that they too can see what we are capable of seeing. This is not just a lucky coincidence, but an inevitable feature of Darwinian evolution due to the fact that those individuals who happened to perceive the surrounding world in a distorted way would have been frequently injured by natural accidents and would not have been able to survive.

8.6   Testing the quantum information theory of consciousness

8.6.1   Tests for logical consistency

One way to test the quantum information theory of consciousness is to investigate whether its predictions contain logical inconsistencies. Since from logical contradiction follows anything, a scientific theory can be meaningful only if it does not prove both some result and its negation.

One may wonder whether it is possible to prove once and for all that a given scientific theory is logically consistent given that the axioms of the theory are true. The answer is negative and follows from *Gödel's incompleteness theorems* according to which if a consistent formal axiomatic theory is strong enough to capture the arithmetic of the natural numbers, then it cannot prove its own consistency [218, 194, 445]. Thus, not only are we unable to verify the empirical adequacy of a scientific theory with the use of a finite amount of experimental data, but we are also unable to guarantee that a proposed theory is not just self-contradictory nonsense. The theoretical research cannot end with the formulation of a scientific theory, but has to continue scrutinizing the theory for unnoticed contradictions. The search for logical inconsistencies will never end if the theory is consistent, but it may end in a finite amount of time if an inconsistency is discovered.

In Chapter 5, we have discussed multiple inconsistencies in classical theories of consciousness that are based on the axioms of classical physics. The most entertaining one is provided by eliminative materialism according to which consciousness does not exist at all, hence we are hallucinating having experiences while in fact we have none. A remarkable contradiction also occurs in the functional approach according to which consciousness is a product of brain function that has been selected for by natural selection in the course of millions of years of evolution. From Theorem 22, however, it follows that in classical functionalism consciousness is a causally ineffective epiphenomenon that cannot evolve as there is nothing to be selected for. Yet another contradiction occurs in the reductive approach according to which the conscious mind is the brain, but some brains such as anesthetized or crushed brains are unconscious. Since discarding massive amounts of knowledge from disciplines such as Darwin's theory of evolution or anesthesiology is a heavy price to be paid, one is forced to conclude that these classical approaches to consciousness are flawed beyond repair.

In regard to the quantum information theory of consciousness, the search for possible problems should not be considered over. We have shown that previously calculated decoherence times do rule out quantum theories that require quantum coherence at a millisecond timescale in the brain (Section 7.3), but do not rule the quantum information theory of consciousness (Section 4.15) that uses quantum entanglement for binding of conscious experiences at a picosecond timescale or faster. We have also been able to apply some of the fundamental quantum information theorems from Section 4.20 to address various problems related to consciousness raised by ingenious thought experiments (Chapter 6). Even though the quantum information theory of consciousness provides a huge leap forward in our understanding of ourselves as conscious minds, there are still some open problems that need concrete mathematical solutions. The quest for quantum gravity, aimed at reconciling quantum mechanics and Einstein's theory of general relativity, is one such currently ongoing project that has the potential to shed new light on the mechanism of objective reductions.

8.6.2   Tests for empirical adequacy

A second way to test the quantum information theory of consciousness is to experimentally compare it with other rival theories using Bayes' theorem. For meaningful calculation of Bayes factors, the rival theories should not be inconsistent and should agree on some shared knowledge expressed in protocol sentences that describe how the world appears to us through our conscious experiences. As a precaution against possible dishonest introspective reports from others, one could perform personally all of the empirical tests and then draw one's own conclusions.

The unsuccessful attempts of Giovanni Aldini to resurrect a hanged criminal in 1803 provide an example of how experimental evidence is able to rule out the reductive classical theory of consciousness according to which the electric currents propagating in your brain are what your conscious mind is. Experimentally, one is also able to easily rule out the classical binding of conscious experiences through synchronous electrical firing of neurons. What is needed is simultaneous monitoring of the electrical firing of neurons in two human subjects. Then, after waiting long enough, some of the neurons will fire simultaneously in the two brains and one can obtain verbal reports whether each of the two human minds had been able to take a look into the conscious experiences of the other human mind at the time of the simultaneous electric neuronal firing. Even better, you do not really need to monitor the electric firing of the neurons in your or someone else's brain since by the sheer number of living people on Earth, it is highly probable that at any moment of time, some of the neurons in your brain will fire synchronously with someone else's neurons. Therefore, if you have never been able to look into someone else's conscious mind in your life, then it is rational to conclude that simultaneity of neuronal firing is not in itself responsible for binding of conscious experiences. Similarly, one can also discard the proposed quantum binding of conscious experiences due to quantum coherence in the brain. It suffices to note that the collection of two quantum coherent systems is also quantum coherent, in order to conclude that if quantum coherence were responsible for the binding of conscious experiences, then the collection of all quantum coherent human minds should have been bound into a single global mind, which is not the case.

The apparent strong Bayesian support for the quantum information theory of consciousness, in comparison with other rival theories of consciousness, is not accidental. To address the main problems of consciousness (Chapter 1) using the specific tools of quantum information theory, we had to construct an axiomatization that outputs results consistent with our introspective viewpoint of what our conscious minds are. Thus, if the quantum information theory of consciousness fares better than its rivals, then this is due to the fine theoretical construction that has been done. To really challenge the current theory one has to find novel problems that were not covered here.

8.6.3 Tests for empirical corroboration

A third way to test the quantum information theory of consciousness is to look for empirical corroboration of the novel predictions that are made by the theory out of logical necessity rather than by deliberate design. In this respect, the theory offers interesting experimental tests both to biologists and to physicists.

In Section 7.4, we have discussed the importance of quantum tunneling in the zipping of SNARE protein complexes that control the release of neurotransmitter molecules through exocytosis of synaptic vesicles at electrically active synapses. Previously, we have proposed to experimentally test the possible quantum tunneling in exocytosis using the so-called *kinetic isotope effect*, as in the studies of dehydrogenase enzymes [30, 31, 429, 469]. The key idea is to replace some of the *protium* hydrogen atoms, ^{1}H, in –C=O$\cdots$H–N– complexes that form the α-helix spines in the SNARE four-α-helix bundle with the heavier hydrogen isotope, *deuterium*, ^{2}H. Because quantum tunneling is strongly dependent on the particle mass that enters into the Schrödinger equation, such an isotope replacement is predicted to have a detrimental effect on the propagation of the Davydov soliton along the α-helix spines and thus should effectively inhibit exocytosis. In contrast, if the SNARE zipping is just a classical process, the isotope replacement would not have any effect on exocytosis because the chemical properties of the common hydrogen isotope, protium, are identical to the chemical properties of the heavier hydrogen isotope, deuterium. Such an experimental test is well within the capabilities of present-day biochemistry.

Further experimental tests searching for quantum tunneling of electrons in the electrically charged protein α-helices that act as voltage sensors of voltage-gated ion channels [208] could be designed with the use of quantum resonances [501]. For example, electrophysiological recordings from single channels in excised patches of neuronal membrane are routinely done for studying the biophysical properties of voltage-gated ion channels. If an external alternating high frequency driving force is applied to the recording pipette, one might be able to induce resonances in the conductance of the voltage-gated ion channel. Because such resonances do not occur in systems exhibiting classical rate equation dynamics, their existence would confirm the quantum tunneling mechanism in the gating of the ion channels [501].

The existence of a physical process leading to objective reduction and quantum disentanglement is one of the definitive predictions by the quantum information theory of consciousness. Experimental tests sensitive enough to detect the occurrence of objective reductions and capable of determining the energy threshold $\mathcal{E}$ at which these reductions occur have already been proposed [334, 379]. The basic idea is to arrange a Mach–Zehnder interferometer with highly energetic X-ray photons such that in one of the interferometer arms the mirror is small enough to be kicked into a macroscopic superposition of two different locations in space, yet massive enough so that the gravitationally induced objective reduction can decohere the photon before it is registered at the detectors. If the photon does not form an interference pattern, then objective reduction should have occurred. In order to be sure, however, that the loss of the interference pattern is not caused by decoherence due to environmental interaction, the experiment needs to be conducted in ultra-high vacuum. One version of the experiment called Free-orbit Experiment with Laser Interferometry X-rays (FELIX) is suggested to be performed in space, where the photon is reflected between mirrors on two space platforms of perhaps an Earth-diameter separation [379, pp. 856–860]. Another version of the experiment, using less energetic photons and a concave mirror so that the photon can kick the mirror millions of times before exiting toward the interferometer arm, is suggested to be performed on the ground in a vacuum chamber cooled down to temperature near absolute zero [334]. In either case, the experiments appear to be within the reach of present-day technology.

8.6.4 Tests for explanatory power

The fourth and most important test, addressing the explanatory power of quantum information theory of consciousness, I leave to you, the reader. Because the main purpose of every scientific theory is to explain natural phenomena and provide a conceptual framework for approaching the physical world, you could ask yourself whether you are now seeing the world through new eyes. If the answer is yes, the quantum information theory of consciousness would have passed its most significant test.

To assess your progress in understanding consciousness, you may contemplate the following 15 quiz questions and contrast what you think now with what you were thinking when you first opened this book.

Quiz on Consciousness

1. What is the physical fabric from which your mind is made?

2. Why does your mind appear to be localized in physical space?

3. Where is the physical boundary that delimits your mind from other minds?

4. Why is not your mind a constituent part of another larger conscious mind?

5. What is the physical mechanism that binds your conscious experiences together?

6. Why do you lack direct access to the conscious experiences of other minds?

7. Why do you have privileged access to your own conscious experiences?

8. Why you are unable to communicate to others what your conscious experiences feel like?

9. How does your mind causally affect the physical world?

10. How does your mind exercise genuine free will and make choices?

11. How does your mind override the free will of its constituent physical particles?

12. How do different conscious minds evolve through natural selection?

13. Why does your brain not perform its physiological function in a mindless mode?

14. How does your brain generate your conscious experiences?

15. How do your brain and your mind interact with each other?

References

[1] L. F. Abbott, B. DePasquale, and R.-M. Memmesheimer. Building functional networks of spiking model neurons. *Nature Neuroscience*, 19(3):350–355, 2016.

[2] C. Acuna, Q. Guo, J. Burré et al. Microsecond dissection of neurotransmitter release: SNARE-complex assembly dictates speed and Ca^{2+} sensitivity. *Neuron*, 82(5):1088–1100, 2014.

[3] J. E. Alcock. Give the null hypothesis a chance: reasons to remain doubtful about the existence of Psi. *Journal of Consciousness Studies*, 10(6–7):29–50, 2003.

[4] M. T. Alkire, R. J. Haier, J. H. Fallon, and L. Cahill. Hippocampal, but not amygdala, activity at encoding correlates with long-term, free recall of nonemotional information. *Proceedings of the National Academy of Sciences*, 95(24):14506–14510, 1998.

[5] R. K. Allemann and N. S. Scrutton. *Quantum Tunnelling in Enzyme-Catalysed Reactions*. RSC Biomolecular Sciences. Royal Society of Chemistry, Cambridge, 2009.

[6] J. L. Alquist, S. E. Ainsworth, R. F. Baumeister et al. The making of might-have-beens: effects of free will belief on counterfactual thinking. *Personality and Social Psychology Bulletin*, 41(2):268–283, 2015.

[7] O. Alter and Y. Yamamoto. *Quantum Measurement of a Single System*. Wiley, New York, 2001.

[8] H. Amann and J. Escher. *Analysis I*. Birkhäuser, Basel, 2009.

[9] T. Aoyama, M. Hayakawa, T. Kinoshita, and M. Nio. Tenth-order qed contribution to the electron $g-2$ and an improved value of the fine structure constant. *Physical Review Letters*, 109(11):111807, 2012.

[10] P. Apian. *Cosmographicus Liber*. Landshutae, impensis P. Apiani, 1524.

[11] T. M. Apostol. *Calculus. Vol. 1: One-Variable Calculus, with an Introduction to Linear Algebra*. John Wiley & Sons, New York, 2nd edition, 1967.

[12] R. Appignanesi and O. Zarate. *Freud for Beginners*. Pantheon Books, New York, 1979.

[13] Aristotle. *The Works of Aristotle. Vol. 5: De Partibus Animalium. De Motu Animalium. De Incessu Animalium. De Generatione Animalium.* Clarendon Press, Oxford, 1912.

[14] Aristotle. *The Works of Aristotle. Vol. 8: Metaphysica.* Clarendon Press, Oxford, 1928.

[15] Aristotle. *The Works of Aristotle. Vol. 3: Meteorologica. De Mundo. De Anima. Parva Naturalia. De Spiritu.* Clarendon Press, Oxford, 1931.

[16] Aristotle. *The Works of Aristotle. Vol. 4: Historia Animalium.* Clarendon Press, Oxford, 1949.

[17] J. Ashmead. Morlet wavelets in quantum mechanics. *Quanta*, 1(1):58–70, 2012.

[18] A. Aspect. Bell's inequality test: more ideal than ever. *Nature*, 398(6724):189–190, 1999.

[19] A. Aspect, J. Dalibard, and G. Roger. Experimental test of Bell's inequalities using time-varying analyzers. *Physical Review Letters*, 49(25):1804–1807, 1982.

[20] A. Aspect, P. Grangier, and G. Roger. Experimental realization of Einstein–Podolsky–Rosen–Bohm Gedankenexperiment: a new violation of Bell's inequalities. *Physical Review Letters*, 49(2):91–94, 1982.

[21] F. A. C. Azevedo, L. R. B. Carvalho, L. T. Grinberg et al. Equal numbers of neuronal and nonneuronal cells make the human brain an isometrically scaled-up primate brain. *Journal of Comparative Neurology*, 513(5):532–541, 2009.

[22] I. Bahrini, J.-H. Song, D. Diez, and R. Hanayama. Neuronal exosomes facilitate synaptic pruning by up-regulating complement factors in microglia. *Scientific Reports*, 5:7989, 2015.

[23] H. Bai, R. Xue, H. Bao et al. Different states of synaptotagmin regulate evoked versus spontaneous release. *Nature Communications*, 7:10971, 2016.

[24] Z. Bai and S. Du. Maximally coherent states. *Quantum Information & Computation*, 15(15–16):1355–1364, 2015.

[25] P. F. Baker, A. L. Hodgkin, and T. I. Shaw. Replacement of the protoplasm of a giant nerve fibre with artificial solutions. *Nature*, 190(4779):885–887, 1961.

[26] P. F. Baker, A. L. Hodgkin, and T. I. Shaw. Replacement of the axoplasm of giant nerve fibres with artificial solutions. *Journal of Physiology*, 164:330–354, 1962.

[27] J. M. Bardeen, B. Carter, and S. W. Hawking. The four laws of black hole mechanics. *Communications in Mathematical Physics*, 31(2):161–170, 1973.

[28] C. N. Barnard. A human cardiac transplant: an interim report of a successful operation performed at Groote Schuur Hospital, Cape Town. *South African Medical Journal*, 41(48):1271–1274, 1967.

[29] H. Barnum, C. M. Caves, C. A. Fuchs et al. Noncommuting mixed states cannot be broadcast. *Physical Review Letters*, 76(15):2818–2821, 1996.

[30] J. Basran, S. Patel, M. J. Sutcliffe, and N. S. Scrutton. Importance of barrier shape in enzyme-catalyzed reactions. Vibrationally assisted hydrogen tunneling in tryptophan tryptophylquinone-dependent amine dehydrogenases. *Journal of Biological Chemistry*, 276(9):6234–6242, 2001.

[31] J. Basran, M. J. Sutcliffe, and N. S. Scrutton. Enzymatic H-transfer requires vibration driven extreme tunneling. *Biochemistry*, 38(10):3218–3222, 1999.

[32] R. F. Baumeister, A. W. Crescioni, and J. L. Alquist. Free will as advanced action control for human social life and culture. *Neuroethics*, 4(1):1–11, 2011.

[33] T. Baumgratz, M. Cramer, and M. B. Plenio. Quantifying coherence. *Physical Review Letters*, 113(14):140401, 2014.

[34] B. P. Bean. The action potential in mammalian central neurons. *Nature Reviews Neuroscience*, 8(6):451–465, 2007.

[35] C. Beaulieu and M. Colonnier. A laminar analysis of the number of round-asymmetrical and flat-symmetrical synapses on spines, dendritic trunks, and cell bodies in area 17 of the cat. *Journal of Comparative Neurology*, 231(2):180–189, 1985.

[36] F. Beck. Can quantum processes control synaptic emission? *International Journal of Neural Systems*, 7(4):343–353, 1996.

[37] F. Beck and J. C. Eccles. Quantum aspects of brain activity and the role of consciousness. *Proceedings of the National Academy of Sciences*, 89(23):11357–11361, 1992.

[38] F. Beck and J. C. Eccles. Quantum processes in the brain: a scientific basis of consciousness. *Cognitive Studies: Bulletin of the Japanese Cognitive Science Society*, 5(2):95–109, 1998.

[39] J. S. Bell. On the Einstein–Podolsky–Rosen paradox. *Physics*, 1(3):195–200, 1964.

[40] C. M. Bender. *Mathematical Physics*. Perimeter Institute Recorded Seminar Archive. Perimeter Institute for Theoretical Physics, Waterloo, Ontario, Canada, 2011.

[41] C. M. Bender and S. A. Orszag. *Advanced Mathematical Methods for Scientists and Engineers*. McGraw-Hill, New York, 1978.

[42] C. H. Bennett, G. Brassard, C. Crépeau et al. Teleporting an unknown quantum state via dual classical and Einstein–Podolsky–Rosen channels. *Physical Review Letters*, 70(13):1895–1899, 1993.

[43] M. R. Bennett and P. M. S. Hacker. *History of Cognitive Neuroscience*. Wiley-Blackwell, Chichester, 2013.

[44] M. Bennett-Smith. Stephen Hawking: brains could be copied to computers to allow life after death. *The Huffington Post*, September 24, 2013.

[45] D. L. Benson, L. M. Schnapp, L. Shapiro, and G. W. Huntley. Making memories stick: cell-adhesion molecules in synaptic plasticity. *Trends in Cell Biology*, 10(11):473–482, 2000.

[46] M. N. Bera, T. Qureshi, M. A. Siddiqui, and A. K. Pati. Duality of quantum coherence and path distinguishability. *Physical Review A*, 92(1):012118, 2015.

[47] G. Berkeley. *A Treatise Concerning the Principles of Human Knowledge*. J. F. Dove, London, 1820.

[48] F. Berna, P. Goldberg, L. K. Horwitz et al. Microstratigraphic evidence of in situ fire in the Acheulean strata of Wonderwerk Cave, Northern Cape province, South Africa. *Proceedings of the National Academy of Sciences*, 109(20):E1215–E1220, 2012.

[49] F. Bezanilla. The voltage sensor in voltage-dependent ion channels. *Physiological Reviews*, 80(2):555–592, 2000.

[50] F. Bezanilla. The voltage-sensor structure in a voltage-gated channel. *Trends in Biochemical Sciences*, 30(4):166–168, 2005.

[51] D. J. Bierman and D. I. Radin. Anomalous anticipatory response on randomized future conditions. *Perceptual and Motor Skills*, 84(2):689–690, 1997.

[52] N. Block. Troubles with functionalism. In C. W. Savage, editor, *Perception & Cognition: Issues in the Foundations of Psychology*, volume 9 of *Minnesota Studies in The Philosophy of Science*, pages 261–325. University of Minnesota Press, Minneapolis, 1978.

[53] D. Boatman, B. Gordon, J. Hart et al. Transcortical sensory aphasia: revisited and revised. *Brain*, 123(8):1634–1642, 2000.

[54] J. Bogousslavsky, J. Miklossy, J. P. Deruaz et al. Lingual and fusiform gyri in visual processing: a clinico-pathologic study of superior altitudinal hemianopia. *Journal of Neurology, Neurosurgery, and Psychiatry*, 50(5):607–614, 1987.

[55] M. Born. Zur Quantenmechanik der Stoßvorgänge. *Zeitschrift für Physik A*, 37(12):863–867, 1926.

[56] M. Born and A. Einstein. *The Born–Einstein Letters: Correspondence Between Albert Einstein and Max and Hedwig Born from 1916–1955, with Commentaries by Max Born*. Macmillan, London, 1971.

[57] D. Boschi, S. Branca, F. De Martini et al. Experimental realization of teleporting an unknown pure quantum state via dual classical and Einstein–Podolsky–Rosen channels. *Physical Review Letters*, 80(6):1121–1125, 1998.

[58] M. A. Brandimonte, N. Bruno, and S. Collina. Cognition. In K. Pawlik and G. d'Ydewalle, editors, *Psychological Concepts: An International Historical Perspective*, pages 11–26. Psychology Press, Hove, 2006.

[59] F. Briggs. Organizing principles of cortical layer 6. *Frontiers in Neural Circuits*, 4:3, 2010.

[60] L. Brizhik, A. Eremko, B. Piette, and W. Zakrzewski. Solitons in α-helical proteins. *Physical Review E*, 70(3):031914, 2004.

[61] C. D. Broad. *The Mind and Its Place in Nature*. Harcourt, Brace & Company, Inc., New York, 1925.

[62] Č. Brukner. Quantum complementarity and logical indeterminacy. *Natural Computing*, 8(3):449–453, 2009.

[63] D. Bucher, L. Guidoni, P. Carloni, and U. Rothlisberger. Coordination numbers of K^+ and Na^+ ions inside the selectivity filter of KcsA potassium channel: insights from first principles molecular dynamics. *Biophysical Journal*, 98(10):L47–L49, 2010.

[64] D. Bucher and U. Rothlisberger. Molecular simulations of ion channels: a quantum chemist's perspective. *Journal of General Physiology*, 135(6):549–554, 2010.

[65] P. Busch. Is the quantum state (an) observable? In R. S. Cohen, M. Horne, and J. Stachel, editors, *Potentiality, Entanglement and Passion-at-a-Distance: Quantum Mechanical Studies for Abner Shimony, Vol. 2*, volume 194 of *Boston Studies in the Philosophy of Science*, pages 61–70. Kluwer, Dordrecht, 1997.

[66] G. Cañas, S. Etcheverry, E. S. Gómez et al. Experimental implementation of an eight-dimensional Kochen–Specker set and observation of its connection with the Greenberger–Horne–Zeilinger theorem. *Physical Review A*, 90(1):012119, 2014.

[67] S. Canavero. HEAVEN: the head anastomosis venture. Project outline for the first human head transplantation with spinal linkage (GEMINI). *Surgical Neurology International*, 4(Supplement 1):S335–S342, 2013.

[68] L. H. Caporale. *Darwin in the Genome: Molecular Strategies in Biological Evolution*. McGraw-Hill, New York, 2003.

[69] J. M. Carmena, M. A. Lebedev, R. E. Crist et al. Learning to control a brain-machine interface for reaching and grasping by primates. *PLoS Biology*, 1(2):e42, 2003.

[70] R. Carnap. On protocol sentences. *Noûs*, 21(4):457–470, 1987.

[71] R. Carnap, H. Hahn, and O. Neurath. The scientific conception of the world: the Vienna circle. In M. Neurath and R. S. Cohen, editors, *Empiricism and Sociology*, volume 1 of *Vienna Circle Collection*, pages 299–318. D. Reidel, Dordrecht, 1973.

[72] M. Cattani, I. L. Caldas, S. L. de Souza, and K. C. Iarosz. Deterministic chaos theory: basic concepts. *Revista Brasileira de Ensino de Física*, 39(1):e1309, 2017.

[73] J. M. Cattell. The time it takes to see and name objects. *Mind*, 11(41):63–65, 1886.

[74] J. M. Cattell. The time taken up by cerebral operations. Parts I & II. *Mind*, 11(42):220–242, 1886.

[75] J. M. Cattell. The time taken up by cerebral operations. Part III. *Mind*, 11(43):377–392, 1886.

[76] J. M. Cattell. The time taken up by cerebral operations. Part IV. *Mind*, 11(44):524–538, 1887.

[77] J. M. Cattell. Experiments on the association of ideas. *Mind*, 12(45):68–74, 1887.

[78] J. M. Cattell and F. Galton. Mental tests and measurements. *Mind*, 15(59):373–381, 1890.

[79] G. J. Chaitin. Information-theoretic limitations of formal systems. *Journal of the Association for Computing Machinery*, 21(3):403–424, 1974.

[80] G. J. Chaitin. The halting probability Ω: irreducible complexity in pure mathematics. *Milan Journal of Mathematics*, 75(1):291–304, 2007.

[81] G. J. Chaitin. *Thinking about Gödel and Turing: Essays on Complexity, 1970–2007*, chapter Leibniz, information, math & physics, pages 227–239. World Scientific, Singapore, 2007.

[82] D. J. Chalmers. Absent qualia, fading qualia, dancing qualia. In T. Metzinger, editor, *Conscious Experience*, pages 309–328. Imprint Academic, Thorverton, 1995.

[83] D. J. Chalmers. Facing up to the problem of consciousness. *Journal of Consciousness Studies*, 2(3):200–219, 1995.

[84] D. J. Chalmers. *The Conscious Mind: In Search of a Fundamental Theory*. Philosophy of Mind. Oxford University Press, Oxford, 1996.

[85] D. J. Chalmers. How can we construct a science of consciousness? *Annals of the New York Academy of Sciences*, 1303(1):25–35, 2013.

[86] C. C. Chancey, S. A. George, and P. J. Marshall. Calculations of quantum tunnelling between closed and open states of sodium channels. *Journal of Biological Physics*, 18(4):307–321, 1992.

[87] E. R. Chapman. Synaptotagmin: a Ca^{2+} sensor that triggers exocytosis? *Nature Reviews Molecular Cell Biology*, 3(7):498–508, 2002.

[88] E. R. Chapman and A. F. Davis. Direct interaction of a Ca^{2+}-binding loop of synaptotagmin with lipid bilayers. *Journal of Biological Chemistry*, 273(22):13995–14001, 1998.

[89] R. Chaudhuri and I. Fiete. Computational principles of memory. *Nature Neuroscience*, 19(3):394–403, 2016.

[90] Y. A. Chen and R. H. Scheller. SNARE-mediated membrane fusion. *Nature Reviews Molecular Cell Biology*, 2(2):98–106, 2001.

[91] T.-P. Cheng. *A College Course on Relativity and Cosmology*. Oxford University Press, Oxford, 2015.

[92] M. C. Chicka, E. Hui, H. Liu, and E. R. Chapman. Synaptotagmin arrests the SNARE complex before triggering fast, efficient membrane fusion in response to Ca^{2+}. *Nature Structural & Molecular Biology*, 15(8):827–835, 2008.

[93] A. M. Childs, R. Cleve, E. Deotto et al. Exponential algorithmic speedup by a quantum walk. In *Proceedings of the 35th annual ACM Symposium on Theory of Computing*, pages 59–68. Association for Computing Machinery, 2003.

[94] F. Cioffi. *Freud and the Question of Pseudoscience*. Open Court, Chicago, 1998.

[95] J. H. Cole, A. D. Greentree, L. C. L. Hollenberg, and S. Das Sarma. Spatial adiabatic passage in a realistic triple well structure. *Physical Review B*, 77(23):235418, 2008.

[96] N. Collins. Hawking: 'in the future brains could be separated from the body'. *The Telegraph*, September 20, 2013.

[97] M. Colombo, A. Colombo, and C. G. Gross. Bartolomeo Panizza's Observations on the optic nerve (1855). *Brain Research Bulletin*, 58(6):529–539, 2002.

[98] J. H. Conway. *Free Will Lecture Series.* Princeton University, Princeton, NJ, 2009.

[99] J. H. Conway and S. B. Kochen. The free will theorem. *Foundations of Physics,* 36(10):1441–1473, 2006.

[100] J. H. Conway and S. B. Kochen. The strong free will theorem. *Notices of the AMS,* 56(2):226–232, 2009.

[101] J. H. Conway and S. B. Kochen. Thou shalt not clone one bit! *Foundations of Physics,* 40(4):430–433, 2010.

[102] F. C. Crews. *Unauthorized Freud: Doubters Confront a Legend.* Viking, New York, 1998.

[103] L. Cruzeiro, J. Halding, P. L. Christiansen et al. Temperature effects on the Davydov soliton. *Physical Review A,* 37(3):880–887, 1988.

[104] J. Cutting and H. Silzer. Psychopathology of time in brain disease and schizophrenia. *Behavioural Neurology,* 3(4):197–215, 1990.

[105] C. S. Darrow. *The Story of My Life.* Charles Scribner's Sons, New York, 1932.

[106] C. Darwin. *On the Origin of Species by Means of Natural Selection, or the Preservation of Favoured Races in the Struggle for Life.* John Murray, London, 1859.

[107] C. Darwin. *From So Simple a Beginning: The Four Great Books of Charles Darwin (The Voyage of the Beagle, On the Origin of Species, The Descent of Man, The Expression of the Emotions in Man and Animals).* W. W. Norton & Company, New York, 2006.

[108] G. M. Davidson. A syndrome of time-agnosia. *Journal of Nervous and Mental Disease,* 94(3):336–343, 1941.

[109] A. S. Davydov. The theory of contraction of proteins under their excitation. *Journal of Theoretical Biology,* 38(3):559–569, 1973.

[110] A. S. Davydov. Solitons and energy transfer along protein molecules. *Journal of Theoretical Biology,* 66(2):379–387, 1977.

[111] A. S. Davydov. *Biology and Quantum Mechanics.* Naukova Dumka, Kiev, 1979.

[112] A. S. Davydov. The lifetime of molecular (Davydov) solitons. *Journal of Biological Physics,* 18(2):111–125, 1991.

[113] R. Dawkins. *The Selfish Gene.* Oxford University Press, Oxford, 1976.

[114] R. Dawkins. *The Extended Phenotype.* Oxford University Press, Oxford, 1982.

[115] R. Dawkins. *The Blind Watchmaker: Why the Evidence of Evolution Reveals a Universe without Design.* W. W. Norton, New York, 1986.

[116] R. Dawkins. *The Ancestor's Tale: A Pilgrimage to the Dawn of Life.* Houghton Mifflin, Boston, 2004.

[117] R. Dawkins. *The Greatest Show on Earth: The Evidence for Evolution.* Free Press, New York, 2009.

[118] P. Dayan and L. F. Abbott. *Theoretical Neuroscience: Computational and Mathematical Modeling of Neural Systems.* MIT Press, Cambridge, MA, 2001.

[119] C. Dean and T. Dresbach. Neuroligins and neurexins: linking cell adhesion, synapse formation and cognitive function. *Trends in Neurosciences*, 29(1):21–29, 2006.

[120] V. P. Demikhov. *Transplantation of Vital Organs in Experiment. Experiments on Transplantion of Heart, Lungs, Head, Kidneys and Other Organs.* Medgiz, Moscow, 1960.

[121] D. C. Dennett. *Elbow Room: The Varieties of Free Will Worth Wanting.* Clarendon Press, Oxford, 1984.

[122] D. C. Dennett. *Consciousness Explained.* Back Bay Books, New York, 1991.

[123] D. C. Dennett. The message is: there is no medium. *Philosophy and Phenomenological Research*, 53(4):919–931, 1993.

[124] D. C. Dennett. Précis of Consciousness Explained. *Philosophy and Phenomenological Research*, 53(4):889–892, 1993.

[125] D. C. Dennett. *Freedom Evolves.* Viking, New York, 2003.

[126] D. C. Dennett. *Sweet Dreams: Philosophical Obstacles to a Science of Consciousness.* MIT Press, Cambridge, MA, 2005.

[127] R. Descartes. *The Method, Meditations and Philosophy of Descartes. Translated from the original texts, with a new introductory essay, historical and critical by John Vietch and a special introduction by Frank Sewall.* Tudor Publishing Co., New York, 1901.

[128] M. Detlefsen. *Hilbert's Program: An Essay on Mathematical Instrumentalism.* Synthese Library. D. Reidel, Dordrecht, 1986.

[129] V. Detlovs and K. Podnieks. *Introduction to Mathematical Logic.* University of Latvia, Riga, 2014.

[130] B. S. DeWitt and N. Graham. *The Many Worlds Interpretation of Quantum Mechanics.* Princeton Series in Physics. Princeton University Press, Princeton, NJ, 1973.

[131] D. Dieks. Communication by EPR devices. *Physics Letters A*, 92(6):271–272, 1982.

[132] T. L. Dimitrova and A. Weis. The wave-particle duality of light: a demonstration experiment. *American Journal of Physics*, 76(2):137–142, 2008.

[133] T. L. Dimitrova and A. Weis. Lecture demonstrations of interference and quantum erasing with single photons. *Physica Scripta*, 2009(T135):014003, 2009.

[134] T. L. Dimitrova and A. Weis. Single photon quantum erasing: a demonstration experiment. *European Journal of Physics*, 31(3):625–637, 2010.

[135] L. Diósi. Gravitation and quantum-mechanical localization of macro-objects. *Physics Letters A*, 105(4–5):199–202, 1984.

[136] L. Diósi. A universal master equation for the gravitational violation of quantum mechanics. *Physics Letters A*, 120(8):377–381, 1987.

[137] L. Diósi. Gravity-related wave function collapse: mass density resolution. *Journal of Physics: Conference Series*, 442(1):012001, 2013.

[138] L. Diósi. Gravity-related wave function collapse. Is superfluid He exceptional? *Foundations of Physics*, 44(5):483–491, 2014.

[139] P. A. M. Dirac. A new notation for quantum mechanics. *Mathematical Proceedings of the Cambridge Philosophical Society*, 35(3):416–418, 1939.

[140] P. A. M. Dirac. *The Principles of Quantum Mechanics*. Oxford University Press, Oxford, 4th edition, 1967.

[141] W. H. Dobelle. Artificial vision for the blind by connecting a television camera to the visual cortex. *ASAIO Journal*, 46(1):3–9, 2000.

[142] L. E. Dobrunz and C. F. Stevens. Heterogeneity of release probability, facilitation, and depletion at central synapses. *Neuron*, 18(6):995–1008, 1997.

[143] F. A. Dodge and R. Rahamimoff. Co-operative action of calcium ions in transmitter release at the neuromuscular junction. *Journal of Physiology*, 193(2):419–432, 1967.

[144] W. H. Donovan. Spinal cord injury—past, present, and future. *Journal of Spinal Cord Medicine*, 30(2):85–100, 2007.

[145] N. F. Dronkers, O. Plaisant, M. T. Iba-Zizen, and E. A. Cabanis. Paul Broca's historic cases: high resolution MR imaging of the brains of Leborgne and Lelong. *Brain*, 130(5):1432–1441, 2007.

[146] T. Dufresne. *Killing Freud: Twentieth Century Culture and the Death of Psychoanalysis*. Continuum, London, 2003.

[147] D. Eagleman. *The Brain with David Eagleman, 6 episodes*. Public Broadcasting Service, Arlington, Virginia, 2015.

[148] J. C. Eccles. Hypotheses relating to the brain-mind problem. *Nature*, 168(4263):53–57, 1951.

[149] J. C. Eccles. The synapse: from electrical to chemical transmission. *Annual Review of Neuroscience*, 5:325–339, 1982.

[150] J. C. Eccles. Do mental events cause neural events analogously to the probability fields of quantum mechanics? *Proceedings of the Royal Society of London B*, 227(1249):411–428, 1986.

[151] J. C. Eccles. A unitary hypothesis of mind-brain interaction in the cerebral cortex. *Proceedings of the Royal Society of London. Series B, Biological Sciences*, 240(1299):433–451, 1990.

[152] J. C. Eccles. *How the Self Controls Its Brain*. Springer, Berlin, 1994.

[153] E. Eggermann, I. Bucurenciu, S. P. Goswami, and P. Jonas. Nanodomain coupling between Ca^{2+} channels and sensors of exocytosis at fast mammalian synapses. *Nature Reviews Neuroscience*, 13(1):7–21, 2012.

[154] N. Y. Eidelman. *Looking for My Ancestors*. Eureka. Molodaya Gvardiya, Moscow, 1967.

[155] A. Einstein. *Relativity: The Special and General Theory*. Henry Holt and Company, New York, 1921.

[156] A. Einstein. *My Credo*. German League of Human Rights, Berlin, 1932.

[157] A. Einstein, H. A. Lorentz, H. Weyl, and H. Minkowski. *The Principle of Relativity. A Collection of Original Memoirs on the Special and General Theory of Relativity with Notes by A. Sommerfeld*. Dover Publications, New York, 1952.

[158] A. Einstein, B. Podolsky, and N. Rosen. Can quantum-mechanical description of physical reality be considered complete? *Physical Review*, 47(10):777–780, 1935.

[159] D. J. S. Elliott, E. J. Neale, Q. Aziz et al. Molecular mechanism of voltage sensor movements in a potassium channel. *EMBO Journal*, 23(24):4717–4726, 2004.

[160] F. Engels. *Dialectics of Nature*. International Publishers, New York, 1940.

[161] H. Everett III. "Relative state" formulation of quantum mechanics. *Reviews of Modern Physics*, 29(3):454–462, 1957.

[162] H. Everett III. *The Everett Interpretation of Quantum Mechanics: Collected Works 1955–1980 with Commentary*. Princeton University Press, Princeton, NJ, 2012.

[163] H. Eves. *Elementary Matrix Theory*. Dover Books on Mathematics. Dover Publications, New York, 1980.

[164] W. Ewald and W. Sieg. *David Hilbert's Lectures on the Foundations of Arithmetic and Logic 1917–1933*. Springer, Berlin, 2013.

[165] U. B. Eyo and M. E. Dailey. Microglia: key elements in neural development, plasticity, and pathology. *Journal of Neuroimmune Pharmacology*, 8(3):494–509, 2013.

[166] J. Farquhar, H. Bao, and M. Thiemens. Atmospheric influence of Earth's earliest sulfur cycle. *Science*, 289(5480):756–758, 2000.

[167] J. Feng. *Computational Neuroscience: A Comprehensive Approach*. Mathematical & Computational Biology. Chapman & Hall/CRC, Boca Raton, FL, 2004.

[168] D. Ferrier. Experimental researches in cerebral physiology and pathology. *West Riding Lunatic Asylum Medical Reports*, 3:30–96, 1873.

[169] D. Ferrier. Experiments on the brain of monkeys.—No. I. *Proceedings of the Royal Society of London*, 23:409–430, 1874.

[170] D. Ferrier. The Croonian Lecture: Experiments on the brain of monkeys (second series). *Philosophical Transactions of the Royal Society of London*, 165:433–488, 1875.

[171] P. K. Feyerabend. *Against Method*. Verso, London, 1984.

[172] R. P. Feynman. Space-time approach to non-relativistic quantum mechanics. *Reviews of Modern Physics*, 20(2):367–387, 1948.

[173] R. P. Feynman. Space-time approach to quantum electrodynamics. *Physical Review*, 76(6):769–789, 1949.

[174] R. P. Feynman. The concept of probability in quantum mechanics. In J. Neyman, editor, *Proceedings of the Second Berkeley Symposium on Mathematical Statistics and Probability, July 31–August 12, 1950*, pages 533–541. University of California Press, Berkeley, 1951.

[175] R. P. Feynman. *Quantum Electrodynamics*. W. A. Benjamin, New York, 1961.

[176] R. P. Feynman. The development of the space-time view of quantum electrodynamics. *Science*, 153(3737):699–708, 1966.

[177] R. P. Feynman. *QED: The Strange Theory of Light and Matter*. Princeton University Press, Princeton, NJ, 1983.

[178] R. P. Feynman and A. R. Hibbs. *Quantum Mechanics and Path Integrals.* McGraw-Hill Companies, New York, 1965.

[179] R. P. Feynman, R. B. Leighton, and M. Sands. *The Feynman Lectures on Physics, Volume III.* California Institute of Technology, Pasadena, CA, 2013.

[180] M. P. A. Fisher. Quantum cognition: the possibility of processing with nuclear spins in the brain. *Annals of Physics*, 362:593–602, 2015.

[181] M. P. A. Fisher. Are we quantum computers, or merely clever robots? *International Journal of Modern Physics B*, 31(7):1743001, 2017.

[182] W. Förner. Quantum and temperature effects on Davydov soliton dynamics: averaged Hamiltonian method. *Journal of Physics: Condensed Matter*, 4(8):1915–1923, 1992.

[183] W. Förner. Quantum and temperature effects on Davydov soliton dynamics. II. The partial dressing state and comparisons between different methods. *Journal of Physics: Condensed Matter*, 5(7):803–822, 1993.

[184] W. Förner. Quantum and temperature effects on Davydov soliton dynamics. III. Interchain coupling. *Journal of Physics: Condensed Matter*, 5(7):823–840, 1993.

[185] W. Förner. Quantum and temperature effects on Davydov soliton dynamics. IV. Lattice with a thermal phonon distribution. *Journal of Physics: Condensed Matter*, 5(23):3883–3896, 1993.

[186] W. Förner. Quantum and temperature effects on Davydov soliton dynamics. V. Numerical estimate of the errors introduced by the $|D_1\rangle$ ansatz. *Journal of Physics: Condensed Matter*, 5(23):3897–3916, 1993.

[187] W. Förner. Davydov soliton dynamics in proteins: I. Initial states and exactly solvable special cases. *Molecular Modeling Annual*, 2(5):70–102, 1996.

[188] W. Förner. Davydov soliton dynamics in proteins: II. The general case. *Molecular Modeling Annual*, 2(5):103–135, 1996.

[189] W. Förner. Davydov soliton dynamics in proteins: III. Applications and calculation of vibrational spectra. *Molecular Modeling Annual*, 3(2):78–116, 1997.

[190] W. Förner. Davydov solitons in proteins. *International Journal of Quantum Chemistry*, 64(3):351–377, 1997.

[191] M. Foster. *A Textbook of Physiology. Part III. The Central Nervous System.* Macmillan & Co., London, 1897.

[192] A. Fraenkel. Zu den Grundlagen der Cantor–Zermeloschen Mengenlehre. *Mathematische Annalen*, 86:230–237, 1922.

[193] H. Frankfurt. Alternate possibilities and moral responsibility. *Journal of Philosophy*, 66(23):829–839, 1969.

[194] T. Franzen. *Gödel's Theorem: An Incomplete Guide to Its Use and Abuse*. A. K. Peters, Wellesley, MA, 2005.

[195] M. Fréchet. Généralisation du théorème des probabilités totales. *Fundamenta Mathematicae*, 25(1):379–387, 1935.

[196] S. J. Freedman and J. F. Clauser. Experiment test of local hidden-variable theories. *Physical Review Letters*, 28(14):938–941, 1972.

[197] S. Freud. *Das Ich und das Es*. Internationaler Psychoanalytischer Verlag, Vienna, 1923.

[198] S. Freud. *The Ego and the Id*. W. W. Norton & Company, New York, 1990.

[199] I. Fried, C. L. Wilson, K. A. MacDonald, and E. J. Behnke. Electric current stimulates laughter. *Nature*, 391(6668):650, 1998.

[200] K. Friston. The free-energy principle: a unified brain theory? *Nature Reviews Neuroscience*, 11(2):127–138, 2010.

[201] M. Fu and Y. Zuo. Experience-dependent structural plasticity in the cortex. *Trends in Neurosciences*, 34(4):177–187, 2011.

[202] T. W. Gamelin. *Complex Analysis*. Undergraduate Texts in Mathematics. Springer, New York, 2001.

[203] G. Geddes, A. Ehlers, and D. Freeman. Hallucinations in the months after a trauma: An investigation of the role of cognitive processing of a physical assault in the occurrence of hallucinatory experiences. *Psychiatry Research*, 246:601–605, 2016.

[204] A. Gelman, J. B. Carlin, H. S. Stern et al. *Bayesian Data Analysis*. Texts in Statistical Science. CRC Press, Boca Raton, FL, 3rd edition, 2014.

[205] A. Gelman and C. R. Shalizi. Philosophy and the practice of Bayesian statistics. *British Journal of Mathematical and Statistical Psychology*, 66(1):8–38, 2013.

[206] D. Georgiev. Falsifications of Hameroff–Penrose Orch OR model of consciousness and novel avenues for development of quantum mind theory. *NeuroQuantology*, 5(1):145–174, 2007.

[207] D. Georgiev. Quantum no-go theorems and consciousness. *Axiomathes*, 23(4):683–695, 2013.

[208] D. Georgiev. Monte Carlo simulation of quantum Zeno effect in the brain. *International Journal of Modern Physics B*, 29(7):1550039, 2015.

[209] D. Georgiev, D. Arion, J. F. Enwright et al. Lower gene expression for KCNS3 potassium channel subunit in parvalbumin-containing neurons in the prefrontal cortex in schizophrenia. *American Journal of Psychiatry*, 171(1):62–71, 2014.

[210] D. Georgiev and J. F. Glazebrook. Subneuronal processing of information by solitary waves and stochastic processes. In S. E. Lyshevski, editor, *Nano and Molecular Electronics Handbook*, pages 17–1–17–41. CRC Press, Boca Raton, FL, 2007.

[211] D. Georgiev and J. F. Glazebrook. Quasiparticle tunneling in neurotransmitter release. In W. A. Goddard III, D. Brenner, S. E. Lyshevski, and G. J. Iafrate, editors, *Handbook of Nanoscience, Engineering, and Technology*, pages 983–1016. CRC Press, Boca Raton, FL, 3rd edition, 2012.

[212] D. Georgiev, H. Taniura, Y. Kambe et al. A critical importance of polyamine site in NMDA receptors for neurite outgrowth and fasciculation at early stages of P19 neuronal differentiation. *Experimental Cell Research*, 314(14):2603–2617, 2008.

[213] J. K. Ghosh, M. Delampady, and T. Samanta. *An Introduction to Bayesian Analysis: Theory and Methods*. Springer Texts in Statistics. Springer, New York, 2006.

[214] F. Giacosa. On unitary evolution and collapse in quantum mechanics. *Quanta*, 3(1):156–170, 2014.

[215] J. F. Glazebrook and R. Wallace. Rate distortion manifolds as model spaces for cognitive information. *Informatica (Slovenia)*, 33(3):309–345, 2009.

[216] J. F. Glazebrook and R. Wallace. Small worlds and Red Queens in the Global Workspace: an information-theoretic approach. *Cognitive Systems Research*, 10(4):333–365, 2009.

[217] I. Gloning, K. Jellinger, W. Sluga, and K. Weingarten. Über Uhrzeitagnosie. *Archiv für Psychiatrie und Nervenkrankheiten*, 198(1):85–95, 1958.

[218] K. Gödel. On formally undecidable propositions of Principia Mathematica and related systems I. In S. Feferman, J. W. Dawson Jr., S. C. Kleene, G. H. Moore, R. M. Solovay, and J. van Heijenoort, editors, *Collected Works, Vol. I: Publications 1929–1936*, pages 144–195. Oxford University Press, New York, 1986.

[219] I. I. Goldberg, M. Harel, and R. Malach. When the brain loses its self: prefrontal inactivation during sensorimotor processing. *Neuron*, 50(2):329–339, 2006.

[220] H. W. Gordon, J. E. Bogen, and R. W. Sperry. Absence of deconnexion syndrome in two patients with partial section of the neocommissures. *Brain*, 94(2):327–336, 1971.

[221] P. Gottlieb. Aristotle on non-contradiction. In E. N. Zalta, U. Nodelman, and C. Allen, editors, *Stanford Encyclopedia of Philosophy*. Stanford University, Stanford, CA, 2017.

[222] D. M. Greenberger, M. A. Horne, A. Shimony, and A. Zeilinger. Bell's theorem without inequalities. *American Journal of Physics*, 58(12):1131–1143, 1990.

[223] V. K. Gribkoff and L. K. Kaczmarek. *Structure, Function and Modulation of Neuronal Voltage-Gated Ion Channels*. Wiley, Hoboken, NJ, 2009.

[224] P. Grim. *Philosophy of Mind: Brains, Consciousness, and Thinking Machines*. The Teaching Company, Chantilly, VA, 2008.

[225] C. G. Gross. Aristotle on the brain. *The Neuroscientist*, 1(4):245–250, 1995.

[226] C. G. Gross. Galen and the squealing pig. *The Neuroscientist*, 4(3):216–221, 1998.

[227] C. G. Gross. The discovery of motor cortex and its background. *Journal of the History of the Neurosciences*, 16(3):320–331, 2007.

[228] C. G. Gross. *A Hole in the Head: More Tales in the History of Neuroscience*. MIT Press, Cambridge, MA, 2009.

[229] S. Hagan, S. R. Hameroff, and J. A. Tuszyński. Quantum computation in brain microtubules: decoherence and biological feasibility. *Physical Review E*, 65(6):061901, 2002.

[230] J. Hall, C. Kim, B. McElroy, and A. Shimony. Wave-packet reduction as a medium of communication. *Foundations of Physics*, 7(9–10):759–767, 1977.

[231] S. R. Hameroff. Quantum computation in brain microtubules? The Penrose–Hameroff 'Orch OR' model of consciousness. *Philosophical Transactions of the Royal Society of London A*, 356(1743):1869–1896, 1998.

[232] S. R. Hameroff. Consciousness, the brain, and spacetime geometry. *Annals of the New York Academy of Sciences*, 929(1):74–104, 2001.

[233] S. R. Hameroff and R. Penrose. Orchestrated reduction of quantum coherence in brain microtubules: a model for consciousness. *Mathematics and Computers in Simulation*, 40(3–4):453–480, 1996.

[234] X. Han, C.-T. Wang, J. Bai et al. Transmembrane segments of syntaxin line the fusion pore of Ca^{2+}-triggered exocytosis. *Science*, 304(5668):289–292, 2004.

[235] D. Hanneke, S. Fogwell, and G. Gabrielse. New measurement of the electron magnetic moment and the fine structure constant. *Physical Review Letters*, 100(12):120801, 2008.

[236] N. R. Hanson. *Patterns of Discovery: An Inquiry into the Conceptual Foundations of Science*. Cambridge University Press, Cambridge, 1958.

[237] J. D. Hardy, W. R. Webb, M. L. Dalton Jr., and G. R. Walker Jr. Lung homotransplantation in man: report of the initial case. *Journal of the American Medical Association*, 186(12):1065–1074, 1963.

[238] K. M. Harris and R. J. Weinberg. Ultrastructure of synapses in the mammalian brain. *Cold Spring Harbor Perspectives in Biology*, 4(5):a005587, 2012.

[239] S. Harris. *Free Will*. Free Press, New York, 2012.

[240] G. K. Harrison. A moral argument for substance dualism. *Journal of the American Philosophical Association*, 2(1):21–35, 2016.

[241] S. W. Hawking. Stephen Hawking: brain could exist outside body. *The Guardian*, September 21, 2013.

[242] S. W. Hawking and L. Mlodinow. *The Grand Design: New Answers to the Ultimate Questions of Life*. Bantam Press, New York, 2010.

[243] M. Hayakawa. Theory of anomalous magnetic dipole moments of the electron. In W. Quint and M. Vogel, editors, *Fundamental Physics in Particle Traps*, pages 41–71. Springer, Berlin, 2014.

[244] M. Hayashi, S. Ishizaka, A. Kawachi et al. *Introduction to Quantum Information Science*. Graduate Texts in Physics. Springer, Berlin, 2015.

[245] W. Heisenberg. Über den anschaulichen Inhalt der quantentheoretischen Kinematik und Mechanik. *Zeitschrift für Physik A*, 43(3–4):172–198, 1927.

[246] W. Heisenberg. *The Physical Principles of the Quantum Theory*. Dover, New York, 1949.

[247] D. Heiss. *Fundamentals of Quantum Information. Quantum Computation, Communication, Decoherence and All That*, volume 587 of *Lecture Notes in Physics*. Springer, Berlin, 2002.

[248] C. G. Hempel and P. Oppenheim. Studies in the logic of explanation. *Philosophy of Science*, 15(2):135–175, 1948.

[249] D. Hilbert and P. Bernays. *Grundlagen der Mathematik I*. Die Grundlehren der mathematischen Wissenschaften. Springer, Berlin, 1968.

[250] D. Hilbert and P. Bernays. *Grundlagen der Mathematik II*. Die Grundlehren der mathematischen Wissenschaften. Springer, Berlin, 1970.

[251] A. R. Hildebrand, G. T. Penfield, D. A. Kring et al. Chicxulub Crater: a possible Cretaceous/Tertiary boundary impact crater on the Yucatán Peninsula, Mexico. *Geology*, 19(9):867–871, 1991.

[252] J. Hilgevoord and J. Uffink. The uncertainty principle. In E. N. Zalta, U. Nodelman, and C. Allen, editors, *Stanford Encyclopedia of Philosophy*. Stanford University, Stanford, CA, 2017.

[253] H. Hilmisdóttir and J. Kozlowski. *Beginner's Icelandic*. Hippocrene Beginner's Series. Hippocrene Books, New York, 2009.

[254] I. J. Hirsh and C. E. Sherrick Jr. Perceived order in different sense modalities. *Journal of Experimental Psychology*, 62(5):423–432, 1961.

[255] W. His. *Die Neuroblasten und deren Entstehung im embryonalen Mark.* Abhandlungen der mathematisch-physischen Classe der Königlich Sächsischen Gesellschaft der Wissenschaften, 15. Bd, no 4. S. Hirzel, Leipzig, 1889.

[256] T. Hobbes. *The English Works of Thomas Hobbes of Malmesbury. Vol. IV: Tripos, In Three Discourses: The First, Humane Nature; The Second, De Corpore Politico; The Third, Of Liberty And Necessity.* John Bohn, London, 1839.

[257] L. R. Hochberg, M. D. Serruya, G. M. Friehs et al. Neuronal ensemble control of prosthetic devices by a human with tetraplegia. *Nature*, 442(7099):164–171, 2006.

[258] A. L. Hodgkin and A. F. Huxley. A quantitative description of membrane current and its application to conduction and excitation in nerve. *Journal of Physiology*, 117(4):500–544, 1952.

[259] J. F. Hoffecker. *Landscape of the Mind: Human Evolution and the Archaeology of Thought.* Columbia University Press, New York, 2011.

[260] A. S. Holevo. Bounds for the quantity of information transmitted by a quantum communication channel. *Problemy Peredachi Informatsii*, 9(3):3–11, 1973.

[261] A. S. Holevo. Bounds for the quantity of information transmitted by a quantum communication channel. *Problems of Information Transmission*, 9(3):177–183, 1973.

[262] A. Holtmaat and K. Svoboda. Experience-dependent structural synaptic plasticity in the mammalian brain. *Nature Reviews Neuroscience*, 10(9):647–658, 2009.

[263] D. Home and A. Robinson. Einstein and Tagore: man, nature and mysticism. *Journal of Consciousness Studies*, 2(2):167–179, 1995.

[264] M. Horky. *Brevissima Peregrinatio Contra Nuncium Sidereum.* Iulianum Cassianum, Modena, 1610.

[265] L. R. Horn. Contradiction. In E. N. Zalta, U. Nodelman, and C. Allen, editors, *Stanford Encyclopedia of Philosophy*. Stanford University, Stanford, CA, 2017.

[266] D. H. Hubel. *Eye, Brain, and Vision*. W. H. Freeman, New York, 2nd edition, 1995.

[267] D. Hume. *An Enquiry Concerning Human Understanding*. The Open Court Publishing Company, Chicago, 1900.

[268] D. Hume. *A Treatise of Human Nature*. Clarendon Press, Oxford, 1960.

[269] T. H. Huxley. On the hypothesis that animals are automata, and its history. *The Fortnightly Review*, 16:555–580, 1874.

[270] T. H. Huxley. *Darwiniana*. D. Appleton and Company, New York, 1896.

[271] J. M. Hyman, D. W. McLaughlin, and A. C. Scott. On Davydov's alpha-helix solitons. *Physica D: Nonlinear Phenomena*, 3(1):23–44, 1981.

[272] O. A. Imas, K. M. Ropella, B. D. Ward et al. Volatile anesthetics enhance flash-induced γ oscillations in rat visual cortex. *Anesthesiology*, 102(5):937–947, 2005.

[273] W. Israel. Event horizons in static vacuum space-times. *Physical Review*, 164(5):1776–1779, 1967.

[274] F. Jackson. Epiphenomenal qualia. *Philosophical Quarterly*, 32(127):127–136, 1982.

[275] F. Jackson. What Mary didn't know. *Journal of Philosophy*, 83(5):291–295, 1986.

[276] R. G. Jahn and B. J. Dunne. On the quantum mechanics of consciousness, with application to anomalous phenomena. *Foundations of Physics*, 16(8):721–772, 1986.

[277] S. R. James, R. W. Dennell, A. S. Gilbert et al. Hominid use of fire in the lower and middle pleistocene: a review of the evidence. *Current Anthropology*, 30(1):1–26, 1989.

[278] W. James. Are we automata? *Mind*, 4(13):1–22, 1879.

[279] W. James. *The Principles of Psychology*, volume 1. Henry Holt and Company, New York, 1890.

[280] S. Jeffers. Physics and claims for anomalous effects related to consciousness. *Journal of Consciousness Studies*, 10(6–7):135–152, 2003.

[281] J. S. Johansson, D. Scharf, L. A. Davies et al. A designed four-α-helix bundle that binds the volatile general anesthetic halothane with high affinity. *Biophysical Journal*, 78(2):982–993, 2000.

[282] D. Johnston and S. M.-S. Wu. *Foundations of Cellular Neurophysiology*. Bradford Books. MIT Press, Cambridge, MA, 1995.

[283] S. Jordan. *Quantum Algorithm Zoo*. National Institute of Standards and Technology, Gaithersburg, MD, 2016.

[284] M. Kanabus, E. Szelag, E. Rojek, and E. Pöppel. Temporal order judgement for auditory and visual stimuli. *Acta Neurobiologiae Experimentalis*, 62(4):263–270, 2002.

[285] E. R. Kandel, J. H. Schwartz, T. M. Jessell et al. *Principles of Neural Science*. McGraw-Hill Professional, New York, 5th edition, 2012.

[286] A. M. Kariev and M. E. Green. Quantum calculations on water in the KcsA channel cavity with permeant and non-permeant ions. *Biochimica et Biophysica Acta (BBA) - Biomembranes*, 1788(5):1188–1192, 2009.

[287] A. M. Kariev and M. E. Green. Voltage gated ion channel function: gating, conduction, and the role of water and protons. *International Journal of Molecular Sciences*, 13(2):1680–1709, 2012.

[288] A. M. Kariev, P. Njau, and M. E. Green. The open gate of the Kv1.2 channel: quantum calculations show the key role of hydration. *Biophysical Journal*, 106(3):548–555, 2014.

[289] A. M. Kariev, V. S. Znamenskiy, and M. E. Green. Quantum mechanical calculations of charge effects on gating the KcsA channel. *Biochimica et Biophysica Acta (BBA) - Biomembranes*, 1768(5):1218–1229, 2007.

[290] R. F. Keep, Y. Hua, and G. Xi. Brain water content: a misunderstood measurement? *Translational Stroke Research*, 3(2):263–265, 2012.

[291] E. H. Kennard. Zur Quantenmechanik einfacher Bewegungstypen. *Zeitschrift für Physik*, 44(4–5):326–352, 1927.

[292] A. Kent. Quanta and qualia. In *FQXi's 5th International Conference "If a Tree Falls: The Physics of What Happens and Who Is Listening"*, Banff, Canada, August 17–22, 2016.

[293] R. Kirk. Zombies. In E. N. Zalta, U. Nodelman, and C. Allen, editors, *Stanford Encyclopedia of Philosophy*. Stanford University, Stanford, CA, 2017.

[294] M. Kneussel and W. Wagner. Myosin motors at neuronal synapses: drivers of membrane transport and actin dynamics. *Nature Reviews Neuroscience*, 14(4):233–247, 2013.

[295] S. B. Kochen and E. P. Specker. The problem of hidden variables in quantum mechanics. *Journal of Mathematics and Mechanics*, 17(1):59–87, 1967.

[296] A. N. Kolmogorov. *Foundations of the Theory of Probability*. Chelsea Publishing Company, New York, 1950.

[297] M. Korte and D. Schmitz. Cellular and system biology of memory: timing, molecules, and beyond. *Physiological Reviews*, 96(2):647–693, 2016.

[298] J. Kortelainen, X. Jia, T. Seppänen, and N. Thakor. Increased electroencephalographic gamma activity reveals awakening from isoflurane anaesthesia in rats. *British Journal of Anaesthesia*, 109(5):782–789, 2012.

[299] A. Korzybski. *Science and Sanity: An Introduction to Non-Aristotelian Systems and General Semantics*. Institute of General Semantics, New York, 5th edition, 2000.

[300] I. Lakatos. Criticism and the methodology of scientific research programmes. *Proceedings of the Aristotelian Society*, 69:149–186, 1968.

[301] I. Lakatos. Falsification and the methodology of scientific research programmes. In I. Lakatos and A. Musgrave, editors, *Criticism and the Growth of Knowledge: Proceedings of the International Colloquium in the Philosophy of Science, London, 1965*, volume 4, pages 91–196. Cambridge University Press, Cambridge, 1970.

[302] I. Lakatos. History of science and its rational reconstructions. In R. C. Buck and R. S. Cohen, editors, *PSA 1970: In Memory of Rudolf Carnap; Proceedings of the 1970 Biennial Meeting Philosophy of Science Association*, volume 8 of *Boston Studies in the Philosophy of Science*, pages 91–136. D. Reidel, Dordrecht, 1971.

[303] I. Lakatos. The role of crucial experiments in science. *Studies In History and Philosophy of Science Part A*, 4(4):309–325, 1974.

[304] I. Lakatos. *The Methodology of Scientific Research Programmes*, volume 1 of *Philosophical Papers*. Cambridge University Press, Cambridge, 1978.

[305] I. Lakatos. *Proofs and Refutations: The Logic of Mathematical Discovery*. Cambridge University Press, Cambridge, 1978.

[306] I. Lakatos and P. K. Feyerabend. *For and Against Method: Including Lakatos's Lectures on Scientific Method and the Lakatos–Feyerabend Correspondence*. University of Chicago Press, Chicago, 1999.

[307] V. A. F. Lamme, K. Zipser, and H. Spekreijse. Figure-ground activity in primary visual cortex is suppressed by anesthesia. *Proceedings of the National Academy of Sciences*, 95(6):3263–3268, 1998.

[308] R. Land, G. Engler, A. Kral, and A. K. Engel. Auditory evoked bursts in mouse visual cortex during isoflurane anesthesia. *PLoS ONE*, 7(11):e49855, 2012.

[309] L. D. Landau and E. M. Lifshitz. *Quantum Mechanics*. Course of Theoretical Physics. Pergamon Press, Oxford, 1965.

[310] P.-S. Laplace. *A Philosophical Essay on Probabilities*. John Wiley & Sons, New York, 1902.

[311] A. Lawrence, J. McDaniel, D. Chang et al. Dynamics of the Davydov model in α-helical proteins: effects of the coupling parameter and temperature. *Physical Review A*, 33(2):1188–1201, 1986.

[312] G. W. Leibniz. *New Essays Concerning Human Understanding*. Macmillan, London, 1896.

[313] S. Lem. *Solaris*. MON, Warsaw, 1961.

[314] V. I. Lenin. *Lenin Collected Works, Vol. 14: Materialism and Empirio-Criticism (1908)*. Progress Publishers, Moscow, 1977.

[315] H. R. Leuchtag. *Voltage-Sensitive Ion Channels: Biophysics of Molecular Excitability*. Springer, Berlin, 2008.

[316] P. J. Lewis. Conspiracy theories of quantum mechanics. *British Journal for the Philosophy of Science*, 57(2):359–381, 2006.

[317] C. Li, S. Wang, Y. Zhao et al. The freedom to pursue happiness: belief in free will predicts life satisfaction and positive affect among Chinese adolescents. *Frontiers in Psychology*, 7:2027, 2016.

[318] B. Libet. Reflections on the interaction of the mind and brain. *Progress in Neurobiology*, 78(3–5):322–326, 2006.

[319] J. T. Littleton, M. Stern, M. Perin, and H. J. Bellen. Calcium dependence of neurotransmitter release and rate of spontaneous vesicle fusions are altered in Drosophila synaptotagmin mutants. *Proceedings of the National Academy of Sciences*, 91(23):10888–10892, 1994.

[320] J. Locke. *An Essay Concerning Human Understanding. And a Treatise on the Conduct of the Understanding*. Troutman & Hayes, Philadelphia, 1850.

[321] P. S. Lomdahl and W. C. Kerr. Do Davydov solitons exist at 300 K? *Physical Review Letters*, 55(11):1235–1238, 1985.

[322] P. S. Lomdahl, S. P. Layne, and I. J. Bigio. Solitons in biology. *Los Alamos Science*, 10:2–31, 1984.

[323] E. N. Lorenz. Deterministic nonperiodic flow. *Journal of the Atmospheric Sciences*, 20(2):130–141, 1963.

[324] E. N. Lorenz. Predictability: does the flap of a butterfly's wings in Brazil set off a tornado in Texas? In *139th Annual Meeting of the American Association for the Advancement of Science*, Washington, DC, December 29, 1972.

[325] J. P. Lowe and K. Peterson. *Quantum Chemistry*. Academic Press, Amsterdam, 3rd edition, 2005.

[326] J. H. Lowenstein. *Essentials of Hamiltonian Dynamics*. Cambridge University Press, Cambridge, 2012.

[327] X. Lü and F. Lin. Soliton excitations and shape-changing collisions in alpha helical proteins with interspine coupling at higher order. *Communications in Nonlinear Science and Numerical Simulation*, 32:241–261, 2016.

[328] P. Ludlow, Y. Nagasawa, and D. Stoljar. *There's Something About Mary: Essays on Phenomenal Consciousness and Frank Jackson's Knowledge Argument*. MIT Press, Cambridge, MA, 2004.

[329] M. B. MacIver, A. A. Mikulec, S. M. Amagasu, and F. A. Monroe. Volatile anesthetics depress glutamate transmission via presynaptic actions. *Anesthesiology*, 85(4):823–834, 1996.

[330] M. J. MacKenzie, K. D. Vohs, and R. F. Baumeister. You didn't have to do that: belief in free will promotes gratitude. *Personality and Social Psychology Bulletin*, 40(11):1423–1434, 2014.

[331] H. A. Mallot. *Computational Neuroscience: A First Course*, volume 2 of *Springer Series in Bio-/Neuroinformatics*. Springer, Cham, Switzerland, 2013.

[332] S. Mangini, B. R. Alves, O. M. Silvestre et al. Heart transplantation: review. *Einstein (São Paulo)*, 13(2):310–318, 2015.

[333] J. Markovits. Frankfurt on moral responsibility. In *24.231 Ethics*, MIT OpenCourseWare. Massachusetts Institute of Technology, Cambridge, MA, 2009.

[334] W. Marshall, C. Simon, R. Penrose, and D. Bouwmeester. Towards quantum superpositions of a mirror. *Physical Review Letters*, 91(13):130401, 2003.

[335] K. C. Martin, Y. Hu, B. A. Armitage et al. Evidence for synaptotagmin as an inhibitory clamp on synaptic vesicle release in Aplysia neurons. *Proceedings of the National Academy of Sciences*, 92(24):11307–11311, 1995.

[336] G. A. Mashour. The cognitive binding problem: from Kant to quantum neurodynamics. *NeuroQuantology*, 2(1):29–38, 2004.

[337] E. Matevossian, H. Kern, N. Hüser et al. Surgeon Yurii Voronoy (1895–1961)—a pioneer in the history of clinical transplantation: in Memoriam at the 75th Anniversary of the First Human Kidney Transplantation. *Transplant International*, 22(12):1132–1139, 2009.

[338] J. L. McGaugh. Memory—a century of consolidation. *Science*, 287(5451):248–251, 2000.

[339] B. McGuinness. *Unified Science: The Vienna Circle Monograph Series originally edited by Otto Neurath, now in an English edition*. Vienna Circle Collection. D. Reidel, Dordrecht, 1987.

[340] N. D. Mermin. Hidden variables and the two theorems of John Bell. *Reviews of Modern Physics*, 65(3):803–815, 1993.

[341] A. Messiah. *Quantum Mechanics*, volume 1. North Holland Publishing Company, Amsterdam, 1967.

[342] M. M. Mesulam. From sensation to cognition. *Brain*, 121(6):1013–1052, 1998.

[343] B. J. Molyneaux, P. Arlotta, R. M. Fame et al. Novel subtype-specific genes identify distinct subpopulations of callosal projection neurons. *Journal of Neuroscience*, 29(39):12343–12354, 2009.

[344] B. J. Molyneaux, P. Arlotta, J. R. L. Menezes, and J. D. Macklis. Neuronal subtype specification in the cerebral cortex. *Nature Reviews Neuroscience*, 8(6):427–437, 2007.

[345] H. Morris and J. P. McMurrich. *Morris's Human Anatomy: A Complete Systematic Treatise by English and American Authors. Part III. Nervous System. Organs of Special Sense*. P. Blakiston's Son & Co., Philadelphia, 4th edition, 1907.

[346] H. G. J. Moseley. The high-frequency spectra of the elements. *Philosophical Magazine, Series 6*, 26(156):1024–1034, 1913.

[347] H. G. J. Moseley. The high-frequency spectra of the elements. Part II. *Philosophical Magazine, Series 6*, 27(160):703–713, 1914.

[348] J. M. Musacchio. Why do qualia and the mind seem nonphysical? *Synthese*, 147(3):425–460, 2005.

[349] A. Musgrave. The ultimate argument for scientific realism. In R. Nola, editor, *Relativism and Realism in Science*, volume 6 of *Australasian Studies in History and Philosophy of Science*, pages 229–252. Springer, Dordrecht, 1988.

[350] A. Musgrave. The 'miracle argument' for scientific realism. *The Rutherford Journal*, 2:020108, 2007.

[351] T. Nagel. What is it like to be a bat? *Philosophical Review*, 83(4):435–450, 1974.

[352] T. Nagel. *What Does It All Mean? A Very Short Introduction to Philosophy*. Oxford University Press, New York, 1987.

[353] P. Nagele, J. B. Mendel, W. J. Placzek et al. Volatile anesthetics bind rat synaptic SNARE proteins. *Anesthesiology*, 103(4):768–778, 2005.

[354] S. Nasr and R. B. H. Tootell. Role of fusiform and anterior temporal cortical areas in facial recognition. *NeuroImage*, 63(3):1743–1753, 2012.

[355] O. Neurath. *Philosophical Papers 1913–1946*, chapter The unity of science as a task, pages 115–120. Vienna Circle Collection. D. Reidel, Dordrecht, 1983.

[356] O. Neurath. *Philosophical Papers 1913–1946*, chapter Protocol statements, pages 91–99. Vienna Circle Collection. D. Reidel, Dordrecht, 1983.

[357] I. Newton. *Newton's Principia: The Mathematical Principles of Natural Philosophy; to which is added Newton's System of the World*. Daniel Adee, New York, 1846.

[358] M. A. Nielsen and I. L. Chuang. *Quantum Computation and Quantum Information*. Cambridge University Press, Cambridge, 10th anniversary edition, 2010.

[359] R. Nieuwenhuys. The neocortex: an overview of its evolutionary development, structural organization and synaptology. *Anatomy and Embryology*, 190(4):307–337, 1994.

[360] A. Nishida, K. M. Xu, T. Croudace et al. Adolescent self-control predicts midlife hallucinatory experiences: 40-year follow-up of a national birth cohort. *Schizophrenia Bulletin*, 40(6):1543–1551, 2014.

[361] M. Ohya and N. Watanabe. Quantum entropy and its applications to quantum communication and statistical physics. *Entropy*, 12(5):1194–1245, 2010.

[362] M. Oizumi, L. Albantakis, and G. Tononi. From the phenomenology to the mechanisms of consciousness: integrated information theory 3.0. *PLOS Computational Biology*, 10(5):e1003588, 2014.

[363] B. Pakkenberg, D. Pelvig, L. Marner et al. Aging and the human neocortex. *Experimental Gerontology*, 38(1–2):95–99, 2003.

[364] C. Pandarinath, P. Nuyujukian, C. H. Blabe et al. High performance communication by people with paralysis using an intracortical brain-computer interface. *eLife*, 6:e18554, 2017.

[365] Z. P. Pang and T. C. Südhof. Cell biology of Ca^{2+}-triggered exocytosis. *Current Opinion in Cell Biology*, 22(4):496–505, 2010.

[366] A. Pathak. *Elements of Quantum Computation and Quantum Communication*. CRC Press, Boca Raton, FL, 2013.

[367] A. K. Pati and S. L. Braunstein. Impossibility of deleting an unknown quantum state. *Nature*, 404(6774):164–165, 2000.

[368] A. K. Pati and S. L. Braunstein. Quantum-no-deleting principle. *Current Science*, 79(10):1426–1427, 2000.

[369] A. K. Pati and S. L. Braunstein. Quantum deleting and signalling. *Physics Letters A*, 315(3–4):208–212, 2003.

[370] L. Pauling and R. B. Corey. The pleated sheet, a new layer configuration of polypeptide chains. *Proceedings of the National Academy of Sciences*, 37(5):251–256, 1951.

[371] L. Pauling, R. B. Corey, and H. R. Branson. The structure of proteins: two hydrogen-bonded helical configurations of the polypeptide chain. *Proceedings of the National Academy of Sciences*, 37(4):205–211, 1951.

[372] D. T. Pegg. Wave function collapse in atomic physics. *Australian Journal of Physics*, 46(1):77–86, 1993.

[373] C. Pelham. *The Chronicles of Crime; Or The New Newgate Calendar. Being a Series of Memoirs and Anecdotes of Notorious Characters who have outraged the laws of Great Britain from the earliest period to 1841*, volume 1. Thomas Tegg, London, 1886.

[374] W. Penfield. *The Mystery of the Mind: A Critical Study of Consciousness and the Human Brain*. Princeton University Press, Princeton, NJ, 1978.

[375] R. Penrose. Gravitational collapse and space-time singularities. *Physical Review Letters*, 14(3):57–59, 1965.

[376] R. Penrose. On gravity's role in quantum state reduction. *General Relativity and Gravitation*, 28(5):581–600, 1996.

[377] R. Penrose. Quantum computation, entanglement and state reduction. *Philosophical Transactions of the Royal Society of London A*, 356(1743):1927–1939, 1998.

[378] R. Penrose. *The Emperor's New Mind: Concerning Computers, Minds and The Laws of Physics*. Oxford University Press, Oxford, 1999.

[379] R. Penrose. *The Road to Reality: A Complete Guide to the Laws of the Universe*. Jonathan Cape, London, 2004.

[380] R. Penrose. *Shadows of the Mind: A Search for the Missing Science of Consciousness*. Vintage, London, 2005.

[381] R. Penrose. A theory of everything? *Nature*, 433(7023):259, 2005.

[382] R. Penrose. On the gravitization of quantum mechanics 1: quantum state reduction. *Foundations of Physics*, 44(5):557–575, 2014.

[383] A. E. Pereda. Electrical synapses and their functional interactions with chemical synapses. *Nature Reviews Neuroscience*, 15(4):250–263, 2014.

[384] A. Peres. Two simple proofs of the Kochen–Specker theorem. *Journal of Physics A: Mathematical and General*, 24(4):L175–L178, 1991.

[385] V. C. Piatti, M. G. Davies-Sala, M. S. Espósito et al. The timing for neuronal maturation in the adult hippocampus is modulated by local network activity. *Journal of Neuroscience*, 31(21):7715–7728, 2011.

[386] T. Pink. *Free Will: A Very Short Introduction*, volume 110 of *Very Short Introductions*. Oxford University Press, Oxford, 2004.

[387] I. Pitowsky. Quantum mechanics as a theory of probability. In W. Demopoulos and I. Pitowsky, editors, *Physical Theory and Its Interpretation. Essays in Honor of Jeffrey Bub*, volume 72 of *The Western Ontario Series in Philosophy of Science*, pages 213–240. Springer, Dordrecht, 2006.

[388] K. Podnieks. *What Is Mathematics: Gödel's Theorem and Around*. University of Latvia, Riga, 2015.

[389] H. Poincaré. *Science and Hypothesis*. Walter Scott, London, 1905.

[390] J. Polkinghorne. *Quantum Theory: A Very Short Introduction*, volume 69 of *Very Short Introductions*. Oxford University Press, Oxford, 2002.

[391] E. Pöppel. Reconstruction of subjective time on the basis of hierarchically organized processing system. In M. A. Pastor and J. Artieda, editors, *Time, Internal Clocks and Movement*, volume 115 of *Advances in Psychology*, pages 165–185. North-Holland, Amsterdam, 1996.

[392] E. Pöppel. A hierarchical model of temporal perception. *Trends in Cognitive Sciences*, 1(2):56–61, 1997.

[393] E. Pöppel and N. Logothetis. Neuronal oscillations in the human brain. Discontinuous initiations of pursuit eye movements indicate a 30-Hz temporal framework for visual information processing. *Naturwissenschaften*, 73(5):267–268, 1986.

[394] K. R. Popper. The propensity interpretation of probability. *British Journal for the Philosophy of Science*, 10(37):25–42, 1959.

[395] K. R. Popper. *Conjectures and Refutations: The Growth of Scientific Knowledge*. Routledge & Kegan Paul, London, 4th edition, 1981.

[396] K. R. Popper. *Objective Knowledge: An Evolutionary Approach*. Oxford University Press, Oxford, 1983.

[397] K. R. Popper. *The Logic of Scientific Discovery*. Routledge, London, 2002.

[398] K. R. Popper and J. C. Eccles. *The Self and Its Brain: An Argument for Inter-actionism*. Routledge & Kegan Paul, London, 1983.

[399] J. Preskill. *Lecture Notes for Physics 219/Computer Science 219: Quantum Computation*. California Institute of Technology, Pasadena, CA, 2015.

[400] C. J. Price, R. J. S. Wise, E. A. Warburton et al. Hearing and saying: the functional neuro-anatomy of auditory word processing. *Brain*, 119(3):919–931, 1996.

[401] G. Priest and K. Tanaka. Paraconsistent logic. In E. N. Zalta, U. Nodelman, and C. Allen, editors, *Stanford Encyclopedia of Philosophy*. Stanford University, Stanford, CA, 2017.

[402] C. Ptolemaeus. *Almagestum*. Petrus Lichtenstein, Venice, Italy, 1515.

[403] B. Qin, B. Tian, W. Liu et al. Solitonic excitations and interactions in the three-spine α-helical protein with inhomogeneity. *SIAM Journal on Applied Mathematics*, 71(4):1317–1353, 2011.

[404] D. I. Radin, L. Michel, K. Galdamez et al. Consciousness and the double-slit interference pattern: six experiments. *Physics Essays*, 25(2):157–171, 2012.

[405] D. I. Radin and R. D. Nelson. Evidence for consciousness-related anomalies in random physical systems. *Foundations of Physics*, 19(12):1499–1514, 1989.

[406] V. S. Ramachandran. *The Tell-Tale Brain: A Neuroscientist's Quest for What Makes Us Human*. W. W. Norton & Company, New York, 2010.

[407] V. S. Ramachandran and S. Blakeslee. *Phantoms in the Brain: Probing the Mysteries of the Human Mind*. HarperCollins, New York, 1998.

[408] S. Ramón y Cajal. *Textura del sistema nervioso del hombre y de los vertebrados: estudios sobre el plan estructural y composición histológica de los centros nerviosos adicionados de consideraciones fisiológicas fundadas en los nuevos descubrimientos*. Nicolas Moya, Madrid, 1904.

[409] S. Ramón y Cajal. The structure and connexions of neurons. *Nobel Lecture*, 1906.

[410] S. M. Rao, A. R. Mayer, and D. L. Harrington. The evolution of brain activation during temporal processing. *Nature Neuroscience*, 4(3):317–323, 2001.

[411] J. Rawls. Justice as fairness. *Philosophical Review*, 67(2):164–194, 1958.

[412] X.-P. Ren, Y. Song, Y.-J. Ye et al. Allogeneic head and body reconstruction: mouse model. *CNS Neuroscience & Therapeutics*, 20(12):1056–1060, 2014.

[413] F. Rieke, D. Warland, R. de Ruyter van Steveninck, and W. Bialek. *Spikes: Exploring the Neural Code*. MIT Press, Cambridge, MA, 1999.

[414] D. Rigoni, G. Pourtois, and M. Brass. 'Why should I care?' Challenging free will attenuates neural reaction to errors. *Social Cognitive and Affective Neuroscience*, 10(2):262–268, 2015.

[415] H. J. Risselada and H. Grubmüller. How SNARE molecules mediate membrane fusion: recent insights from molecular simulations. *Current Opinion in Structural Biology*, 22(2):187–196, 2012.

[416] J. Rizo and J. Xu. The synaptic vesicle release machinery. *Annual Review of Biophysics*, 44(1):339–367, 2015.

[417] R. D. Rosenkrantz. *Inference, Method and Decision: Towards a Bayesian Philosophy of Science*, volume 115 of *Synthese Library*. D. Reidel, Dordrecht, 1977.

[418] B. Russell. *Why I Am Not a Christian and Other Essays on Religion and Related Subjects*. Routledge, London, 2005.

[419] B. Sakmann and E. Neher. *Single-Channel Recording*. Springer, New York, 2nd edition, 1995.

[420] J.-P. Sartre. *Existentialism Is a Humanism*. Yale University Press, New Haven, CT, 2007.

[421] V. Scarani and A. Suarez. Introducing quantum mechanics: one-particle interferences. *American Journal of Physics*, 66(8):718–721, 1998.

[422] E. Schrödinger. An undulatory theory of the mechanics of atoms and molecules. *Physical Review*, 28(6):1049–1070, 1926.

[423] E. Schrödinger. Die gegenwärtige Situation in der Quantenmechanik. I. *Naturwissenschaften*, 23(48):807–812, 1935.

[424] E. Schrödinger. Indeterminism and free will. *Nature*, 138(3479):13–14, 1936.

[425] H. G. Schuster and W. Just. *Deterministic Chaos: An Introduction*. Wiley-VCH, Weinheim, 4th edition, 2005.

[426] J. W. Schweitzer. Lifetime of the Davydov soliton. *Physical Review A*, 45:8914–8923, 1992.

[427] J. Schwinger. On quantum-electrodynamics and the magnetic moment of the electron. *Physical Review*, 73(4):416–417, 1948.

[428] A. C. Scott. Davydov's soliton. *Physics Reports*, 217(1):1–67, 1992.

[429] N. S. Scrutton, J. Basran, and M. J. Sutcliffe. New insights into enzyme catalysis. Ground state tunnelling driven by protein dynamics. *European Journal of Biochemistry*, 264(3):666–671, 1999.

[430] J. R. Searle. Minds, brains, and programs. *Behavioral and Brain Sciences*, 3(3):417–457, 1980.

[431] J. R. Searle. *The Mystery of Consciousness*. The New York Review of Books, New York, 1997.

[432] M. N. Shadlen and J. A. Movshon. Synchrony unbound: a critical evaluation of the temporal binding hypothesis. *Neuron*, 24(1):67–77, 1999.

[433] V. Shahrezaei and K. R. Delaney. Consequences of molecular-level Ca^{2+} channel and synaptic vesicle colocalization for the Ca^{2+} microdomain and neurotransmitter exocytosis: a Monte Carlo study. *Biophysical Journal*, 87(4):2352–2364, 2004.

[434] V. Shahrezaei and K. R. Delaney. Brevity of the Ca^{2+} microdomain and active zone geometry prevent Ca^{2+}-sensor saturation for neurotransmitter release. *Journal of Neurophysiology*, 94(3):1912–1919, 2005.

[435] C. E. Shannon. A mathematical theory of communication. *The Bell System Technical Journal*, 27(3):379–423, 1948.

[436] C. E. Shannon. A mathematical theory of communication. *The Bell System Technical Journal*, 27(4):623–656, 1948.

[437] Y. Shen, L. Hao, and G.-L. Long. Why can we copy classical information? *Chinese Physics Letters*, 28(1):010306, 2011.

[438] J. J. Shih, D. J. Krusienski, and J. R. Wolpaw. Brain-computer interfaces in medicine. *Mayo Clinic Proceedings*, 87(3):268–279, 2012.

[439] P. Shipman. *The Man Who Found the Missing Link: Eugène Dubois and His Lifelong Quest to Prove Darwin Right*. Simon & Schuster, New York, 2001.

[440] G. O. Sipe, R. L. Lowery, M. E. Tremblay et al. Microglial P2Y12 is necessary for synaptic plasticity in mouse visual cortex. *Nature Communications*, 7:10905, 2016.

[441] B. F. Skinner. *Science and Human Behavior*. The Macmillan Company, New York, 1953.

[442] T. A. Skolem. *Abstract Set Theory*, volume 8 of *Notre Dame Mathematical Lectures*. University of Notre Dame, Notre Dame, IN, 1962.

[443] F. Smeets, T. Lataster, M. Dominguez et al. Evidence that onset of psychosis in the population reflects early hallucinatory experiences that through environmental risks and affective dysregulation become complicated by delusions. *Schizophrenia Bulletin*, 38(3):531–542, 2012.

[444] L. Smith. *Chaos: A Very Short Introduction*. Very Short Introductions. Oxford University Press, Oxford, 2007.

[445] P. Smith. *An Introduction to Gödel's Theorems*. Cambridge Introductions to Philosophy. Cambridge University Press, Cambridge, 2007.

[446] R. Spataro, A. Chella, B. Allison et al. Reaching and grasping a glass of water by locked-in ALS patients through a BCI-controlled humanoid robot. *Frontiers in Human Neuroscience*, 11:68, 2017.

[447] R. Sperry. Some effects of disconnecting the cerebral hemispheres. *Science*, 217(4566):1223–1226, 1982.

[448] J. C. Sprott. Simplifications of the Lorenz attractor. *Nonlinear Dynamics, Psychology, and Life Sciences*, 13(3):271–278, 2009.

[449] S. A. Stamper, M. S. Madhav, N. J. Cowan, and E. S. Fortune. Beyond the jamming avoidance response: weakly electric fish respond to the envelope of social electrosensory signals. *Journal of Experimental Biology*, 215(23):4196–4207, 2012.

[450] G. Stanghellini, Á. I. Langer, A. Ambrosini, and A. J. Cangas. Quality of hallucinatory experiences: differences between a clinical and a non-clinical sample. *World Psychiatry*, 11(2):110–113, 2012.

[451] H. P. Stapp. Quantum theory and the role of mind in nature. *Foundations of Physics*, 31(10):1465–1499, 2001.

[452] H. P. Stapp. *Mindful Universe: Quantum Mechanics and the Participating Observer*. The Frontiers Collection. Springer, Heidelberg, 2nd edition, 2011.

[453] P. Stehle. *Quantum Mechanics*. Holden-Day Series in Physics. Holden-Day, San Francisco, 1966.

[454] V. J. Stenger. The myth of quantum consciousness. *The Humanist*, 53(3):13–15, 1993.

[455] C. F. Stevens and J. M. Sullivan. The synaptotagmin C2A domain is part of the calcium sensor controlling fast synaptic transmission. *Neuron*, 39(2):299–308, 2003.

[456] I. Stewart. *In Pursuit of the Unknown: 17 Equations That Changed the World*. Basic Books, New York, 2012.

[457] C. Stringer. *The Origin of Our Species*. Penguin, London, 2012.

[458] T. C. Südhof. The synaptic vesicle cycle. *Annual Review of Neuroscience*, 27(1):509–547, 2004.

[459] T. C. Südhof. Neuroligins and neurexins link synaptic function to cognitive disease. *Nature*, 455(7215):903–911, 2008.

[460] T. C. Südhof. Calcium control of neurotransmitter release. *Cold Spring Harbor Perspectives in Biology*, 4(1):a011353, 2012.

[461] T. C. Südhof. The presynaptic active zone. *Neuron*, 75(1):11–25, 2012.

[462] T. C. Südhof. A molecular machine for neurotransmitter release: synapto-tagmin and beyond. *Nature Medicine*, 19(10):1227–1231, 2013.

[463] T. C. Südhof. Neurotransmitter release: the last millisecond in the life of a synaptic vesicle. *Neuron*, 80(3):675–690, 2013.

[464] T. C. Südhof and J. E. Rothman. Membrane fusion: grappling with SNARE and SM proteins. *Science*, 323(5913):474–477, 2009.

[465] W.-R. Sun, B. Tian, Y.-F. Wang, and H.-L. Zhen. Soliton excitations and interactions for the three-coupled fourth-order nonlinear Schrödinger equations in the alpha helical proteins. *European Physical Journal D*, 69(6):146, 2015.

[466] W.-R. Sun, B. Tian, H. Zhong, and H.-L. Zhen. Soliton interactions for the three-coupled discrete nonlinear Schrödinger equations in the alpha helical proteins. *Studies in Applied Mathematics*, 132(1):65–80, 2014.

[467] L. Susskind. *The Theoretical Minimum: Quantum Mechanics*. Stanford Continuing Studies, Stanford University, Stanford, CA, 2012.

[468] L. Susskind and A. Friedman. *Quantum Mechanics: The Theoretical Minimum. What You Need to Know to Start Doing Physics*. Basic Books, New York, 2014.

[469] M. J. Sutcliffe and N. S. Scrutton. Enzyme catalysis: over-the-barrier or through-the-barrier? *Trends in Biochemical Sciences*, 25(9):405–408, 2000.

[470] J. Symons, O. Pombo, and J. M. Torres. *Otto Neurath and the Unity of Science*, volume 18 of *Logic, Epistemology, and the Unity of Science*. Springer, Dordrecht, 2011.

[471] E. Takahashi, K. Ohki, and Y. Miyashita. The role of the parahippocampal gyrus in source memory for external and internal events. *NeuroReport*, 13(15):1951–1956, 2002.

[472] C. S. R. Taylor and C. G. Gross. Twitches versus movements: a story of motor cortex. *The Neuroscientist*, 9(5):332–342, 2003.

[473] J. R. Taylor. *Classical Mechanics*. University Science Books, Sausalito, CA, 2005.

[474] M. Tegmark. The interpretation of quantum mechanics: many worlds or many words? *Fortschritte der Physik*, 46(6–8):855–862, 1998.

[475] M. Tegmark. Importance of quantum decoherence in brain processes. *Physical Review E*, 61(4):4194–4206, 2000.

[476] M. Tegmark. Many lives in many worlds. *Nature*, 448(7149):23–24, 2007.

[477] M. Tegmark. Consciousness is a state of matter, like a solid or gas. *New Scientist*, 222(2964):28–31, 2014.

[478] M. Tegmark. Consciousness as a state of matter. *Chaos, Solitons & Fractals*, 76:238–270, 2015.

[479] M. Tegmark. Improved measures of integrated information. *PLOS Computational Biology*, 12(11):e1005123, 2016.

[480] S. Teki, M. Grube, and T. D. Griffiths. A unified model of time perception accounts for duration-based and beat-based timing mechanisms. *Frontiers in Integrative Neuroscience*, 5:90, 2011.

[481] S. Thuret, L. D. F. Moon, and F. H. Gage. Therapeutic interventions after spinal cord injury. *Nature Reviews Neuroscience*, 7(8):628–643, 2006.

[482] G. Tononi. An information integration theory of consciousness. *BMC Neuroscience*, 5:42, 2004.

[483] G. Tononi. Consciousness, information integration, and the brain. *Progress in Brain Research*, 150:109–126, 2005.

[484] G. Tononi. Consciousness as integrated information: a provisional manifesto. *Biological Bulletin*, 215(3):216–242, 2008.

[485] G. Tononi. The integrated information theory of consciousness: an updated account. *Archives Italiennes de Biologie*, 150(4):293–329, 2012.

[486] G. Tononi, M. Boly, M. Massimini, and C. Koch. Integrated information theory: from consciousness to its physical substrate. *Nature Reviews Neuroscience*, 17(7):450–461, 2016.

[487] G. Tononi and C. Koch. Consciousness: here, there and everywhere? *Philosophical Transactions of the Royal Society of London Series B*, 370(1668):20140167, 2015.

[488] G. Tononi and O. Sporns. Measuring information integration. *BMC Neuroscience*, 4:31, 2003.

[489] O. B. Toon, K. Zahnle, D. Morrison et al. Environmental perturbations caused by the impacts of asteroids and comets. *Reviews of Geophysics*, 35(1):41–78, 1997.

[490] J. S. Townsend. *A Modern Approach to Quantum Mechanics*. University Science Books, Sausalito, CA, 2000.

[491] J. S. Townsend. *Quantum Physics: A Fundamental Approach to Modern Physics*. University Science Books, Sausalito, CA, 2010.

[492] T. Trappenberg. *Fundamentals of Computational Neuroscience*. Oxford University Press, Oxford, 2002.

[493] M. Tye. Qualia. In E. N. Zalta, U. Nodelman, and C. Allen, editors, *Stanford Encyclopedia of Philosophy*. Stanford University, Stanford, CA, 2015.

[494] D. Ungar and F. M. Hughson. SNARE protein structure and function. *Annual Review of Cell and Developmental Biology*, 19(1):493–517, 2003.

[495] R. Ursin, T. Jennewein, M. Aspelmeyer et al. Quantum teleportation across the Danube. *Nature*, 430(7002):849, 2004.

[496] G. van den Bogaart, M. G. Holt, G. Bunt et al. One SNARE complex is sufficient for membrane fusion. *Nature Structural & Molecular Biology*, 17(3):358–364, 2010.

[497] B. van Swinderen, L. B. Metz, L. D. Shebester et al. Goα regulates volatile anesthetic action in Caenorhabditis elegans. *Genetics*, 158(2):643–655, 2001.

[498] B. van Swinderen, O. Saifee, L. Shebester et al. A neomorphic syntaxin mutation blocks volatile-anesthetic action in Caenorhabditis elegans. *Proceedings of the National Academy of Sciences*, 96(5):2479–2484, 1999.

[499] S. Vanni, T. Tanskanen, M. Seppä et al. Coinciding early activation of the human primary visual cortex and anteromedial cuneus. *Proceedings of the National Academy of Sciences*, 98(5):2776–2780, 2001.

[500] M. J. Vansteensel, E. G. M. Pels, M. G. Bleichner et al. Fully implanted brain–computer interface in a locked–in patient with ALS. *New England Journal of Medicine*, 375(21):2060–2066, 2016.

[501] A. Vaziri and M. B. Plenio. Quantum coherence in ion channels: resonances, transport and verification. *New Journal of Physics*, 12(8):085001, 2010.

[502] V. Vedral and M. B. Plenio. Basics of quantum computation. *Progress in Quantum Electronics*, 22(1):1–39, 1998.

[503] B. A. Vogt. Pain and emotion interactions in subregions of the cingulate gyrus. *Nature Reviews Neuroscience*, 6(7):533–544, 2005.

[504] K. D. Vohs and J. W. Schooler. The value of believing in free will: encouraging a belief in determinism increases cheating. *Psychological Science*, 19(1):49–54, 2008.

[505] G. von der Emde. Distance and shape: perception of the 3-dimensional world by weakly electric fish. *Journal of Physiology-Paris*, 98(1–3):67–80, 2004.

[506] C. von der Malsburg. Binding in models of perception and brain function. *Current Opinion in Neurobiology*, 5(4):520–526, 1995.

[507] A. von Kölliker. *Handbuch der Gewebelehre des Menschen, Zweiter Band: Nervensystem des Menschen und der Thiere.* Wilhelm Engelmann, Leipzig, 1896.

[508] J. von Neumann. Various techniques used in connection with random digits. In A. S. Householder, G. E. Forsythe, and H. H. Germond, editors, *Monte Carlo Method*, National Bureau of Standards Applied Mathematics Series, pages 36–38. U.S. Government Printing Office, Washington, DC, 1951.

[509] J. Wackermann. On cumulative effects and averaging artefacts in randomised S–R experimental designs. In *Proceedings of the 45th Annual Convention of the Parapsychological Association*, pages 293–305, 2002.

[510] J. Wackermann, C. Seiter, H. Keibel, and H. Walach. Correlations between brain electrical activities of two spatially separated human subjects. *Neuroscience Letters*, 336(1):60–64, 2003.

[511] X. Wang, D. W. Brown, and K. Lindenberg. Quantum Monte Carlo simulation of the Davydov model. *Physical Review Letters*, 62(15):1796–1799, 1989.

[512] T. Weber, B. V. Zemelman, J. A. McNew et al. SNAREpins: minimal machinery for membrane fusion. *Cell*, 92(6):759–772, 1998.

[513] R. Webster. *Why Freud Was Wrong: Sin, Science and Psychoanalysis.* Orwell Press, Oxford, 2005.

[514] D. M. Wegner. *The Illusion of Conscious Will.* MIT Press, Cambridge, MA, 2002.

[515] A. Wehrl. General properties of entropy. *Reviews of Modern Physics*, 50(2):221–260, 1978.

[516] C. P. Weingarten, P. M. Doraiswamy, and M. P. A. Fisher. A new spin on neural processing: quantum cognition. *Frontiers in Human Neuroscience*, 10:541, 2016.

[517] W. Whewell. *The Philosophy of the Inductive Sciences Founded Upon Their History*, volume 2. John W. Parker, London, 1847.

[518] R. J. White, L. R. Wolin, L. C. Massopust Jr. et al. Cephalic exchange transplantation in the monkey. *Surgery*, 70(1):135–139, 1971.

[519] A. N. Whitehead and B. Russell. *Principia Mathematica*, volume 1. Cambridge University Press, Cambridge, 2nd edition, 1963.

[520] A. P. Wickens. *A History of the Brain: From Stone Age Surgery to Modern Neuroscience.* Psychology Press, New York, 2015.

[521] E. P. Wigner. Remarks on the mind-body question. In I. J. Good, editor, *The Scientist Speculates*, pages 284–302. Heinemann, London, 1961.

[522] L. Wittgenstein. *Tractatus Logico-Philosophicus*. Routledge & Kegan Paul, London, 1922.

[523] L. Wittgenstein. *Philosophical Investigations*. Blackwell Publishers, Oxford, 2nd edition, 1999.

[524] M. Wittmann. The inner experience of time. *Philosophical Transactions of the Royal Society B: Biological Sciences*, 364(1525):1955–1967, 2009.

[525] P. Woit. *Quantum Theory, Groups and Representations: An Introduction*. Columbia University, New York, 2016.

[526] B. Wood. *Human Evolution: A Very Short Introduction*, volume 142 of *Very Short Introductions*. Oxford University Press, Oxford, 2005.

[527] N. J. Woolf and S. R. Hameroff. A quantum approach to visual consciousness. *Trends in Cognitive Sciences*, 5(11):472–478, 2001.

[528] W. K. Wootters and W. H. Zurek. A single quantum cannot be cloned. *Nature*, 299(5886):802–803, 1982.

[529] Z. Xi, Y. Li, and H. Fan. Quantum coherence and correlations in quantum system. *Scientific Reports*, 5:10922, 2015.

[530] D. L. K. Yamins and J. J. DiCarlo. Using goal-driven deep learning models to understand sensory cortex. *Nature Neuroscience*, 19(3):356–365, 2016.

[531] H.-D. Zeh. Roots and fruits of decoherence. In B. Duplantier, J.-M. Raimond, and V. Rivasseau, editors, *Quantum Decoherence. Poincaré Seminar 2005*, volume 48 of *Progress in Mathematical Physics*, pages 151–175. Birkhäuser, Basel, 2007.

[532] H.-D. Zeh. Quantum discreteness is an illusion. *Foundations of Physics*, 40(9–10):1476–1493, 2010.

[533] E. Zermelo. Über Grenzzahlen und Mengenbereiche. *Fundamenta Mathematicae*, 16(1):29–47, 1930.

[534] X. Zhao, L. Liu, X.-X. Zhang et al. The effect of belief in free will on prejudice. *PLoS ONE*, 9(3):e91572, 2014.

[535] Q. Zhou, Y. Lai, T. Bacaj et al. Architecture of the synaptotagmin-SNARE machinery for neuronal exocytosis. *Nature*, 525(7567):62–67, 2015.

[536] M. P. Zlatev. *Theoretical Electrotechnics*, volume 1. Tehnika, Sofia, 1972.

[537] W. H. Zurek. Pointer basis of quantum apparatus: into what mixture does the wave packet collapse? *Physical Review D*, 24(6):1516–1525, 1981.

Glossary

acceleration is a physical measure $a = \frac{dv}{dt}$ of how quickly the velocity v of an object changes in time t. 173

action potential of a neuron is a localized rise in the transmembrane electrical potential whose propagation along the neuronal projections is used for the transmission of classical information. 27

analytic function is a function that is expressible in the neighborhood of some point x_0 as a convergent power series $f(x) = \sum_{n=0}^{\infty} a_n (x - x_0)^n$. 51

anesthetized brain is an electrically active brain under general anesthesia that does not generate any conscious experiences. 25

anomalous magnetic moment of the electron $a = \frac{g-2}{2}$ is a contribution of quantum electrodynamic effects, expressed by Feynman diagrams with loops, to the magnetic moment of the electron, which at the zeroth order of perturbation is calculated to be exactly 2. 85

associativity is a mathematical property that allows arbitrary grouping of arguments with the use of parentheses in a repeated operation $\circ$ without changing the final result of the calculation, $x_1 \circ (x_2 \circ x_3) = (x_1 \circ x_2) \circ x_3$. 86

atom is the smallest unit of matter that has the properties of a chemical element; atoms are made of protons, neutrons and electrons. 45

atomic magic wand is a thought experiment showing that emergent consciousness is untenable as it could be switched on or off by the addition or removal of a single atom. 163

atomic number is the number of protons in the nucleus of an atom, which determines its chemical identity and properties. 45

axiom is an empirically motivated statement that is taken to be true without proof; axioms are the building blocks of scientific theories since axioms can be used for the logical deduction of other true statements called theorems. 33

axon is a long thin neuronal projection capable of generating electric action potentials that output information from a given neuron to other target neurons, or to effector organs such as muscles or glands. 25

Bayesian inference is a method of statistical inference in which Bayes' theorem is used to update the probability or odds for a hypothesis T being true as more evidence E becomes available. 270

behaviorism is a philosophical stance that all statements about mental states and conscious processes should be equivalent in meaning to statements about behavioral dispositions. 161

belief in free will is our common sense viewpoint that we are free to make genuine choices among alternative future courses of action; since our introspective testimony confirms the existence of free will, believing in free will is heavily evidence-based, unlike believing in fairies in the forest. 175

biomolecule is any substance that is produced by living organisms. 178

bra is a row vector $\langle \psi | = \sum_i a_i^* \langle \psi_i |$ in a dual Hilbert space $\mathcal{H}^*$; a bra combined with a ket on the right forms a bra-ket that denotes the inner product of two vectors. 96

brain is an anatomical organ that is part of the central nervous system; the brain controls our sensations, thoughts, bodily functions and movements. 6

brain cortex is the outer layer of gray matter of the cerebrum of the brain; the brain cortex is the seat of our consciousness. 8

brain in a vat is a thought experiment of a disembodied brain kept alive in a vat filled with electrolyte solution; instead of a body at the other end of peripheral nerves is connected a classical computer that feeds the brain with sensory information and reads out the brain motor responses. 228

butterfly effect is the sensitive dependence of deterministic nonlinear systems on the initial conditions in which a small change in the current state could lead to large differences in future states beyond a short period of time. 172

canonical momentum is one of $3n$ coordinates in phase space that are used to describe the components of momentum p_i of a classical physical system consisting of n particles. 68

canonical position is one of $3n$ coordinates in phase space that are used to describe the components of position q_i of a classical physical system consisting of n particles. 68

Cartesian coordinates specify the location in n-dimensional space using a list of n distances to each of n mutually perpendicular coordinate axes. 87

central nervous system is a part of the nervous system that consists of the brain and spinal cord. 6

chaos is a property of deterministic nonlinear systems to exhibit behavior that is effectively unpredictable for extended periods of time due to great sensitivity to small changes in initial conditions. 48, 172

chaotic system is a deterministic nonlinear system whose behavior is effectively unpredictable for extended periods of time due to great sensitivity to small changes in initial conditions. 172

chemical element is a material substance that cannot be chemically interconverted or broken down into simpler substances. 45

choice is the act of picking or deciding between two or more possibilities. 16

circular proof is a logically fallacious chain of reasoning that assumes, either implicitly or explicitly, what it is attempting to prove. 34

classical bit is a basic unit of classical information contained in the answer of a single yes-or-no question. 35

classical computer is a physical device that performs computational tasks with the use of classical bits of information. 221

classical information is information stored in classical physical states; it is observable, local, clonable, broadcastable, and erasable. 35

classical state is a point (p, q) in phase space that undergoes deterministic evolution in time given by Hamilton's equations. 68

classical theory of consciousness is any extension of classical physics with new laws that introduce consciousness into the physical world. 143

cloning is the process of creating an exact copy of a physical system. 125

closed system is a physical system that does not interact with its environment. 48, 111

cognition is the mental act of knowing or understanding through conscious experience, reasoning, or intuition. 146

cognitive process is the performance of any cognitive operation that affects mental contents including, but not limited to, sensory perception, learning, attention, memorization, memory recall, language, reasoning, problem solving, prediction, decision making, or error correction. 146

collection of minds is a group of conscious entities, each with its own private conscious experiences that are inaccessible to the other minds in the group. 10

collective consciousness is a metaphorical concept that refers to the set of shared beliefs, ideas, attitudes, and knowledge that are common to a society; going beyond the metaphor entails paradoxical occurrence of minds within minds. 11

color blindness is a genetically inherited condition that affects the color-sensing pigments in the eye and causes difficulty or inability to distinguish colors. 220

communicability is the property of being able to be communicated to others. 177

commutativity is a mathematical property that allows changing the order of arguments in an operation $\circ$ without changing the final result of the calculation, $x_1 \circ x_2 = x_2 \circ x_1$. 86

compatibilism is a philosophical stance that free will and determinism are not mutually exclusive. 167

complex conjugate of a complex number $z = x + \imath y$ is the number $z^* = x - \imath y$ formed by changing the sign of the imaginary part. 93

complex conjugation is the operation of changing the sign of the imaginary part of a complex number. 86

complex number is a number that can be expressed in the form $z = x + \imath y$, where x and y are real numbers and $\imath = \sqrt{-1}$. 85

complex plane provides a geometric picture of complex numbers $z = x + \imath y$ as pairs of coordinates: x gives the position on the horizontal real axis and y gives the position on the vertical imaginary axis. 86

complexity is the property of not being simple. 39

compression ratio is the ratio between the uncompressed size of the data and the compressed size of the theory; it provides a quantification of the explanatory power of a scientific theory. 39

computationalism is a philosophical stance that the conscious mind is a computer program or software, whereas the brain is hardware that runs the mind software. 161

computer program is a list of instructions that performs a specific task when interpreted and executed by a computer. 39

conditional probability is a measure of the probability $p(A|B)$ of an event A given that another event B has occurred. 64

conscious mind is a sentient phyical entity endowed with consciousness. 4

consciousness refers to the subjective, first-person point of view of our mental states, experiences or feelings; a conscious state is a state of experience. 3

continuity of a function $f(x)$ at a point x_0 means that the limit of the function as the argument x approaches x_0 exists and this limit is equal to $f(x_0)$, or symbolically $\lim_{x \to x_0} f(x) = f(x_0)$. 53

contradiction is the logical conjunction of a statement P with its own negation $\neg P$, or symbolically $P \wedge \neg P$; since from contradiction follows everything, contradictions lead to nonsense. 33

corpus callosum is a wide bundle of neural fibers beneath the brain cortex connecting the left and right brain hemispheres. 13

corticospinal tract is a descending motor pathway which contains bundles of axons that originate in the brain cortex and terminate on motor neurons in the spinal cord; the axons of the motor neurons then convey the motor commands to the muscles whose contraction moves the limbs and trunk. 13

cosmic mind is a mystical concept endorsing the existence of cosmic consciousness of which we are all a part; going beyond the metaphor entails paradoxical occurrence of minds within minds. 11

crucial experiment is an experiment capable of decisively determining whether or not a particular scientific theory is superior to another rival theory or group of theories. 275

crushed brain is a brain apparently lacking conscious experiences, even though its composition in terms of chemical atoms is exactly the same as an intact conscious brain. 160

curl of a vector field $\vec{A}$ is a vector operator $\nabla \times \vec{A}$ that describes the infinitesimal rotation of $\vec{A}$. 76

current density is the flow $J = \frac{di}{ds}$ of electric current di per elementary area of cross section ds. 73

data compression is the process of reducing the number of classical bits needed to store or transmit data. 39

Davydov soliton is a quantum quasiparticle that represents a self-trapped amide I excitation propagating along the protein α-helix. 254

dead brain is an anatomical organ left after the death of a person; the dead brain apparently lacks conscious experiences, which justifies the practice of post-mortem pathologoanatomical examination. 25

decoherence is the loss of quantum coherence by a quantum system due to interaction, and inevitable quantum entanglement, with its environment. 244

definite integral is an integral $\int_a^b f(x)dx = F(b) - F(a)$ expressed as the difference between the values of the indefinite integral $F(x)$ at specified upper limit b and lower limit a of the independent variable x. 60

dendrite is a neuronal projection along which electric impulses received from other neurons at dendritic synapses are transmitted to the cell body. 25

density matrix is a quantum mechanical operator $\hat{\rho} = \sum_i p_i |\psi_i\rangle\langle\psi_i|$ that describes the state of a quantum system as a statistical ensemble, $\sum_i p_i = 1$, of several, not necessarily orthogonal, pure quantum states $|\psi_i\rangle$. 114

dependent events are any two events A and B such that the outcome of one event affects the outcome of the other event; in terms of conditional probabilities $p(A|B) \neq p(A)$ or $p(B|A) \neq p(B)$. 66

determinism is the ability of a theory consisting of a set of physical laws to predict in principle with absolute certainty and arbitrarily high precision the state of any closed physical system at any future moment of time given the current state of the system; complex nonlinear systems exhibiting chaotic behavior are an example of deterministic systems. 48

deuterium is a stable isotope of hydrogen, ^{2}H, containing one proton and one neutron. 45

differentiable function is a function $f(x)$ whose derivative $\frac{df}{dx}$ exists at each point in its domain. 53

differentiation is the process of finding the derivative $\frac{df}{dx}$ of a function $f(x)$. 51

directed edge in a graph G is an ordered pair of graph vertices, $v_1 \rightarrow v_2$. 153

distance is a mathematical measure of how far apart two objects are. 93

domain of a function is the set of all inputs x over which the function $f(x)$ is defined. 51

domino theory of moral nonresponsibility is a consequence of determinism by which no one is ever ultimately responsible for his actions because they are the result of what other people or the physical circumstances have done to him. 170

dualism is a philosophical stance that both the brain and the mind do exist in the physical world but they are made of different substances; brain is material, whereas the mind is mental. 181

easy problems of consciousness are to explain the function, dynamics, and structure of consciousness in terms of how the brain participates in perception, cognition, learning and behavior. 27

eigenstate of a quantum observable $\hat{A}$ is a quantum state $|\psi\rangle$ that has a definite value λ when $\hat{A}$ is measured, $\hat{A}|\psi\rangle = \lambda|\psi\rangle$. 100

eigenvalue of a quantum observable $\hat{A}$ is a possible value λ that can be obtained as a measurement outcome when $\hat{A}$ is measured. 100

eigenvector is a vector $|\psi\rangle$ which when operated on by a given operator $\hat{A}$ gives a scalar multiple of itself, $\hat{A}|\psi\rangle = \lambda|\psi\rangle$); eigenvectors are mathematical objects that can represent anything in the physical world, not just eigenstates of quantum systems. 100

electric current is the flow of electric charge. 72

electric field is a vector field $\vec{E}$ generated by electric charges or time-varying magnetic fields; the direction of the field is taken to be the direction of the force it would exert on a positive test charge q. 71

electrocardiography is a noninvasive procedure for recording the electrical activity of the heart using electrodes placed on the skin. 158

electroencephalography is a noninvasive procedure for recording the electrical activity of the brain using electrodes placed on the scalp. 158

electrolyte is a conducting medium in which the flow of electric current is due to the movement of ions. 145

electromagnetic spectrum is the range of all possible frequencies or wavelengths of electromagnetic radiation. 79

electromagnetic theory of consciousness is any theory that reduces consciousness to an electromagnetic phenomenon. 165

electron is an elementary particle e^- with a negative elementary electric charge; electrons are main constituents of atoms and are responsible for the chemical properties of the elements. 45

elementary electric charge is the electric charge carried by a single proton. 45

eliminativism is a philosophical stance that the brain is built from classical physical particles that obey deterministic physical laws, whereas conscious minds do not exist and are just illusions. 143

emergentism is a philosophical stance that conscious experiences are emergent properties of the brain that are not identical with, reducible to, or deducible from the other physical properties of the brain; emergentism is a disguised form of pseudoscience because conscious experiences are able to miraculously pop into existence where physical laws predict none. 163

epiphenomenalism is a philosophical stance that conscious experiences are caused by physical events in the brain, but have no effects upon any event in the brain and the physical world. 164

epiphenomenon is a causally ineffective secondary phenomenon or byproduct that results from and occurs alongside a causally effective primary phenomenon. 163

equation is a statement that the values of two mathematical expressions are equal. 49

eukaryote is an organism consisting of one or more cells whose genetic material is contained within a membrane-enclosed nucleus. 15

existentialism is a philosophical stance that we are born without a greater purpose or predetermined plan set by something outside us; instead we are free to define ourselves, choose our own purpose of life and become what we want to be within the limits of the physically possible. 17

exocytosis is a process by which the content of an intracellular vesicle is released to the extracellular space through fusion of the vesicle membrane with the plasma membrane of the cell. 252

experience is the subjective, first-person, phenomenal point of view of our mental states, perceptions, or feelings; the totality of all experiences that we have at a single moment of time makes up our consciousness. 3

explanandum is that which needs to be explained. 36

explanans is that which contains the explanation. 36

explanation is a statement about how or why something is the way it is; to be able to explain means to be able to predict the occurrence of the phenomenon under consideration given a set of physical laws and initial conditions. 36

falsification is the act of disproving a proposition, hypothesis, or theory with the use of experimental evidence. 268

feeling is a consciously experienced sensation or emotion. 3

Feynman diagram is a graphical way to represent a series of emissions and absorptions of subatomic particles by other subatomic particles, from which the quantum probability of the whole series can be calculated. 244

force is any influence that has the capacity to change the motion of an object. 56

free will is the capacity of agents to choose a future course of action from among at least two different alternatives. 16

free-will skepticism is a philosophical stance that free will is impossible regardless of whether determinism is true. 209

free will within free will is a paradoxical occurrence of nested free will agents inside other free will agents. 206

frequency is the number of occurrences of a phenomenon over a unit period of time. 78

frontal lobe is an anatomically distinct part of the brain cortex located at the front of the head. 11

function is a mathematical relation from a set of inputs to a set of possible outputs where each input is related to exactly one output; to function is to execute a replacement of an input with an output specified by a functional relation. 50

functionalism is a philosophical stance that the conscious mind is a function or a functional product of the physical brain. 160

fundamental physical law is a physical law that happens to be true for no other simpler reason. 37

Golgi apparatus is a membrane-bound cellular organelle that is involved in protein packaging and trafficking with the use of membrane-bound vesicles. 15

gradient is a change in the magnitude of a property observed in passing from one point in space to another. 57

graph is a network of vertices connected with directed or undirected edges. 153

graph of a function is a pictorial representation showing the relationship between the inputs and the outputs of a function. 51

gyrus is a ridge on the brain cortex. 11

hard determinism is a philosophical stance that because we are living in a universe that is governed by deterministic physical laws, we are agents without free will. 167

hard problem of consciousness is to explain how and why the brain generates any qualia or phenomenal experiences at all. 27

heart is a hollow muscular organ which pumps the blood through the circulatory system by rhythmic contraction and dilation. 5

Hermitian matrix is a matrix $\hat{A}$ that is equal to its own conjugate transpose $\hat{A}^\dagger$; the eigenvalues of a Hermitian matrix are always real. 100

Hermitian operator is an operator represented by a Hermitian matrix $\hat{A} = \hat{A}^\dagger$. 100

Hilbert space is a vector space $\mathcal{H}$ possessing an inner product $\langle \phi | \psi \rangle$ such that the norm defined by $|\psi| = \sqrt{\langle \psi | \psi \rangle}$ turns $\mathcal{H}$ into a complete metric space. 94

hydrogen bond is an electrostatic attraction between two polar groups that occurs when a hydrogen atom bonded to a strongly electronegative atom exists in the vicinity of another electronegative atom with a lone pair of electrons; hydrogen bonds are weaker than covalent bonds. 241

idealism is a philosophical stance that only conscious minds exist, whereas insentient physical matter is a nonentity or just an illusion. 178

imaginary unit is a solution $\imath = \sqrt{-1}$ to the equation $x^2 + 1 = 0$. 86

immortality is the ability to live forever as a conscious mind. 166

improbable is the property of being not likely to be true or to happen. 125

incommunicability is the property of not being able to be communicated. 220

incompatibilism is a philosophical stance that free will and determinism are mutually exclusive. 167

inconsistency is the presence of a logical contradiction. 33

indefinite integral of a function $f(x)$ is a differentiable function $F(x)$ whose derivative $\frac{dF(x)}{dx}$ is equal to the original function $f(x)$. 58

independent events are any two events A and B such that the outcome of one event does not affect the outcome of the other event; in terms of conditional probabilities $p(A|B) = p(A)$ and $p(B|A) = p(B)$. 66

indeterminism is the inability of a theory consisting of a complete set of physical laws to predict in principle with absolute certainty and arbitrarily high precision the future state of a closed physical system given its current state. 199

infinite regress is an infinitely long proof composed of an ever growing list of novel statements; such a proof cannot be actually completed, written down in a finite amount of classical bits of information and communicated to others. 34

information gain also known as the Kullback–Leibler divergence $D_{KL}(P_2\|P_1)$ is the amount of classical bits of information gained when one revises one's beliefs from the prior probability distribution P_1 to the posterior probability distribution P_2; for discrete distributions $D_{KL}(P_2\|P_1) = \sum_i P_2(i)\log_2 \frac{P_2(i)}{P_1(i)}$, whereas for continuous distributions $D_{KL}(P_2\|P_1) = \int_{-\infty}^{\infty} P_2(x)\log_2 \frac{P_2(x)}{P_1(x)}dx$. 153

inner monologue is our inner voice, internal thinking in words, or verbal stream of consciousness. 248

inner privacy of consciousness is the private character of our experiences that are accessible through introspection from a first-person, subjective, phenomenal perspective, but are unobservable from a third-person, objective perspective. 20

input is the argument x that is operated on by a mathematical function $f(x)$. 36

instability is the capacity of a physical system not to return to its initial physical state after an infinitesimal perturbation. 171

integer is a whole number that can be written without a fractional component. 49

intelligent amoeba is a metaphor used by Hugh Everett III to illustrate his proposal of conscious minds that split into parallel universes. 235

interactionism is a philosophical stance that material events in the brain can cause conscious experiences and those experiences can subsequently cause material events in the brain. 181

irrational number is a number that has an infinite aperiodic decimal expansion. 50

isotope is a variant of a chemical element with specified number of neutrons. 45

jerk is a physical measure $j = \frac{da}{dt}$ of how quickly the acceleration a of an object changes in time t. 173

ket is a column vector $|\psi\rangle = \sum_i a_i |\psi_i\rangle$ in a Hilbert space $\mathcal{H}$; a ket combined with a bra on the left forms a bra-ket that denotes the inner product of two vectors. 95

kinetic isotope effect is the change in the rate of a chemical reaction when one of the atoms in the reactants is substituted with one of its isotopes; substitution of protium ^{1}H with deuterium ^{2}H is commonly used in biochemical studies. 279

label is a string of symbols used to describe people, activities or things. 23

lateral spinothalamic tract is an ascending pathway in the spinal cord that carries sensory information from the body to the thalamus; thalamic neurons in turn relay processed sensory information to the brain cortex. 151

lawlessness is a state not restrained or controlled by a law. 19

learning is the act of gaining knowledge or skill by studying, practicing, or experiencing something. 146

libertarianism is a philosophical stance that because we are agents with free will, we cannot be living in a universe that is governed by deterministic physical laws. 167

light is an electromagnetic wave in classical physics or a beam of photons exhibiting wave-particle duality in quantum physics. 45

magnetic field is a vector field $\vec{B}$ generated by moving electric charges or time-varying electric fields; the magnetic field acts only on moving charges. 73

mammal is a warm-blooded vertebrate animal of which the female secretes milk for the nourishment of her young. 15

mass is a physical measure $m = \frac{F}{a}$ of the resistance to acceleration a of an object when a net force F is applied. 67

matter is any physical substance that has mass and takes up space; massive particles such as electrons are particles of matter, whereas massless particles such as photons of light are particles of energy. 45

meaning of life is a personal understanding of the significance of living or existence. 17

mind is a sentient physical entity endowed with consciousness. 3

mind duplication is a paradoxical existence of a conscious mind at two different locations in space at the same instant of time. 143

mind dust is a metaphor used by William James to denote the raw material upon which natural selection acts thereby leading to evolution of complex minds from simpler minds, without necessity of miraculous popping into existence of conscious experiences where the physical laws predict none. 181

mind–brain identity is a philosophical stance that the mind is the same as the brain; in such a case the evolution of the mind is equivalent to the evolution of the brain, hence epiphenomenalism is avoided. 164

mindless brain is an electrically active brain that does not generate any conscious experiences. 27

mindless machine is a machine that simulates the electrical activity of a conscious brain, but itself does not generate any conscious experiences. 182

minds within minds is a paradoxical occurrence of nested conscious minds inside other conscious minds. 11

mitochondrion is a membrane-bound cellular organelle that is involved in cellular respiration providing the cell with ready-to-use biochemical energy in the form of adenosine triphosphate (ATP). 15

mixed density matrix is a quantum operator $\hat{\rho} = \sum_i p_i |\psi_i\rangle\langle\psi_i|$ where $\{|\psi_i\rangle\}$ is some set of pure states, not necessarily orthogonal, and the probabilities $0 \leq p_i < 1$ sum up to unity $\sum_i p_i = 1$; mixed density matrices can be used to describe the quantum states of open physical systems that are quantum entangled with their environment. 114

momentum is the quantity of motion $p = mv$ of a moving body given by the product of the body's mass and its velocity. 67

moral responsibility is the obligation of an agent with free will to account for the consequences of his own choices. 17

multi-valued function is a relation that assumes two or more distinct values in its range for at least one point in its domain; each multi-valued function of a single complex variable can be thought of as a single-valued function on a Riemann surface consisting of several sheets. 88

multiple personality disorder is a condition in which the identity of a person is fragmented into at least two distinct personalities; each personality may inhabit the person's conscious awareness for a period of time and remain unaware of the existence of the other personalities. 265

nature is the physical world and everything in it. 43

natural number is any non-negative integer. 49

neurite is any cable-like projection from the cell body of a neuron. 25

neuron is an electrically excitable cell in the nervous system that inputs, processes and transmits information through electrical and chemical signals. 25

neurotransmitter is a chemical substance released by neurons to stimulate other neurons, muscle cells, or gland cells. 250

neutron is a subatomic particle n^0 with no net electric charge; neutrons are main constituents of atomic nuclei. 45

norm is a mathematical function $|\vec{x}| = \sqrt{\vec{x} \cdot \vec{x}}$ that assigns a positive length to each vector $\vec{x}$ in a vector space, except for the zero vector $\vec{0}$ whose length is zero. 93

nucleus is a membrane-enclosed organelle storing the genetic material in eukaryotic cells. 15

objective knowledge is knowledge that can be shared with others in the form of classical bits of information; an example of objective knowledge is the content of a book, a song or a movie. 32

objective reduction is a discontinuous jump in the dynamics of the quantum state vector $|\psi\rangle$ of a sufficiently large quantum entangled system that has reached a certain energy threshold $\mathcal{E}$; in the Diósi–Penrose gravitational model for objective reductions, the energy threshold $\mathcal{E} = G\frac{m^2}{r}$ is expressed in terms of the gravitational interaction energy between two displaced macroscopic superpositions with equal mass m, where G is the gravitational constant, and r is a short distance cutoff. 185

observability is the property of being observable or measurable. 45

observable is a physical variable that can be measured. 45

observable brain is the anatomical organ that can be observed inside one's skull. 227

occipital lobe is an anatomically distinct part of the brain cortex located at the back of the head. 11

odds for an event A are given by the ratio $O(A) = \frac{p(A)}{1-p(A)}$, where $p(A)$ is the probability of A occurring, and $1 - p(A)$ is the probability of A not occurring. 271

open system is a physical system that interacts with its environment. 111

operator in functional analysis is a mapping that takes as an input a function and produces as an output another function; in logic the term operator is used for denoting the symbol of a mathematical operation. 99

ordered pair is a pair of numbers (x,y) written in a particular order; the ordered pair (x,y) is not the same as the ordered pair (y,x) unless $x = y$. 62

output is the value $y = f(x)$ that is generated by a mathematical function $f(x)$ when the argument x is operated on. 36

panpsychism is a philosophical stance that all physical particles possess primordial mental features or psyche. 181

paradox is a logically contradictory statement or proposition; since from contradiction follows everything, paradoxes lead to nonsense. 158

parietal lobe is an anatomically distinct part of the brain cortex located in the middle upper part of the head above the temporal lobe. 11

peripheral nervous system is a part of the nervous system that consists of nerves and ganglia outside of the brain and spinal cord. 6

phase space is a multidimensional space in which every degree of freedom of a physical system is represented as an axis. 68

phosphene is a consciously experienced flash of light generated through direct electrical stimulation of the visual brain cortex. 9

physical law is a universal statement about the physical world based on empirical observations and scientific generalization. 32

physical world is everything that exists. 32

physics is the natural science that studies everything in existence. 32

plane wave is a wave whose wave fronts are plane surfaces corresponding to parallel rays. 105

polar coordinates is a pair of coordinates (r, θ) locating the position of a point in the plane, where r is the length of the radius vector $\vec{r}$ of the point, and θ is the counterclockwise angle between $\vec{r}$ and a fixed ray for which $\theta = 0$. 87

position is the location of an object in space. 67

posterior probability of a proposition T is the conditional probability $p(T|E)$ that is assigned after the relevant evidence E is taken into account. 270

potential field is any physical field φ that obeys Laplace's equation given by $\nabla^2 \varphi = \frac{\partial^2 \varphi}{\partial x^2} + \frac{\partial^2 \varphi}{\partial y^2} + \frac{\partial^2 \varphi}{\partial z^2} = 0$. 71

power set of any set S is the set 2^S of all subsets of S, including the empty set $\emptyset$ and S itself. 63

preferred basis problem is a known deficiency of Everett's many worlds interpretation of quantum mechanics due to the fact that maximally mixed states are decoherent in every basis, hence the splitting into parallel universes cannot avoid the observation of Schrödinger's cats around us. 235

principal value of a multi-valued function is the value on the principal sheet of the Riemann surface of that function. 88

prior probability of a proposition T is the probability $p(T)$ that is assigned before any evidence E is taken into account. 38

problem shift is the claim that a theory T has not been falsified by an experimental result E due to hypothesized uncontrolled factors $C_{n+1}, C_{n+2}, \ldots$ acting alongside the factors $C_1, C_2 \ldots, C_n$ controlled by the experimenter. 274

projection operator is a linear transformation $\hat{P}$ from a vector space to itself such that applying $\hat{P}$ twice gives the same output as applying $\hat{P}$ once, $\hat{P}^2 = \hat{P}$. 100

prokaryote is a unicellular organism whose genetic material is not contained within a membrane-enclosed nucleus or other specialized organelles. 15

proof is a logically valid inference of the truth of a statement, based on axioms and theorems derived from those axioms. 36

propensity is a natural inclination or tendency to behave in a particular way. 85

protium is the common, stable isotope of hydrogen, ^{1}H, containing one proton and zero neutrons. 45

protocol sentence is a communicable description of regularities in one's conscious experiences; protocol sentences describe what we think, how we feel, or how the surrounding world appears to us as perceived through our senses. 276

proton is a subatomic particle p^+ with a positive elementary electric charge; protons are main constituents of atomic nuclei. 45

psyche is the conscious mind, or the center of one's thoughts, feelings and motivation. 181

psychoanalysis is a theory of the human mind originated by Sigmund Freud, according to whom the mind could be divided into id, ego and super-ego; id is a completely unconscious, impulsive component; super-ego is only a partially conscious, moral component; ego is the conscious, rational component. 4

psycho-physical parallelism is a philosophical stance that brain states and mental states do not need to interact at all because they are set in pre-established harmony by a divine creator. 181

pure density matrix is a quantum operator $\hat{\rho} = |\psi\rangle\langle\psi|$ that has a single nonzero eigenvalue equal to unity whose eigenvector is $|\psi\rangle$; since the pure density matrix is the projection operator onto the one-dimensional space spanned by $|\psi\rangle$, it has the property $\hat{\rho}^2 = \hat{\rho}$. 114

purpose of life is the personal aim of one's own living or existence. 17

quale is the subjective, first-person, phenomenal, qualitative property of how a certain conscious experience feels. 3

quantum bit is the basic unit of quantum information contained in the quantum state $|\psi\rangle$ of a two-level quantum system. 123

quantum brain is the quantum state of the physically existing anatomical organ inside one's skull. 227

quantum coherence is a basis-dependent quantum measure that depends on the existence of non-zero off-diagonal entries in the density matrix $\hat{\rho}$ when $\hat{\rho}$ is expressed in the given basis. 215

quantum computer is a physical device that performs computational tasks with the use of quantum bits of information. 221

quantum entanglement is a physical phenomenon that occurs when two or more quantum particles are created or interact in ways such that the quantum state of each particle cannot be described independently of the others, even when the particles are spatially separated. 189

quantum entropy is a basis-independent measure $S = -\text{Tr}\left(\hat{\rho}\log_2\hat{\rho}\right)$ of how close a quantum system is to having its own quantum state vector; quantum states described by a pure density matrix have zero entropy, whereas mixed density matrix states have strictly positive entropy. 214

quantum history is a possible trajectory of the quantum state of a quantum physical system in time. 196

quantum information is information stored in quantum physical states; it is unobservable, non-local, non-clonable, non-broadcastable, and non-erasable. 123

quantum jump is an abrupt transition of a quantum physical system from one quantum state to another $|\psi\rangle \rightarrow |\psi'\rangle$. 203

quantum probability is the probability for a physical event to occur at a given point in space and time given by the Born rule as the squared modulus of the quantum probability amplitude $|\psi(x,y,z,t)|^2$. 85

quantum probability amplitude is the fabric of quantum physical states; the squared modulus of the quantum probability amplitude $|\psi(x,y,z,t)|^2$ gives the probability for a physical event to occur at a given point in space and time. 85

quantum purity is a basis-independent measure $\gamma = \text{Tr}\left(\hat{\rho}^2\right)$ of how close a quantum system is to a pure quantum state; states with a pure density matrix have a unit purity, whereas states with a mixed density matrix have a purity $\gamma < 1$. 213

quantum state is a ket vector $|\psi\rangle$ in Hilbert space $\mathcal{H}$ that undergoes deterministic evolution in time according to the Schrödinger equation; the observed quantum indeterminism arises from objective reductions of macroscopic quantum entangled states that have reached a certain energy threshold $\mathcal{E}$ predicted by Diósi–Penrose models. 102

quantum superposition is a linear combination of two or more distinct quantum states $|\psi_1\rangle, |\psi_2\rangle, \ldots, |\psi_k\rangle$ given by $|\psi\rangle = \sum_{i=1}^{k} a_i|\psi_i\rangle$ where the coefficients $a_1, a_2, \ldots, a_k$ are non-zero; because quantum superpositions are basis-dependent, every quantum state $|\psi\rangle$ is superposed if $|\psi\rangle$ is not among the basis states. 112

quantum theory of consciousness is any extension of quantum physics with new laws that introduce consciousness into the physical world. 246

quantum wave function is a complex-valued probability amplitude distribution in space and time $\psi(x, y, z, t)$ that describes the physical state of a quantum system. 85

radius vector of a point p is a vector $\vec{r}$ from the origin to the point p. 87

range of a function is the set of all output values $y = f(x)$ that the function $f(x)$ takes. 51

rational number is any number expressible as a fraction $\frac{p}{q}$ of two integers. 49

reaction time is the amount of time it takes to respond to a stimulus. 247

real number is a number expressible as a limit of rational numbers. 50

reality is the world or the state of things as they actually exist. 44

reductionism is a philosophical stance that the mental events can be grouped into types which can be identified with types of physical events in the brain. 164

reflex is an involuntary motor response caused by applied stimulus. 10

representationalism is a philosophical stance that representation of a certain kind suffices for the generation of conscious experiences, where the kind of representation needs to be specified as a physical function without any recourse to fundamental mental properties. 163

research program is a sequence of adjustments made to a scientific theory aimed at improving the empirical adequacy or explanatory power of the theory. 274

rough endoplasmic reticulum is a membrane-bound cellular organelle that is involved in protein production. 15

rules of inference are logically valid forms of reasoning that allow derivation of theorems from the list of axioms. 42

scalar is an element of a mathematical field; both real numbers and complex numbers are scalars. 50

scientific theory is a logically consistent, communicable, empirically adequate knowledge about the physical world that describes how the world is and explains why the world appears to be the way it is. 32

sentience is the capacity to feel, perceive, or experience subjectively. 3

set is a collection of objects that is considered as an object in its own right. 61

shared knowledge is a set of beliefs, ideas, attitudes, and knowledge, expressible in classical bits of information that are shared between two or more people. 269

signal transduction is the physical process by which a neuron converts one kind of signal into another. 241

single mind is an entity whose conscious experiences are unified into a single mental picture. 10

singleton is a set with exactly one element. 62

slope of a line is the ratio $m = \frac{\Delta y}{\Delta x}$ between the vertical change $\Delta y = y_2 - y_1$ and the horizontal change $\Delta x = x_2 - x_1$. 52

soma of a neuron is the cell body without the neuronal projections; electric signals received from other neurons are summated at the soma. 25

space is the physical realm or expanse in which all physical objects are located and all events occur. 45

spinal cord is a long, thin, tubular bundle of nervous tissue located inside the vertebral column. 6

splenium is the thick posterior part of the corpus callosum of the brain. 157

split-brain patient is a patient whose corpus callosum is surgically severed as a therapeutic procedure for refractory epilepsy. 13

stability is the capacity of a physical system to return to its initial physical state after a perturbation. 226

subjective knowledge is knowledge that cannot be shared with others in the form of classical bits of information; an example of subjective knowledge is knowing what is it like to have certain conscious experiences. 32

subset is a set whose elements are contained in another set. 61

sulcus is a groove in the brain cortex. 11

superdeterminism is the conspirative property of a deterministic physical theory to mimic the indeterministic appearance of quantum phenomena. 205

superstition is a belief or practice that is not based on reason or knowledge, but arises from ignorance of natural causes and effects. 49

synapse is a specialized intercellular junction through which a neuron sends signals to other neurons or to non-neuronal cells in muscles or glands. 25

system of equations is a set of equations that need to be solved simultaneously. 88

teleportation is the transfer of a physical system from one place to another without traversing the physical space between the two locations. 139

temporal lobe is an anatomically distinct part of the brain cortex located on the side of the head above the ear. 11

theorem is a statement that has been proven with the use of logical reasoning from a set of axioms; if the proof of the theorem is correct, it is guaranteed that the theorem is true given that the axioms are true. 33

theory is a set of axioms devised to explain a collection of facts or phenomena. 36

time is the continued progress of existence and events in the past, present, and future regarded as a whole. 45

time agnosia is a pathological loss of the ability to subjectively experience the flow of time or to comprehend the succession and duration of events. 245

time complexity of an algorithm quantifies the amount of time taken by the algorithm to run as a function of the length of the input string of symbols; if all steps in the algorithm take a unit of time to perform, then the time complexity counts the number of steps needed to complete the algorithm. 222

trace of an $n \times n$ square matrix $\hat{A}$ is the sum of all entries on the main diagonal $\text{Tr}\left(\hat{A}\right) = \sum_{i=1}^{n} a_{ii}$; the trace of the matrix is invariant with respect to a change of basis, and is equal to the sum of the matrix eigenvalues. 114

tritium is a radioactive isotope of hydrogen, ^{3}H, containing one proton and two neutrons. 45

unconscious brain is a brain that does not generate any conscious experiences. 4

unconscious experience is a self-contradictory expression equivalent to unconscious consciousness. 4

unconscious mind is a self-contradictory expression equivalent to unconscious consciousness. 4

unit operator on a Hilbert space $\mathcal{H}$ is the identity operator $\hat{I} = \sum_{i} |i\rangle\langle i|$, where the set $\{|i\rangle\}$ is an orthonormal basis of $\mathcal{H}$. 102

unit vector is a vector with unit length. 101

unity of mind is the introspectively perceived wholeness of our conscious experiences, all of which appear to be unified into a single mental picture. 11

universal mind is a mystical concept endorsing the existence of universal consciousness of which we are all a part; going beyond the metaphor entails paradoxical occurrence of minds within minds. 11

unobservable is the property of not being observable or measurable. 19

unordered pair is a set of numbers $\{x,y\}$; reordering of the elements does not change the identity of the set $\{x,y\} = \{y,x\}$. 62

unstable state of a physical system is a short lived state that is sensitive to infinitesimal perturbations. 171

vacuum is a physical space entirely devoid of matter. 80

valence is the maximal number of chemical bonds that an element can form. 46

vector is a set that contains all directed line segments of the same length and direction. 55

vector field is a function of a space whose value at each point is a vector quantity. 57

velocity is a physical vector quantity $\vec{v}$ measuring the speed of a moving body in a given direction. 173

Venn diagram is a diagram of all possible logical relations between a finite collection of different sets, where set elements are represented as points in the plane and sets as regions inside closed curves. 61

verification is the process of establishing the truth, accuracy, or validity of something. 268

vicious circle is a logically fallacious chain of reasoning that assumes, either implicitly or explicitly, what it is attempting to prove. 34

voltage is the difference in the electric potential $V(x)$ between two points x_1 and x_2 defined by $\Delta V = V(x_2) - V(x_1)$. 71

voluntary action is an action that is executed under conscious control. 167

wavelength of a sinusoidal wave is the distance between two consecutive peaks or troughs. 78

weakly electric fish are electric fishes that use weak, typically less than one volt, electric discharges in the water for navigation and communication. 145

zero ket is the zero vector in Hilbert space $\mathcal{H}$; it is the only vector that is not represented by a ket, but is written as 0 similarly to the scalar zero 0 from which it should be distinguished from the context; the zero ket has a zero inner product, $0 \cdot 0 = 0$, and should not be confused with one of the unit vectors from the computational basis $|0\rangle$ whose inner product is $\langle 0|0 \rangle = 1$. 97

Index